COOPERATIVE

OCCUPATIONAL

EDUCATION

Including Internships, Apprenticeships, and Tech-Prep

Sixth Edition

COOPERATIVE

OCCUPATIONAL

EDUCATION

Stewart W. Husted, Ph.D.

Dean
School of Business and Economics
Lynchburg College
Lynchburg, Virginia

Ralph E. Mason, Ph.D.

Former Instructor
Business–Industry and Related
 Technologies Division
Maysville Community College
Maysville, Kentucky

Including Internships,
Apprenticeships,
and Tech-Prep

Elaine Adams, Ph.D.

Associate Professor
Career & Occupational Education
The University of Georgia
Athens, Georgia

Prentice
Hall

Upper Saddle River, New Jersey 07458

Pearson Education Ltd.

Pearson Education Australia PTY, Limited

Pearson Education Singapore, Pte. Ltd.

Pearson Education North Asia, Ltd.

Pearson Education Canada, Ltd.

Pearson Educación de Mexico, S.A. de C.V.

Pearson Education—Japan

Pearson Education Malaysia, Pte. Ltd.

Library of Congress Control Number: 2001094382

10 9 8 7 6 5 4 3 2 1

ISBN 0-13-110412-8

About the Authors

Stewart W. Husted is Dean of the School of Business and Economics and the Donaldson Brown Distinguished Professor of Marketing at Lynchburg College in Virginia. Dr. Husted is a former high school marketing education teacher-coordinator and a former community college instructor and vocational counselor. For 13 years he was a teacher-educator at Indiana State University. In addition, Dr. Husted has served as the liaison for the Cooperative Professional Practices Office to the Indiana State University School of Business. He has also served on the MarkED Board of Trustees and as marketing editor of the *Business Education Forum.*

Dr. Husted has written over 60 articles and cases and has co-authored three collegiate business and marketing texts and one secondary marketing text. He is a member of the Marketing Education Association, a member of the American Marketing Association, and a former board member of the Sales and Marketing Division of the American Society of Training and Development.

His degrees are B.S., M.Ed., and Ph.D. from Virginia Polytechnic Institute and State University, the University of Georgia, and Michigan State University, respectively.

Ralph E. Mason has had over 50 years of professional experience in Illinois, Indiana, and Kentucky. He has been a teacher-coordinator of marketing education, a director of vocational education, a teacher-educator, and a department chair at the university level. He recently retired as an instructor in the Business–Industry and Related Technologies Division of Maysville Community College in Kentucky.

His writing and consulting contributions to vocational education are extensive. Dr. Mason has served as a consultant on

vocational education to high schools, area vocational schools, and state boards of vocational education. He has held offices in professional associations at the local, state, and national levels, including the presidencies of the Illinois Vocational Association, the Illinois Business Education Association, the Indiana Business Education Association, and the National Council for Distributive Teacher Education, then an affiliate of the American Vocational Association.

His degrees are B.Ed., M.Ed., and Ph.D. from Illinois State University, Northwestern University, and the University of Illinois, respectively.

Elaine Adams is an associate professor in the Department of Occupational Studies at The University of Georgia. Dr. Adams is a former high school marketing education teacher-coordinator and a former program director of the merchandise management program at Chowan College. Before entering education, Dr. Adams held various managerial positions with several retail businesses.

Dr. Adams holds a bachelor of arts degree in marketing, with a minor in statistics, from Radford University. Her master of science and doctor of philosophy degrees were granted at Virginia Polytechnic Institute and State University and are in vocational and technical education. Both these degrees emphasize a concentration in marketing education and a cognate in the areas of marketing and management.

Her research focuses on career and technical teacher preparation, induction, and stress. Dr. Adams' research has earned three national awards: Omicron Tau Theta's National Outstanding Doctoral Dissertation Award for 1997, the American Vocational Research Association's Outstanding Beginning Scholar Award for 2000, and the Outstanding Journal Article Award for 2000 for the *Journal of Vocational Education Research*. In 2001 she was recognized as an outstanding academic advisor and mentor in the College of Education at The University of Georgia.

Preface

The first edition of *Cooperative Occupational Education* was published at a pivotal point in the history of cooperative and work experience education. Prior to that time (1965), the cooperative plan of instruction had been extensively used in some states, but not in all, and in some fields, notably marketing education and industrial education. Although the cooperative plan was being used at the post-secondary level, it was not in widespread evidence. Soon after the passage of the Vocational Education Act of 1963, many districts began to introduce the cooperative plan, especially in business occupations, home economics occupations, health occupations, and agricultural occupations. Thus, the first edition met a felt need—it was adopted for many pre-service collegiate courses and many in-service workshops and conferences.

The second edition was reorganized to serve all vocational areas using the cooperative plan. Like the first edition, it was widely used as a reference handbook by curriculum directors, local administrators and consultants for vocational education, and graduate students.

The third edition continued to have the practical directness of a handbook. Emphasis remained on the everyday details of organizing and operating the cooperative plan. Much more theoretical substance was added to work-study and general work experience plans, and much more detail was given to the concepts behind the career development of the individual. The conceptual base behind educational decision-making was re-emphasized, with the topics presented in a logical sequence of educational planning—**strategy–structure–system**.

The modified edition of 1986 incorporated the legislative changes in vocational education stemming from the Carl D. Perkins Vocational Education Act of 1984 and the Job Training Partnership Act (JTPA) of 1982.

The fourth edition retained the logical sequence of educational planning (strategy–structure–system), and it continued to stress the impact of current vocational legislation. The legal aspects of cooperative education were discussed in considerable detail. The application of the plan at the post-secondary, collegiate, and adult levels and to agricultural occupations, home economics occupations, and health occupations was expanded.

The fifth edition contained material dealing with the Carl D. Perkins Vocational and Applied Technology Education Act of 1990, with developments in new legislation, and with the integration of tech-prep and apprenticeship provisions as they apply to cooperative education.

The sixth edition discusses the key points of the Carl D. Perkins Vocational and Technical Education Act of 1998. Other major developments are covered, and current terminology is used throughout. The Bibliography has been updated to include the latest references.

Stewart W. Husted
Ralph E. Mason
Elaine Adams

Acknowledgments

We would like to acknowledge everyone who helped with the preparation of this edition, but space limitations prevent us from doing so. However, we are especially grateful to:

- The many students, administrators, teacher-coordinators, and teacher-educators who have given suggestions for this edition and previous editions.
- The many people and organizations who contributed illustrations. We have identified them by courtesy lines where appropriate.
- The administrations of Lynchburg College and the University of Georgia for their support and encouragement.

To everyone, we say, "Thank you."

Stewart W. Husted
Ralph E. Mason
Elaine Adams

Contents

ABOUT THE AUTHORS   v

PREFACE   vii

ACKNOWLEDGMENTS   ix

OVERVIEW: A MODEL FOR PROGRAM PLANNING   xv

SECTION ONE • The Strategy of Aims and Goal Inputs

Introduction to Section One   3

1. The Scope of Instructional Programs
   Using the Work Environment   5

2. The Development of Human Resources   27

3. Public Policy Goals and Institutional Roles   41

SECTION TWO • The Structure of Curriculum Patterns— How the Work Environment Is Used at Various Levels

Introduction to Section Two   69

4. Cooperative Education Models   71

5. Coordinators and Their Roles   83

6. Initiating the Plan   101

7. Coordinator Responsibilities at the
    Secondary Level   115

8. Coordinator Responsibilities for
    Adult Workforce Development   133

9. Coordinator Responsibilities at the
    Post-secondary and Collegiate Levels   147

SECTION THREE • The System of Instruction and Coordination

Introduction to Section Three   167

10. Planning and Carrying Out Effective
    In-school Instruction   169

11. Developing Training Stations as
    Instructional Laboratories   191

12. Correlating Instruction Between
    School and Job Laboratories   217

13. The Maturing of the Cooperative Plan   239

14. Student Organizations as an Integral
Part of Instruction   251

15. Accountability Through Evaluation   273

16. Legal and Regulatory Aspects of
Cooperative Education   291

SECTION FOUR • Application of the Systems Approach

Introduction to Section Four   313

17. The Plan in Agricultural Occupations   315

18. The Plan in Business Occupations   335

19. The Plan in Health Occupations   371

20. The Plan in Family and Consumer Sciences Occupations   397

21. The Plan in Marketing Occupations   411

22. The Plan in Trade and Industrial Occupations   447

GLOSSARY OF KEY TERMS   471

READING RESOURCES   483

INDEX   497

Overview: A Model for Program Planning

Cooperative Occupational Education uses the **strategy–structure–system** model as a means of implementing an instructional plan that effectively brings about what is best for students. The use of the work environment is a plan of education—it is a means to an educational end. In no way can the plan be considered a subject to be taught, nor can it be conceived of as an extracurricular area.

Everything that is done and every choice that is made involve a goal and the resultant determination of the process and means of goal attainment. Thus, one's time, energy, and inputs of other resources must be directed toward attaining the values sought, even if the end in sight is not completely understood or delineated.

Without goals, any activity is at best aimless—or worse, misdirected. So it is also with the educational enterprise and its decision-making process. Goals must be established, and the outcomes sought must be identified and described. When these have been accomplished, the enterprise has a strategy, and the operational decisions about curriculum and instruction follow from that strategy. Thus, the appropriate learning experiences can be devised and put to work in the implementation of that strategy.

In Figure 1, the model for decision-making about the use of the work environment in the educational plan consists of three major parts: the **strategy**, the **structure**, and the **system**. Each depends on the others, because within each is a series of decisions that relate to and feed back to the others. Goals form the strategy of the educational endeavor; the length and timing of the learning experiences of the curriculum become the structure of the endeavor; the utilization of human resources (teacher), media, and materials follows in the instructional system.

The Strategy

The strategy emerges from consideration of a series of inputs representing the needs, expressed or implied, of the groups to whom the educational plan will relate. As Figure 1 shows, these inputs are first related to the human resource needs of the community. For instance, what is the school being asked to do in providing a trained labor force, which affects greatly the economic well-being of the community? The goal is one of service to society through helping individuals reach their economic potential. But the human resource element as an input has another dimension that relates to program planning—that is, the recognition of the performance levels expected of the school's product when it enters the labor market. The demands are general as they relate to over-all employee characteristics, but they are specific in the skills and knowledge required for performance within a particular job title. Student occupational aspirations must also be perceived, since a program that meets community goals is of little value if the students are not interested. Student aspirations should be viewed not only in their short-range dimensions of getting the first job for whatever reason but also in their longer range of routes for advancement and promotion, or even shifts to other and more rewarding occupations.

A second category of strategy input rests with the student. In addition to the student's learning style, the capacity to learn—including affinity for cognitive learning, for example—must be considered. The approach or strategy to program building must take into account the student's prior experiences. How much does the student know about the occupation? Has the student acquired skills and knowledge in prior occupational classes, or is the student starting from scratch? Has the student the necessary basic skills of reading, communications, computation, and human relations to profit from the occupation's program immediately, or will the program call for remediation time? The strategy must also examine whether the student has special needs arising from physical disabilities or socio-economic disadvantages, which not only cause learning difficulties but also present problems in the employability demanded of the student who is to work part time under the direction of the school.

The last part of the strategy decision calls for determination of what other educational institutions are doing or can do for the student in terms of training programs of community agencies, other state and federal agencies, and local associations, such as labor unions.

The Structure

The structure that emerges from the plan using the work environment consists of what is popularly known as the curriculum, as well as the supporting staff, physical facilities and equipment, and ancillary or supporting services

A SYSTEMS APPROACH TO IMPROVING INSTRUCTION

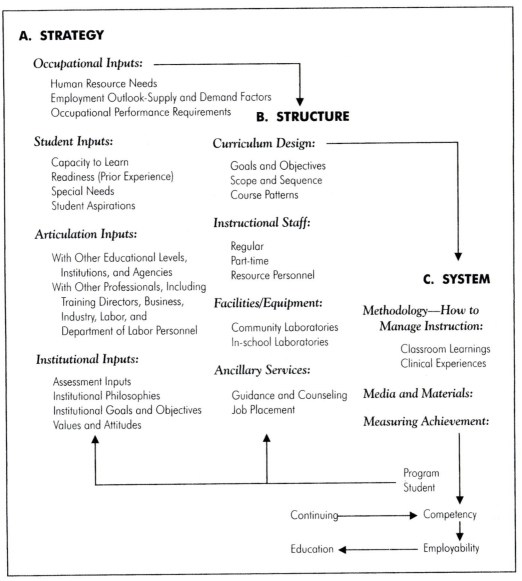

FIGURE 1

such as counseling and placement. For program-planning purposes, the structure should be viewed first as a definition of the behavioral goals to be accomplished, along with those performance goals that are intermediate steps to the facilitation of the end behaviors.

Both the behavioral goals and the performance goals can be measured against the probable student input of readiness and the needed learning experiences seen in broad outline. The learning experiences can then be plotted into a scope-and-sequence chart from which will emerge the final decision regarding the time needed for instruction and the division into whatever courses, modules, or other appropriate units fit the institution's pattern.

The System

The final determination in the decision-making process should appear early in broad outline when the instructional system can be seen in broad perspective. For example, it may be concluded early that because of student inputs and occupational skills demanded, the system will rely heavily upon multimedia-tutorial instruction or upon laboratory simulations. The system in detail emerges when the basic decisions of strategy and structure have been made and when the program is finally about to get underway. The system decisions involve determination of methodology appropriate to each learning experience and to each student's learning capacity and learning style. Questions are answered about what media and materials should be employed, what clinical experiences should be obtained on the job, and how achievement will be measured. The system can readily be seen as a series of daily decisions that depend upon feedback from events as they happen. The system is dynamic and flexible as instruction proceeds.

SECTION ONE

The Strategy of Aims and Goal Inputs

You are here

A SYSTEMS APPROACH TO IMPROVING INSTRUCTION

A. STRATEGY

Occupational Inputs:

Human Resource Needs
Employment Outlook-Supply and Demand Factors
Occupational Performance Requirements

B. STRUCTURE

Student Inputs:

Capacity to Learn
Readiness (Prior Experience)
Special Needs
Student Aspirations

Curriculum Design:

Goals and Objectives
Scope and Sequence
Course Patterns

Articulation Inputs:

With Other Educational Levels,
  Institutions, and Agencies
With Other Professionals, Including
  Training Directors, Business,
  Industry, Labor, and
  Department of Labor Personnel

Instructional Staff:

Regular
Part-time
Resource Personnel

C. SYSTEM

Institutional Inputs:

Assessment Inputs
Institutional Philosophies
Institutional Goals and Objectives
Values and Attitudes

Facilities/Equipment:

Community Laboratories
In-school Laboratories

Ancillary Services:

Guidance and Counseling
Job Placement

*Methodology—How to
Manage Instruction:*

Classroom Learnings
Clinical Experiences

Media and Materials:

Measuring Achievement:

Program
Student

Continuing ⟶ Competency

Education ⟵ Employability

INTRODUCTION TO SECTION ONE

Universal opportunity for human self-realization through public education remains the ultimate goal of society in the United States. There is little doubt that real opportunity exists only when educational programming is sufficiently flexible to provide these learning experiences needed at a time and place in which the individual can profit from them. The provision of such experiences, particularly those related to the world of work, must be available to all young people if they are to become fully participating and contributing members of society.

The United States is truly a dynamic and ever-changing society, characterized by technological innovation and economic growth, compounded by rapid sociological changes. All these changes are bound up in the world of work and in the crucial nature of the job as the central feature of the daily life of adults. Consequently, educators need to provide students with a continuing set of occupational and career development opportunities that will ready them for the transition from school to work. Educators need to help adjust to the occupational situation, to contribute to rising national productivity, and to obtain the style of living desired.

Section One begins with the first step in educational decision-making—determining the **strategy**, which is made up of the goals of those who are involved in the educational enterprise of occupational and career development education—the students, their families, the employers, and the educational institutions. From this amalgam comes the strategy of using the work environment as an instructional plan.

1

The Scope of Instructional Programs Using the Work Environment

An on-the-job sponsor observes the student-learner applying his skills.

KEY CONCEPT: Although many programs use the work environment, only the cooperative occupational education plan has a career-technical objective—the development of occupational competence.

GOALS

After successfully completing the study of this chapter, answering the questions, and carrying out the activities, the student should be able to:

- Distinguish between general work experience, apprenticeship, and cooperative occupational education.

- Explain what the term "experiential learning" means.

- Identify and describe the basic types of plans that use the world of work as an educational experience.

- Distinguish between work-study and internship plans.

- Identify and explain the chief components/characteristics of cooperative education.

KEY TERMS

apprentice
apprenticeship
career and technical education (CTE)
cooperative occupational education plans
correlated instruction
experiential learning

exploratory work experience plans
general work experience plans
internships or practicums
work observation plans
work-study plans

Schools and colleges have used the work environment as an educational experience for many, many years. Professional literature in education for decades has contained a great many articles dealing with such programs. Schools use the work environment for many reasons and in many different ways. These range from career exploration to general education adjustment for the adult world of work and from job-entry training to career development education. The uses are many and varied, and the terminology of the various programs is very diverse.

Colleges and universities use experience in a work environment as an integral part of many professional curricula. Usually, these experiences occur after or near the completion of regular course work and are called *"internships"* or *"practicums."* The main purposes are to allow the student to view acquired theory in actual practice and to give the neophyte the opportunity to practice prior to professional employment. For example, the future teacher interns as a student teacher, the criminal justice student interns as a police officer, the accountant engages in a practicum with a CPA firm, or the student dietician works in a nursing home.

Post–high school vocational-technical schools, career-technical schools, and community–junior colleges use the work environment in their career specialist curricula. Students in curricula such as engineering technology, health care, office administration, and marketing are placed in cooperating firms, where they learn certain job skills and how to use highly specialized equipment. They also have the opportunity to apply classroom theory to practical problems.

Secondary schools use the work environment for learning activities extensively. In recent years as more and more schools have become involved in expanding "schooling" into the work environment, the term "experiential education" has developed in popularity. Whatever the term used to describe learning situations that involve world-of-work experiences, the scope of the programs available for secondary students is far-reaching. Students may be placed as workers (or observers) to aid them in exploring occupations as they make their career choice. In other cases, junior and senior high students who may be having difficulties in achieving academically are released from school part time in order to take jobs that perhaps will provide them with not only earnings but also motivation to continue their general education. To acquire general education work experience, students may be employed in schools as cafeteria helpers, office workers, and library assistants. Some districts have created work-study plans for youth with academic and/or socio-economic disadvantages.

What Is Experiential Learning?

The concept of *experiential learning* is important in elementary and secondary schools. The term is often used generically to describe programs that use the work environment as an integral part of the curriculum. At times experiential learning includes cooperative career and technical education, apprenticeship, experience-based career education (general work experience programs in the academic curricula), JTPA (Job Training Partnership Act) tryout employment, and cooperative education plans. Many projects have been funded by state and federal agencies, including the U.S. Department of Education, the National Institute of Education (NIE), and the Department of Labor, to study how the educational environment can be expanded beyond schooling—expanded into the work environment. Experiential learning includes those experiences that teach individuals about work by studying about it, observing it, and doing it.

The Purposes of Using the Work Environment as a Learning Experience

The range of purposes of using work as a learning experience in education is extensive. So also is the range of operational patterns of such programs. Sometimes the work experience is of short duration—an hour or so a day for two weeks; but, at other times, it is much longer—half days for a year or two or full time for a semester or two. The experience may be obtained during normal school hours; at other times it may be undertaken after school or in the summers, although in both cases it is considered part of the school curriculum. Sometimes students are paid for the work, sometimes they are not. In some cases, the school selects only jobs that provide the desired learning experiences, closely supervises the experiences, and provides instruction in school that relates directly to the jobs. In other cases, the experience is loosely supervised, sometimes amounting to only keeping a record of where the students work.

In general, the work environment is used by educational institutions to accomplish one or more of six major purposes or goals. They are:

1. To keep over-age students (or underachievers and/or potential dropouts) in school part time while they obtain needed general education.

2. To help students explore the world of work and to assist them in making an occupational choice.

3. To help students overcome personality and behavioral problems.

4. To help students earn money who otherwise would need to drop out of school.

5. To provide practice in what has been learned in the classroom and to assist in the transition from school to work.

6. To help students develop general and specific occupational skills, knowledges, and attitudes, particularly those not readily available in the school's laboratories.

CONSIDERING A RANGE OF GOALS

Another way of viewing the possible outcomes of using the work environment is to review a career development scheme. The following scheme is intended to represent the needs of individuals at various stages of their career development. The school should decide at what stage each student is and whether a work-situation–based program will be of assistance.

1. **Career orientation**—Helping students understand themselves and giving them an awareness of their interests and aptitudes and an appreciation for the demands of various occupations.

2. **Work exploration**—Providing an opportunity for students to explore clusters of occupations and to gain experiences related to one or more occupational areas to facilitate more intelligent career selection.

3. **Economic awareness**—Helping students to understand the economy and their place in it as workers and consumers.

4. **Work adjustment and/or personal life adjustment**—Providing an opportunity for students to adjust to a work situation and to resolve problems, especially those in self-realization and human relationships.

5. **Skill development**—Preparing students in those skills needed in their occupational choice to provide entry-level competence.

6. **Skill application**—Easing the transition from institutional training to full-time employment by providing an opportunity for students to apply and test their acquired skills under real conditions.

7. **Upgrading skills**—Providing additional training so that students can upgrade their skills in

order to prepare themselves for increased responsibility and/or advancement.

8. **Job placement**—Helping students to find and hold jobs through a school-directed initial placement and try-out, with special reference given to helping students who are physically and/or mentally disabled and those who are economically and/or academically disadvantaged to adjust.

"FORM FOLLOWS FUNCTION"

Typically, educational decision-making needs to begin with a delineation of goals (strategy), which in turn dictates the curriculum structure as well as the instructional system that is needed. But, before discussing the goals (strategy) of the work environment as an educational device, the decision-maker must perceive how the work environment is used in educational institutions to achieve many diverse results.

Contemporary architecture has popularized the "form-follows-function" concept. Oversimplified, this means that a building is designed according to the purposes it will serve. This principle has great merit for curriculum builders who wish to use a work situation as an instructional laboratory. Paraphrasing, you could say, "The form of the work experience plan, including counseling and in-school instruction, should follow the functions or purposes (educational outcomes) desired." For example, if occupational competence is the goal, the *length* of the experience at work and the intensity of school course work related to the training objective would probably be longer than that of the experience required in a program of occupational exploration. If income for the student is the primary goal, then *almost any legal employment* might suffice. On the other hand, a specific career goal by the student would require *a carefully selected training station* complete with appropriate course work.

GENERIC USE OF THE TERM "WORK EXPERIENCE"

The term "work experience" is widely used to describe any curriculum plan that employs experience in a productive work setting to derive desired educational outcomes. However, what many people call work experience is not occupational education. Furthermore, work experience plans are general in nature, while supervised occupational experience plans are career-technical in nature, even at the college level. This fundamental distinction, when properly understood, will enable educational personnel to determine the organizational and operational techniques necessary to devise the type of plan that will bring about desired educational outcomes.

Six Basic Types of Plans Using Experience in Work Situations

The world of work is the scene of many school-sponsored learning experiences. Although there are hundreds of programs in educational institutions, there are only six types of plans.

For general education purposes:

1. Work observation plans
2. General work experience plans

For occupational education purposes:

3. Work-study plans and work exploration plans
4. Cooperative occupational education plans
5. Internships and practicums
6. Apprenticeship plans

Six Basic Types of Plans That Use the World of Work as an Educational Experience

Work Observation	General Work Experience	Work-Study and Internships	Cooperative Occupational Education	Apprenticeship
Student observes work, does not perform tasks except to understand them. Is unpaid. Plan is usually few weeks in length at most and may be tied with a class in which occupational information is discussed.	Student performs tasks of the actual job. May or may not be paid. Is typically engaged in for general education values, including exploratory. Is usually one semester or less. Limited school supervision; usually no related class.	Student performs in approved job situation. Usually paid and given credit. In-school instruction usually given before work period and seldom tied in directly with job experiences. Typically one semester or more. "Internship" or "practicum" is a term used for collegiate experiences.	Occupational goals based on student's career objective. Work situation is an occupational laboratory for classroom instruction. Selected training stations. Correlated instruction in school. Pay and credit. Consistent school supervision. Typically at least one year.	Student learns trade under guidance of a skilled master crafts worker. The student must be at least 16 years old and no more than 24. Normally phased in periods of 1,000 credit-hours and includes a minimum of 144 hours of related classroom instruction.

The types of plans using the work environment and the significant characteristics of each are illustrated above.

The work plans described have several characteristics in common: (1) each uses a work situation as the source of learning experiences; (2) each is part of the institution's curriculum, regardless of whether or not the experience is during school hours; (3) each involves school approval of the job and at least minimal school supervision; and (4) each includes an observance of local, state, and federal labor laws. However, further analysis of the plans will reveal that the major differences occur first in purpose and second in their organizational and instructional characteristics, such as the length and depth of the experience, the relationship of the training firm to the student's career goal, the amount and type of supervision, and the provision for instruction in school related to the experience.

With these six types of plans broadly differentiated, the differences between general work experience and cooperative occupational education can be compared.

General Work Experience Plans

Curriculum planners, administrators, and teachers who are concerned with helping students develop their abilities to their individual limits have seized upon the work environment as a useful adjunct to the curriculum. Indeed, schools display an amazing variety of curriculum patterns in which work experience is used as a way to learn. The variety of uses is such that classification criteria emerge. These are: (1) the major outcomes expected from the plan and (2) the type of student for whom the plan is intended.

Exploratory and *general work experience plans* found in the curricula of secondary schools are typically of four basic types. These are:

1. **Out-of-school non-remunerative exploratory work experience plans**—As a phase of the guidance program, students in some schools are "observers" or "student-learners" in physicians' or dentists' offices, in architects' studios, in hospitals,

with city and county officials, in banking concerns, or in other business and professional situations. In this plan the principal objective is guidance, an exploration of occupations in which the students believe they have a career interest.

2. **Out-of-school non-remunerative community service work experience education**—In this type of work experience education, students accept assignments to local social service agencies, elementary schools, libraries, and other public service groups. They work on community service projects. The Job Training Partnership Act (JTPA) placed students in public service, as well as private, employment situations.

3. **In-school non-remunerative general work experience plans**—Students can profit from performing regular work tasks. In many schools, students are released from regular classes to act as helping teachers, as assistants and clerks for teachers and administrators, as maintenance workers, etc. This work is generally done without remuneration, although in some cases an outside agency provides funds for paying for work when it is performed by students who are economically disadvantaged.

4. **Remunerative general work experience plans in junior high or middle schools**—Here the general education values of productive work experiences are recognized by well-planned programs, including guidance activities, placement services, and adequate coordination by a special teacher, and in many cases by scheduling for these work experience students a special class dealing with the problems of job success. The students are engaged in any acceptable and legal work assignment and are paid a reasonable wage for their work.

The discussion thus far has shown that work experience plans are general rather than career-technical in nature. Work experience plans involve students in the environment of real jobs, either as (1) observers or as (2) workers. From the descriptions of the four types of plans, it is apparent that among the purposes of general work experience are these:

1. To help students explore occupations, experiencing their demands and testing their interests.

2. To provide youth with opportunities to earn money to help with school or family expenses.

3. To motivate potential dropouts to remain in school, at least to the extent of getting further general education.

4. To provide students with experience at working so as to develop general personal characteristics.

While it is conceivable and demonstrable that a student who works may develop some occupational competencies, it should be clear that work in and of itself does not necessarily develop employable skills. Thus, general work experience plans should not be considered as career and technical education programs. As the personnel director once said to the applicant who claimed five years of experience, "Have you had five years of experience or one year five times?"

OPERATIONAL DESCRIPTIONS OF WORK EXPERIENCE PLANS

The term "work experience plans" has been used generically to describe varying general education plans that use the work situation as a teaching–learning device. Let's now look at some of these basic types of general education plans to learn about their operational characteristics and, hence, their similarities and differences.

Exploratory work experience plans.—Secondary schools have used work experience to help young people learn firsthand of an occupation by either observing it or engaging in productive work. In general, the work experience plan for career exploration purposes has been of limited length, such as six weeks, nine weeks, or perhaps a full semester. In such plans the amount of each school day devoted to the work experience is no more than a half-day, and in most cases, it is usually only one or two periods plus some time after school. In some schools, only one exploratory work experience is allowed; in others, students may schedule several short-term experiences during a semester or a year. The apparent rationale for the short length of the experience is that the student need not invest a great number of hours in order to see, at least in broad outline, the nature of the occupation and the demands it makes upon abilities. The student can conduct a self-assessment against the occupation in a relatively short time.

Most schools that report exploratory plans have a class to which the experience is attached. Usually the class is one in which career guidance is the primary content, with the teacher being either a counselor or an academic teacher who has had some professional training in guidance and counseling. Some schools report that the class meets for one period daily; others report but one hourly session per week. Generally, students are not paid for the work if the work experience is for a short term, such as six weeks. On the other hand, in longer, more intensive internship-like placements, such as a semester for half-days, students actually engage in productive work and may be paid. Most schools attempt to be quite selective regarding the employing firms or the professional persons with whom the students are placed, recognizing the intimate role that can be played in guidance by identification and by conversations between student and sponsor.

There are some secondary schools that have used other approaches to fulfill the exploratory function of work experience. Some have simply released students the last period of the day to work when and where they wished, with little or no supervision by the school and no related class in school. However, if the students do not have the guidance of learning activities, at least related to working in general, the school cannot assure itself that desired learning outcomes occur. Other schools have more formal programs, lasting half-days for a full year, with placement approved or obtained by school supervisors, but these programs are conducted without any related classes. In such cases, the schools rely on the initiative of the students to find jobs appropriate to their career inter-

ests. This can create a problem because some occupations are closed to part-time teenage workers and because many students will choose jobs on the basis of the income to be earned rather than the learning to be acquired. There appears to be an unjustified faith in the ability and initiative of the students to draw occupational meanings from the experience. Students need some organized instructional time in school in which they can discuss their work experiences with other students, have the help of a trained counselor-teacher, and engage in learning activities that provide information and understanding about occupations.

Exploratory work experience plans have in common the following characteristics: (1) placements carefully selected to meet the career interests of each student, (2) experiences with sponsoring firms whose management and supervising employees understand their guidance role, (3) classes in school taught by credentialed career-technical teachers or career-technical counselors who can help young people interpret their experiences, (4) no intent to teach job skills or knowledges, and (5) providing career-technical guidance so that each student gains a systematic sampling of a variety of conditions of work.

Released-time work experience (work placement) plans.—Some secondary and postsecondary schools have the policy of releasing students from their last-period study halls or from some classes in order for them to engage in paid employment. Typically, the school issues a work permit and keeps a record of where the students are employed. Some schools go further and assign personnel to assist students in finding jobs and to act as community placement service officers for part-time positions. A few schools provide personnel who make routine follow-up visitations about once a semester. A student only coincidentally may be taking any courses in school that relate to the job, and only coincidentally may the job relate to that student's occupational goal. These plans are, in effect, not a part of the curriculum, and no credit is given. Therefore it seems quite logical not to classify them as being work experience plans. However, because school administrators sometimes call them "work experience plans," they have been so classified in this text, even though the purpose seems to be either to allow students to earn money or to relieve some students (particularly over-age students and low achievers) from the "burden" of attending school full time. Certainly, without careful supervision or the organization of related learning experiences, such work placement programs cannot be presumed to be part of the curriculum of the school.

General work experience plans.—The chief criterion that a work experience plan must meet to be classified as a general work experience plan is: "Are the outcomes derived from the plan general in nature rather than career-technical, or are they contributing to the solution of special education problems?" Although many types of work experience plans are considered general work experience plans, they are not, according to this text, because the plans are: (1) not part of the school's curriculum, (2) not carefully supervising students, (3) not using a related class, and (4) not carefully selecting work stations.

Outcomes.—The basic outcome of the general work experience plan is an understanding of the world of work and the development of acceptable work habits, attitudes, and personality traits. While the student may learn some career-technical skills and information relevant to a particular occupation or job, these are clearly perceived as being incidental to the main purpose. The primary outcome is to help students develop those habits, abilities, and characteristics that are of value in any occupational field and on any job, from unskilled to professional. Quite obviously, the plan also provides opportunities for occupational exploration and the development of career interests.

Operational aspects.—The general work experience plan is typically offered to eleventh-and-twelfth-grade students, although tenth-grade students are sometimes enrolled in special circum-

stances. The plan is part of the curriculum, even though the actual working hours are primarily or even exclusively after school. Credit is given, but the students are not always paid. A class in school is required of all students, and the content of the course that is taught by the program supervisor relates to general occupational needs and individual personal needs. The school selects training stations that will match the students' interests and that will provide job supervision and attention to the students' needs. The following is a checklist of conditions of operation that a general work experience plan should meet at least minimally.

1. **Selection of students**—Counselors should be involved in determining the students' interests, aptitudes, and motivations for enrolling in the program. Students wishing specific career-technical preparation should be referred to career and technical education (CTE) programs.

2. **Selection of work stations and placement**—The work station need not be related to the student's occupational goal, because the purpose of the program is not vocational preparation. For the student who wishes to try out a given occupation, placement in a training station relating to the student's career aspirations is obviously important. The supervisory personnel at the work stations should understand the purposes of the program and be able to help students understand the world of work. Students may "find" their own stations, but no placements should be made until the program supervisor has made an evaluation visit to each station and has explained the purposes of the plan to the various employers.

3. **Written agreement of employment**—A placement should be approved only after the agreement has been approved as a memorandum of employment signed by employer, school, student, and parent. The agreement should state the conditions of employment and the responsibilities of each party to the others.

4. **Evaluation of students**—The student's performance on the job must be evaluated. The employer

should prepare a written evaluation of the student's performance and schedule a conference to review the student's performance on the job. In addition, the supervisor should evaluate the student's progress during visits to the station. It is recommended that the supervisor make job visitations once a month, or at least twice a semester, exclusive of rating conferences.

5. **Supervision of the program**—To prevent the program from becoming strictly a released-time learning situation, consistent supervision by the school is necessary. Depending on travel distances and other factors, such as newness of the program, the supervisor (coordinator) should be allotted supervision time in the ratio of one period for every 15 students. But, if the supervisor is paid on an extended-day contract, the ratio can easily be 1:20, or 60 students for a three-period allocation for program supervision.

6. **Employment of trained supervisor**—The supervisor need not be a CTE teacher because the program is of a general education nature. Rather, the supervisor needs some training in counseling and sufficient recent experience in the world of work to be able to work with businesspersons. The supervisor will be in the community making contacts with business and industry on a regular basis; therefore, that person must present a favorable image of the school, or the school's credibility will deteriorate.

7. **Use of related class**—Every student of general work experience should be enrolled in a course that promotes general occupational understandings and provides for individual counseling. The related class might meet daily for one period, although weekly or monthly arrangements may fulfill the need. The course should be taught by the teacher-coordinator because that person knows the student and the work situation. The related class does *not* teach specific vocational content.

8. **Interpretation of the plan**—One of the prime responsibilities of the supervisor is that of interpreting the plan to other teachers, the employers, the students, and their parents. The supervisor should make use of public relations channels established at the school and provide for face-to-face interpretation of the plan at every opportunity.

9. **Maintenance of adequate records**—Adequate records regarding items such as work assignments and hours worked, work permits, individual student personnel folders, ratings, and follow-up studies should be kept by the supervisor.

Work-Study Plans

The term "work-study" has been used in many ways, and in recent years, it has taken on some very diversified meanings, which have caused considerable confusion as well as misinterpretation of program administration guidelines. If the various local work-study plans are classified according to goals and relationship of work experiences to occupational goals and in-school learning experiences, three shapes emerge. They are:

1. A plan wherein students who are enrolled in a career and technical education curriculum find jobs in the same general line of work or cluster of occupations. The students work with school approval, but the jobs and the school instruction are not correlated in the sense that the in-school classes are tied daily and directly to job needs. What is assumed in such situations is that not only will the students learn career-technical skills on the job, but they will also relate them to the classes.

2. A scheme in which the students who are enrolled in a CTE curriculum are also employed part time during school hours. The jobs are usually not related to the CTE curriculum, but the income and motivation are believed to enable the students to continue and to complete the vocational studies. Often these positions are in non-profit agencies. Such a program can be correctly termed "career and technical" only if the students are enrolled in a CTE curriculum. But, in most cases, the work situation normally is not

correlated with the in-school vocational instruction. Therefore, the term "cooperative education" does not apply by strict definition.

3. A situation in which students are given jobs regardless of their school curricula, in order to furnish them on school time with a situation that provides both income and, hopefully, the motivation to continue their education and to finish high school. Some believe these programs are effective with some students who need the will (motivation) in their value systems to continue their education. These types of programs do not necessarily provide students with occupational education because they cannot be classified as CTE. However, some students do learn a skill from their income job or from the CTE curriculum in which they are coincidentally enrolled in the school.

As shown by the preceding discussion, the term "work-study" is found in practice to have several meanings, none of which can truly be called cooperative career and technical education, since the essential correlation of school and job learnings in tandem with a defined career objective is missing. Does this mean that none of these approaches is occupationally useful? Not at all. Look at each, and you will see that even coincidental career and technical education has merit. But a program, to be classified as career-technical, must have the essential ingredients that make it accountable.

The previous description of the work-study plan and programs was related to the secondary school. However, the term "work-study" can also be used at the collegiate level and, in such cases, applies to situations mainly where students in one term (or several) learn the theory and principles of the occupation and in subsequent (one or several alternating) terms are on the job in a full-time capacity, presumably applying what has been learned. Here again, the value of the student experience on the job need not be questioned, nor should it be assumed that the students are not applying what has been learned. What is to be questioned is the role of the school in directly and accountably scheduling the work experiences according to each student's needs and in providing in some way a means for insuring that the students do indeed apply and profit from what is learned. If such plans rely upon the goodwill of the employers to provide needed training and concomitant explanation, as well as upon the assumption that the students have the maturity to meet their own learning needs, these plans are not directed. And this absence of school-directed learning on the job, which is correlated in a time sequence with classroom instruction, means that so-called collegiate work-study plans cannot be described as cooperative occupational education.

Although work-study plans and cooperative occupational education plans have some common goals and similar characteristics, there is a fundamental difference between them. The major difference is the basic purpose and, therefore, the provision of related instruction. In work-study plans the purpose is general occupational education, and the instruction in school is only generally related to the work of the training station. There is no effort to teach topics in the order in which they are needed on the student's job. Individual learning needs stemming from the job are not usually a focus of instruction. In addition, the instruction in school is often given before the job experience, rather than concurrently with it. Lastly, the occupational experience may be only generally related to the student's career goal, rather than contributing directly to it.

In contrast, in cooperative occupational education plans, the goal is both general and specific occupational education. The instruction is said to be correlated; that is, there is a direct relationship between the study in school and the activities of the training job, both of which are based on a career objective. This correlation involves both (1) the sequence of learning (what is studied when) and (2) the application of learning (what is learned in school and then applied on the job, with the results being reported in the classroom). In addition, the students will have some individual instruction in school; that is, each student will examine some things not studied by other students, because of individual job needs and individual career goals. Besides the basic difference in instruction between work-study and cooperative education, there are other differences. However, they are more a matter of degree than of contrast. For example, the supervision of the occupational experience in the training firm is likely to be more consistent in cooperative education than in work-study.

An illustration of a work-study plan will serve to point out the indirectness of the related instruction as compared to that in a cooperative plan. The illustrative program comes from the field of business education. At Westport High School (a fictitious institution), seniors who are preparing for administrative assistant careers may elect the work-study plan. In the eleventh grade the students complete one year of computer applications and one year of accounting. If they elect the work-study plan, they will be placed in an office during the second semester of the twelfth grade, working each afternoon from 1 to 5. Any student who has a C average or better is accepted. The firms that provide the training are to pay the trainees the regular beginning wage and are expected to give them as many varieties of experience as possible. The school supervisor (not the teacher) visits the training stations only once during the semester to check on the students' progress. In the twelfth grade the students take second-year accounting and advanced computer applications. In the accounting and computer applications classes are other students who are not working.

While the work-study students may apply what they are learning to their jobs, the teacher does not discuss job problems with them, nor does the sequence of instruction change because of their job needs. Indeed, if one of the working students needed instruction, say, desktop publishing, the student would not receive it because it is not usually taught in computer applications and accounting classes. Thus, from this description of the work-study plan at Westport High School, you can see that the job is only indirectly related to the instruction in school. What is learned on the job depends on what the employer wishes to teach by on-the-job coaching.

However, this example also reveals that the plan, in addition to having an indirect vocational purpose, does indeed provide general occupational experience.

Cooperative Occupational Education and Career Development Plans

Cooperative occupational education plans have as their central purpose the development of occupational competence, using employment in real jobs as a source of learning. The school selects as training agencies firms that will provide the occupational experiences needed by the students, and the school supervises the students' experiences. Class work in school provides those learnings basic to employment and to the occupation sought. The occupational experience is expected to be the source for gaining knowledges as well as a vehicle for applying and testing what has been learned in school.

KEY ELEMENTS OF THE COOPERATIVE OCCUPATIONAL EDUCATION PLAN

Not only does this text describe various uses of the work environment in plans that fulfill the school's purposes but it also endorses such situations in which the school has carefully outlined objectives and made the necessary inputs of resources to insure attainment of those objectives. There is no intent to decry general work experience plans as not being vocational—rather, all educationally sound plans are endorsed. But, it is the goal of the authors to help educators distinguish between general education plans using the work environment and those that are truly career and technical education. All too often school administrators have been led to believe that if a student was released from school and was working, the student was indeed in a CTE program. This is not the case because the true measure of distinguishing between a cooperative occupational education plan and a general work experience plan consists of several key conditions. These are:

1. **The primary and overriding purpose is to provide occupational competence at a defined entry level**—Instruction must be geared to a set of definable performance objectives, and providing students with financial assistance, or employment, or even the inducement to stay in school must be a secondary consideration.

2. **The instruction both in school and at the training stations is based upon the student's**

Comparison of Four Plans Using the Work Environment

Components or Characteristics	Cooperative Education	Work-Study	Work Experience	Apprenticeship
1. Established career objective by student	Yes, primary objective is entry employment training toward a career	Sometimes yes, established in some vocational field. Program objective is usually earning power and motivation for student	No, but sometimes has general education values	Yes, primary objective is entry-level employment training toward a career
2. Classroom instruction related to career objective	Yes	Not necessarily	No	Yes
3. Established training station and close supervision by school	Yes	Not usually	No	Uses established training station but usually not supervised by school
4. On-the-job training plan	Yes	No	No	Yes
5. Paid employment	Yes, usually by profit-making business	Yes (in some plans through government subsidy)	Not necessarily	Yes
6. Advisory committee used	Yes	No	No	Yes
7. Career-technical student organization correlated with instruction	Yes	No	No	No
8. Certified teacher-coordinator in occupational field	Yes	No	No	No
9. Planned home visitations	Yes	No	No	No

career goal—It is this interest that provides the student's motivation for learning.

3. **The kind, extent, and sequence of the training station learning experiences correlate closely with the kind, extent, and sequence of the inschool learning experience**—This correlation is maintained by a coordinator who also teaches or by a coordinator who has sufficient time to work closely with the instructors of the students.

4. **Students may elect the cooperative plan only when they possess the employability characteristics acceptable in the market place, as well as the necessary basic knowledges and skills prerequisite to employment**—To use a quotation attributed to Charles A. Prosser, the student who enters cooperative instruction "must want it, need it, and be able to profit from it."

5. **The employment situation must be truly a training station where the firm understands and accepts its teaching responsibility and where an individual is given time to act as a training sponsor,** one who can fulfill the role of the downtown laboratory teacher.

6. **Not only are the employment conditions legal employment but they are also acceptable by all other standards of the school.**

7. **The coordinator has sufficient time to carry out coordination responsibilities and to be accountable for quality education.**

8. **Individualized learning activities, remediation as needed, and interaction with the program of a career-technical student organization (CTSO) characterize instruction.**

Above all, the key element is the visible and real support by all functioning units of the school in close interaction and communication with the community being served.

Internships/Practicums

Generally, world-of-work experiences that are referred to as internships or practicums are offered by secondary schools that use the work environment in their curricula for two purposes: (1) to further the aims of general education by acquainting their graduates firsthand with the dignity of work and with the significance of work in a contemporary society and (2) to provide their students with the opportunity to sample areas of professional service and to test their own abilities and interests against the demands of these occupations. (Chapter 9 presents an in-depth discussion of cooperative education at the post-secondary and collegiate levels.)

The internship is well known to college personnel because of its widespread inclusion in baccalaureate curricula preparing individuals for professional positions. In some institutions it may be called a practicum. It is frequently used in educating teachers, physicians, social workers, and counselors. In a number of situations, it is used to educate school administrators, labor economists, engineers, police administrators, and various business personnel, such as accountants, sales representatives, and retail managers.

The term "internship" or "practicum" generally describes a technique within a curriculum with the following characteristics:

1. It is used in professional or sub-professional curricula.

2. It is undertaken typically as a culminating experience prior to the student's graduation but after preliminary classroom work.

3. It occurs in an actual job situation in which the intern experiences the requirements of employment.

4. It is conceived primarily as a way of enabling the student to apply the concepts and skills learned in the classroom rather than as a way of learning new skills and knowledges.

5. It is often a full-time resident experience and typically at least three months in length in order to provide a complete experience.

6. It places the intern usually in a firm or an agency that is selected by the school or college for its progressive method of operation.

7. It has a professional person, who has been selected because of his/her ability, professionalism, and competence as a trainer to supervise the student at the internship agency.

8. It may or may not pay the intern a salary, usually at a reduced rate, because the person is a trainee, who is not fully productive all of the work time.

It is important to recognize that internships have as their major purpose the development of occupational (professional) competence through practice after theory education has been completed. Because they provide directed occupational experience tied to related course work, such situations are akin to the secondary cooperative occupational training programs. However, a growing number of secondary schools are using the term "internship" to describe exploratory work experience programs.

Some colleges have freshman- or sophomore-year experiences wherein the student is placed on a job, part or full time. These, however useful, cannot be properly referred to as "internships," because the goal is usually one of having the student explore the field and experience its requirements before undertaking the major portion of the professional course work. Internships have become more widely included in college curricula. Such arrangements have demonstrated their value in aiding students in the transition from formal schooling to the actual professional situation. Employing firms and agencies like the arrangement because it gives them, first, the opportunity to "screen" interns for future employment and, second, the chance to "sell" the interns on employment with them. However, there is a more important reason for the prediction of increasing numbers of internships. Collegiate-level curriculum makers seem to be embarked on a program to eliminate "CTE" courses

from their curricula, preferring instead to emphasize liberal arts and a core of professional theory-oriented courses.[1] Faculties are on record as opposing "how-to-do-it" courses, stressing preparation for a career rather than for the initial job. It is quite likely that the internship will serve to provide much of the job-entry operational training that is removed from the curriculum.

Apprenticeship

Apprenticeship has traditionally been considered to be training in an art, a trade, or a craft, under the guidance of a skilled master worker. Examples of common apprenticeships are found in the construction trades such as plumbing, painting, bricklaying, metalworking, electrical work, and carpentry. Thanks to tech-prep, youth apprenticeships are being created in untraditional fields such as banking, television production and management, appliance repair, and others. Today, about 900 occupations are recognized as apprenticable.

To date, over 200 diverse companies have established programs to address their need for skilled workers. For example, Sears, Roebuck and Co. established school–business partnerships that train students at its 750 service centers nationwide and in CTE classrooms.[2] Other companies include corporate giants such as IBM, Boeing Aircraft, and Blue Cross–Blue Shield of Maine and New York. However, according to a 1993 National Alliance of Business survey of apprenticeship businesses, 75 percent have fewer than 500 employees and 60 percent have fewer than 100.

To create an American-style youth apprenticeship, an analysis was conducted of European apprenticeship programs. Several basic premises based on

---

[1]This trend is particularly evident in engineering and business administration curricula.

[2]Rick Wartzman, "A Boost Up the Job Ladder," *The Wall Street Journal,* Classroom Edition, January 1993, p. 10.

the unique cultural, economic, and social conditions in the United States were developed. The apprenticeship program must:

- Recognize and accommodate the diversity inherent in the American populace;
- Be part of a broad effort to improve the linages between the world of work and the world of high school, and not just for those who are not college bound;
- Provide early exposure to work experience and genuine opportunities for workplace learning, with training wages paid by employers;
- Result in formal, universally recognized credentials that meet nationwide standards that are the product of the collaboration of government, education, and labor agencies, union representatives, and business associations;
- Assure apprentices opportunities to go on to further education, should they seek it, after receiving their apprenticeship credentials.[3]

The apprenticeship is created through a joint legal agreement between employers, unions, and in some cases a school system. Most *apprentices* earn wages and work regular hours. People who complete apprenticeships become *journeypersons*. The lowest age for entry into apprenticeship is 16, but in many trades the requirement for a high school diploma makes the effective age 18. The upper age limit is normally 24, but it is higher in some trades. The training period varies from two to five years, but it is usually three or four. Much of the training takes place at technical-vocational schools or in related instruction supervised by the local school system.

The Bureau of Apprenticeship, established in 1937 as a part of the Department of Labor, allows the National Apprenticeship and Training Council

to promote schemes for individual industries. Most apprenticeships are phased in periods of 1,000 hours, each equivalent to six months and include 144 hours minimum of related classroom instruction. The apprentice advances through the phases after passing qualifying examinations. Pay varying from 60 to 90 percent of the journeyman's rate increases at each phase.[4]

In youth apprenticeship programs, it is the responsibility of a school representative to supervise the related instruction and to administer the program relative to the records required by the state board of vocational or career and technical education. Often the school representative is one of the coordinators of a cooperative education program in the school district. The state of West Virginia has been a pioneer in the use of co-op teacher-coordinators in supervising youth apprentices. This is a natural choice because of the teacher-coordinator's relationships with business and industry. The teacher-coordinator of the trade and industry cooperative program (or a diversified program coordinator) is the logical person to handle these responsibilities because most of the apprenticeship trades are in the technical and industrial occupations. If a teacher-coordinator of a trade and industrial plan is not available, one of the other teacher-coordinators or the director of career and technical education may be given the responsibility of operating the apprenticeship program.

Jobs for the Future, a national advisory group, provides technical assistance to new apprenticeship projects selected by the Department of Labor. This group has identified several elements that distinguish youth apprenticeship programs from other forms of school work programs. The elements include:

- Active participation of employers, not only in providing jobs but also in developing curriculum and industry standards.

---

[3]Stephen F. Hamilton, *Essentials of the European Apprenticeship Systems and American Adaptation,* presentation at the conference on Youth Apprenticeship American Style, December 7, 1990, Washington, D.C.

[4]"Apprenticeship," The New Encyclopedia Britannica, 1987, pp. 494–495.

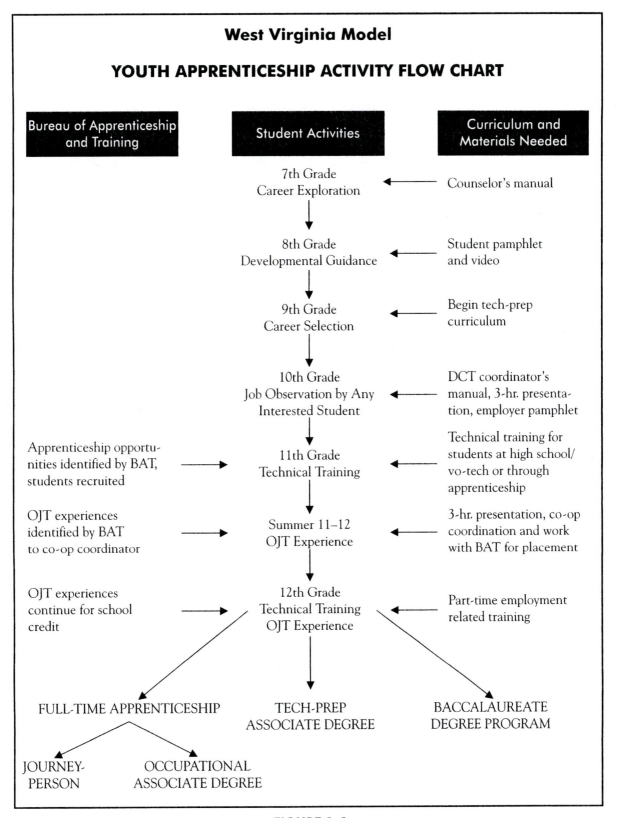

West Virginia Model

YOUTH APPRENTICESHIP ACTIVITY FLOW CHART

Bureau of Apprenticeship and Training	Student Activities	Curriculum and Materials Needed
	7th Grade Career Exploration	Counselor's manual
	8th Grade Developmental Guidance	Student pamphlet and video
	9th Grade Career Selection	Begin tech-prep curriculum
	10th Grade Job Observation by Any Interested Student	DCT coordinator's manual, 3-hr. presentation, employer pamphlet
Apprenticeship opportunities identified by BAT, students recruited	11th Grade Technical Training	Technical training for students at high school/vo-tech or through apprenticeship
OJT experiences identified by BAT to co-op coordinator	Summer 11–12 OJT Experience	3-hr. presentation, co-op coordination and work with BAT for placement
OJT experiences continue for school credit	12th Grade Technical Training OJT Experience	Part-time employment related training

FULL-TIME APPRENTICESHIP   TECH-PREP ASSOCIATE DEGREE   BACCALAUREATE DEGREE PROGRAM

JOURNEY-PERSON   OCCUPATIONAL ASSOCIATE DEGREE

FIGURE 1–1

▌CASE-IN-POINT ▌

Apprenticeship Program Transforms Student's Life

Matt Fleak was a senior at Parkersburg South High School when he became West Virginia's first participant in a federal youth apprentice and training program that is preparing him to become a journeyperson carpenter at about 21 years old (average age 27). Prior to joining the program, Matt was working at a local car wash. Now between classes, he spends three afternoons per week at Dupont's Washington Works learning the tricks of the carpenter trade.

Matt says the wages he earns has had a substantial impact on his grades at his high school. He was told that to keep his job he had to keep his grades up. According to Matt's mom, he is now earning a 4.0 average.

Matt is supervised by his CTE teacher, Craig Frazier, of the Wood County Technical Center. He is employed by M. K. Ferguson, a Belpre contractor. Matt is also registered as an apprentice in the Carpenters and Joiners Local 699 of Parkersburg.

- Integration of work-based and school-based learning so that the learning at one place reinforces learning at the other.
- Integration of academic and career-technical learning, focusing on cognitive as well as technical skill development.
- Structural links between secondary and post-secondary institutions, so that credits may be transferred to four-year colleges and universities.

- Award of a broadly recognized qualification of occupational skill, particularly a certificate of mastery of skills recognized by firms across the industry in which the students train.[5]

A Comparison of Cooperative Occupational Education with Work Experience Education

There is in some schools and among some groups of educators a degree of confusion about the differences between cooperative occupational education and work experience education. In some cases the confusion has been due to the surface resemblance of the plans—with the superficial observers, who, having seen one, believe they have seen the other. In other cases, confusion has arisen from poorly operated cooperative education plans—ones that turned out to be just plain work experience through errors such as inadequate supervision, lack of correlated classroom instruction, and the enrollment of nonqualified students. In still other cases, educators have compounded the problem by a semantics error, that of using the term "work experience" to describe any situation in which a student obtains experience on an actual job.

---

[5]Jobs for the Future, The National Advisory Group for Jobs for the Future's National Youth Apprenticeship Initiative, *Essential Elements of Model Youth Apprenticeship Programs*, Draft, U.S. Department of Education, Washington, D.C., August 8, 1991.

Situation A

The school is concerned with developing outcomes suitable for all students:

1. Occupational exploration and personality development as facets of the education of the whole person.
2. Common learnings—what all workers need to know—as a part of general education.

If these outcomes are to be learned through a job situation, then the method is:

WORK EXPERIENCE EDUCATION

Situation B

The school is concerned with developing outcomes for those who are preparing for the world of work:

1. Learnings that all workers need to know.
2. Knowledges, abilities, and attitudes that only the worker in a given occupational field needs to know.

If these outcomes are to be learned through a job situation and correlated study in school, then the method is:

COOPERATIVE OCCUPATIONAL EDUCATION

In addition, the following differences typically occur between the plans.

Work Experience Education

1. Usually has as its primary goal the improvement of general education, the gaining of exploratory occupational experiences, or the holding of the student in school.
2. Is based on the student's general education needs or need for employment to remain in school.
3. Places student in position as a "producing worker" or as an observer.
4. Often utilizes any available part-time job.
5. Usually relies on the job to provide trainee experiences; the in-school class (not always used) is not usually directly related to the job.
6. Provides job rotation that is usually coincidental rather than part of a planned program.
7. Usually open to any student without restrictions of selection.
8. Utilizes any firm that can provide employment within legal and moral limits.
9. Often makes supervision an "extra" assignment for a person with a full-time teaching load.
10. Usually limits school supervision to job approval.
11. Provides that student may or may not be paid, depending on the type of program operated.

Basic Facts About the Cooperative Plan of Occupational Instruction

1. Has as its primary goal the development of occupational competency.
2. Is based on the student's stated career objectives.
3. Places student in position as a "learning worker."
4. Places the trainee in a job commensurate with his/her ability and career objectives.
5. Provides classroom activities directly related to job activities and to the trainee's occupational goal.
6. Provides the trainee with a variety of job experiences, often involving rotation through different departments of the firm.
7. Selects students who "need the instruction, want it, and can profit from it."
8. Utilizes as training stations those firms that responsibly can and will provide training.
9. Underscores the responsibility of the school by employment of an occupationally competent coordinator with sufficient time to provide consistent program direction.
10. Emphasizes a program of coordinator job visitations and employer conferences to plan learning experiences, evaluate trainee's progress, and aid in solving any problem.
11. Requires that the trainee be employed at the "going" rate of pay for trainees in the occupation.

A number of analyses similar to the preceding one appear in CTE literature. Another description of what cooperative career and technical education is and is not is illustrated as follows: In many cases these analyses appear to create the impression that general work experience is undesirable because it lacks related instruction and employs limited supervision. This point is not valid because general work experience plans can range from poor to excellent, depending on how they are operated. For those in CTE, the important point to be understood is that general work experience can be successful in its general objectives of keeping youth in school, providing general work experience, and giving exploratory experiences; but, it is quite improbable that general work experience plans can develop occupational competence. Developing job competence following career objectives of students is the task of cooperative occupational education.

This discussion has served to underscore an important generalization. Work experience plans and cooperative education plans serve different purposes; in common, each utilizes experience on a job but via a different set of organizational, operational, and instructional practices and for dissimilar purposes. Therefore, each method should be recognized for

---

Basic Facts About the Cooperative Plan of Occupational Instruction

It Is	It Is *Not*
1. An attempt to acquaint trainees with the realities of the world of work and to help them adjust and make the transition from school to work.	1. A way to force students who are hard to handle out of the general classroom.
2. A planned career development program, designed at a minimum to produce entry-level competence.	2. A way for students to make money.
3. An opportunity for employers to assist in training—for themselves or for the occupational world in general.	3. A way to lighten the load of the classroom instructors by putting the education burden on the community.
4. A method whereby instructors of in-school occupational courses can get feedback from potential employers of the trainees.	4. A way to cut job rates for full-time workers.
5. A logical approach for a pre-employment program designed to break the poverty cycle of some youth.	5. A panacea for the student's shortcomings, personal and/or academic, or to overcome a history of failure.
6. A means of providing realistic opportunities to apply and test skills and knowledges learned in school.	6. A "gimmick" to be discarded after a short trial; short-term as well as long-term commitment of resources is necessary.

what it can do most effectively and efficiently. Neither per se is "good" or "poor"; the paramount point is that school personnel should understand the differences between the plans, choose the one that will best meet their objectives, and then provide the resources necessary for the success of the method chosen.

A Summary View

In the past, schools were asked to communicate the basics of math, English, history and other social studies courses, a little art and music, some physical education, perhaps several foreign languages, and, on occasion, home economics and industrial arts. Schools were then asked to assume the responsibility for some additional "teachings" to include sex education and driver education. Schools are now asked to bridge the gap between school and work. The career education movement has more clearly identified the school's role in helping individuals select satisfying careers.

Work and careers cannot, and should not, be separated from other aspects of individuals' lives. On the contrary, the evidence is that more and more people view work as an integral part of their total lifestyle, not as a separate activity. Increasingly, education and other institutions are collaborating and becoming more involved in expanding the educational environment beyond the school into the "real" world of work. These efforts all recognize, to some degree, the need for work experiences as a necessary part of the total school curriculum. Capitalizing upon the learning potential of experiences derived from on-the-job applications has produced many different types of plans that use the work environment in the curriculum.

In this chapter, various plans that use the work environment have been discussed, primarily in terms of the differences and similarities existing between these plans. In addition, the characteristics of cooperative occupational education—what it is and what it is not and how it differs from other work education plans—have been identified. Cooperative education is the one on-the-job learning plan that can directly assist career and technical education in accomplishing its main purpose—to provide future workers and present workers with the skills, knowledges, and attitudes that are realistic in light of future and present job market requirements.

Questions and Activities

1. Do you agree or disagree with this statement: "Some form of experience or observation of the world of work should be included in the curriculum of every senior high school"? Why or why not?

2. Under what circumstances can a school or an educational institution at any level completely turn over the cooperative occupational education of students to an employer?

3. Make arrangements to study some type of school program that uses the work environment as a teaching situation to accomplish educational objectives. Analyze the program studied. What were the program's objectives? How were these objectives being achieved?

4. Which of the three types of work-study plans described is least appropriate as a CTE plan? Why? Which one comes closest to approximating the cooperative plan? Which elements does it lack?

5. What are the basic differences between a high school and a post-secondary cooperative plan of instruction and a collegiate internship at the baccalaureate level?

6. Make a few brief basic statements about the socio-economic situation in which the schools in your area now find themselves. With this as a premise, what is your prediction about the use of work experience and observation as a means of providing education in your area?

7. Give specific illustrations of why it is that "general work experience plans are unlikely to develop CTE competence except on a coincidental basis."

8. Define "experiential learning." Why do you believe the term has become so popular in educational circles?

9. Prepare a one-page report that clearly distinguishes between a cooperative career and technical education plan and a general work experience plan.

10. Explain how apprenticeship training can be integrated into a secondary education program.

For references pertinent to the subject matter of this chapter, see Reading Resources.

2

The Development of Human Resources

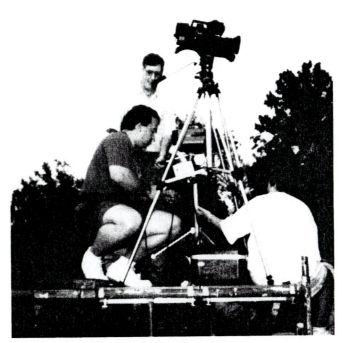

A cooperative student who is receiving on-the-job training in a photo shop gets hands-on experience on location.

KEY CONCEPT: Career and technical education, cooperative education in particular, plays an important role in an individual's own career development and in all human resource development.

GOALS

After successfully completing the study of this chapter, answering the questions, and carrying out the activities, the student should be able to:

- Explain why career and technical education program needs should be projected through labor market forecasting.

- Identify and explain at least four over-all labor force growth patterns having major impact on the development of occupational programs.

- Explain the crucial role work plays in the lives of adults in our society.

- Identify and explain three major occupational competencies needed for employment.

KEY TERMS

"assured jobs" program
at-risk population
career development competency
changing occupation
clusteral concept

demographic trends
job adjustment competency
job intelligence competency
occupational competencies
new occupation

It has long been a tenet of career and technical education (CTE) that training and educational programs have to be geared to the demand of the labor market. The occupations for which training is offered are those in which the trainee can find employment in the local labor market. Further, the demands of the job in effect dictate the content of the instruction. When this principle is violated, career-technical educators are accorded severe criticism for outmoded curricula.

This basic tenet is still one of primary importance to those interested in occupational and career development education. But, the concept in recent years has been broadened considerably beyond what some saw as narrowly conceived job training programs, focusing only on employer needs and not recognizing individual worker needs. The concept of identifying job market needs reaches far beyond the mere compilation of job availability and entry-level requirements to encompass many facets of the world of work, including assisting employees in successful career development. One facet is the realization that occupational and career education is part of a national policy of human resource development. Another is recognition that situations in the labor market, such as the changing nature of jobs, women's roles in the workforce, fair employment practices, and employment upgrading, are vital to program planning. Another critical facet is the recognition that a worker's positive self-concept toward his/her role in the world of work is a key to productivity.

The Labor Market as Part of a Strategy

One of the initial strategies in local program planning, whenever the school uses the work environment in career and technical education programs, is to establish goals that reflect the changing nature of the labor market. In a sense, this is projecting educational needs through labor market forecasting. In a rapidly changing economy, education cannot live in isolation from the realities of labor market demands. But, forecasting labor market demands is no easy task. There are some who believe long-term (5- to 10-year) forecasting is needed if career and technical education is to prepare individuals for tomorrow's jobs. No matter what schemes are used for gathering, processing, and drawing conclusions from the data collected, most career-technical educators would agree that reliable forecasting of labor market needs is critical to program planning for effective delivery of career and technical education.

GROWTH CHANGES IN THE LABOR FORCE PATTERNS

The size and characteristics of the labor force determine the number and types of people competing for jobs. In addition, the size of the labor force affects the quantity of goods and services that can be produced. Growth, alterations in the age structure, and rising educational levels are among the labor force changes that will affect employment opportunities during the next few years.

It is projected that the labor force will grow, but at a slower rate. The civilian labor force—people with jobs and people looking for jobs—is projected to be about 155 million by the year 2008—an increase of about 12 percent from the 1998 level of 138 million.

The chief cause of labor force growth will be the continued rise in the number of women who seek jobs. Women will account for about 63 percent of the new entrants in the labor force. Labor force growth will be slower than in the 1960's and 1970's. However, for the first time in a quarter of a century, the youth labor force (ages 16 to 24) is projected to increase faster than the total labor force. Three other key factors affecting the U.S. labor force are the exporting of U.S. manufacturing jobs, the restructuring and downsizing of U.S. businesses, and the rapid growth of migrant populations, with the Hispanic labor force increasing four times faster than the overall labor force.

PATTERNS OF CHANGE AMONG OCCUPATIONAL CLUSTERS

The over-all labor force growth patterns are one important element counselor-educators should consider when they are setting goals for career and technical education. But, more important, are the differential patterns of growth among occupational clusters and in specific job titles within clusters. According to projections, the fastest growing group of occupations will be in the most skilled occupations—executive, professional, and technical. It is predicted that 10 million more jobs will become available in the skilled category within the next decade, as compared to only about 1 million new jobs in the less-skilled and laborer categories.

The completion of high school education has become standard for U.S. workers. Employers are seeking people with higher levels of education because job performance requirements are more complex, thus necessitating higher levels of skill. The U.S. Bureau of Labor Statistics *Occupational Outlook Handbook,* which indicates future employment opportunities in several hundred important occupations, describes the current trend in job performance requirements. A few illustrations are:

1. Farmers have to know much more about scientific and technical phases of their work, which includes fertilization, selection of feed and breeding stock, and identification of animal and plant diseases.

2. Some workers in the craft occupations have had to shift from participation in the production process per se to maintenance and repair work, which requires greater diagnostic ability and more all-around skills.

3. Accountants must understand not only increasingly complex tax laws and accounting methods but also the potentialities and application of computers.

4. Economists must have a much stronger foundation in mathematics and quantitative theory.

The United States is experiencing a dramatic change in the work environment. Technology is changing faster than today's worker can keep pace with. This demand for advanced-level skills will dictate the content of instruction in career and technical education. In the future an estimated 80 percent of all job openings will require a high school diploma, and many more will require additional post-secondary training. Thus, there is a need for educational programs that will provide for lifetime learning and success in the technological world of the future. Employees will need not only advanced technological skills but also the ability to communicate effectively, solve problems efficiently, and work cooperatively in teams. Furthermore, all jobs will require literacy, dependability, responsibility, and good thinking skills.

AGE GROUPS IN THE LABOR FORCE

Between the years 1998 and 2008, the number of young people ages 16 to 24 in the workforce is projected to increase by 3 million. Still employers may have increasing difficulty in finding young workers. A slower growth number of young people as workers could be particularly important to the Armed Forces—the single largest employer of men in this age group.

The number of people ages 25 to 44 in the labor force is expected to decline slightly, from about 51 percent of the labor force in 1998 to about 44 percent in 2008. The number ages 45 to 54 is projected to increase by approximately 3 percent. The number age 55 and over is projected to increase slightly. By 2008, more than half the labor force will consist of persons over the age of 40.[1]

---

[1]These data were taken from the Monthly Labor Review, Vol. 122, No. 11, November 1999.

The Changing Nature of Occupations

The changing nature of the occupational world, the world of work, is in reality an interaction of many forces. Among the major aspects of the forces for change are: (1) changing the makeup of the labor force, (2) changing employment demands in occupational areas, and (3) changing the requirements of worker competence. Educators must analyze carefully the changing occupational "mix" if curricula and counseling are to be responsive to actual job market conditions in the local service area. Career and technical education programs have often been slow in matching people and jobs. Most of the specific programs developed for outreach purposes have been designed by the affiliated state offices of the U.S. Employment Service. As federal legislation for career and technical education has mandated more aggressive recruiting of *at-risk populations,* outreach programs,[2] which recruit and place certain segments of society, such as minority populations, women, persons who are disabled, economically and/or academically disadvantaged, and persons who are physically and/or mentally disabled, are becoming more popular.

THE PEOPLE WHO WORK

In the next decade, the labor force will grow slower than the population as a whole, mainly because of slower population growth. Of significance to the educator is that the growth in the labor force is only the net increase over this period (2 percent over the present 40 percent). The number of new workers entering the labor force far exceeds this figure. The difference between the total entering or re-entering and the new growth of the labor force represents persons absorbed into the labor force as replacements for those workers who, during the same period, have died, retired, or left for some other reason, such as disability, marriage, or child rearing.

WOMEN IN THE LABOR FORCE

Women continue to play a prominent role in the labor market. Not only have they increased in number (200 percent increase since World War II), but they are also entering occupations at one time considered to be the man's province. The increasing population of women who work will continue to be a major factor contributing to the anticipated increase in the labor force. Through the year 2008, women will account for about 48 percent of the labor force. The highest proportion of working women is found in the 25 to 34 age group, with approximately 50 percent of all women in that age group working.

CHANGES IN THE AGES OF THOSE IN THE LABOR FORCE

The labor market is beginning to be composed of more "older" people. In the next 10 years, our society will be made up of older people. At the same time that there is a decline in the teenage population, there will be a definite increase in the number of adults in their 40's and 50's. These predictions are based on national statistics, and local variations will exist. For example, in 1990, the average worker's age was 36.6. The U.S. Department of Labor predicts that by 2008 the average age will have increased to 41 years. The 45 to 54 age group is expected to have the greatest increase in the future.

The next 10 years will see a "senior citizen boom." The over-55 age group is expected to grow dramatically. Those over 55 years of age are expected to increase by 5 percent.

During this period, there will be a substantial difference in the composition of the younger population. The teenage population of minority groups will

---

[2]Outreach programs purposely seek out those who are underemployed, physically and/or mentally disabled, and/or economically and/or academically disadvantaged, and who can benefit from occupational training or retraining.

have a substantial increase and will represent a larger percentage of all younger people. An estimated 29 percent of the net growth in the workforce during the next 10 years will be in minority groups.

The national unemployment rate for teenagers ages 16 to 19 in 1998 was 16.2 percent, while the overall jobless rate was 4.5 percent.

The unemployment rate for black teenagers was 35 percent in 1998. Some authorities believe, since many black youths are not counted, the actual rate may be even higher than reported. The unemployment rate for Hispanics and Native Americans is not believed to be any better. Because the unemployment rate of all young people is so much higher than for other age groups, changes in the composition of the teenage population have become an even more serious problem.

OCCUPATIONS IN THE POST-INDUSTRIAL ECONOMY

This transformation is characterized by worldwide free trade, downsizing and restructuring of companies, an Asian consumer boom, the spread of democracy and free enterprise, a relative abundance of natural resources, and the containment of inflation and interest rates.

The post-industrial era has seen manufacturing and agriculture decline in relative importance and the service economy become the key to U.S. economic growth. Today, agriculture employs only about 3 percent of the total workforce, while only one of every six U.S. jobs is in manufacturing. Of the 20.3 million new jobs created by the year 2008, approximately 12 million will be in service-producing industries. These industries contain 16 of the 20 fastest growing jobs. More than 2 million jobs have been lost in the manufacturing section since 1980.

The post-industrial economy has created new challenges for career and technical education. Lifelong learning is essential. A significant number of today's workers are employed in knowledge-related

jobs that require the most powerful and modern information technologies.

The Importance of a Job to an Individual's Self-Identity

Most individuals would agree that being employed assumes a crucial role in the lives of people in our society. Work occupies a large part of an individual's waking hours in actual hours spent at the work station and in many additional hours spent in work-related problems and discussions and in commuting. Furthermore, not only does work provide individuals and their families with a living but it also determines and influences a person's social life, attitudes, and lifestyle. Even the choice of friends and leisure-time activities, as well as residence, is conditioned to some degree by the job. After reaching adulthood, almost every individual is involved in earning a living. Almost half of those individuals in the labor market are women. Many senior citizens live largely on the results of earlier work years. According to the U.S. Department of Labor, the average person works for at least 50 years. Allowing for child bearing, the average young woman can look forward to a work span of about 25 years, even though she may leave and reenter the labor market several times.

Since work is a primary influence in adult life, being able to have the optimum occupational situation possible is critical. Therefore, the informed and free selection of an occupation and the proper preparation for it should merit the most careful planning. Educational institutions must give more than lip service to career development and have as a major priority the preparation of students for life. Appropriate resources must be allocated for this priority. No longer can the educational system disregard the fact that probably half or more of our citizens fall into an occupation either by circumstance or by the work available, rather than through their conscious, deliberate,

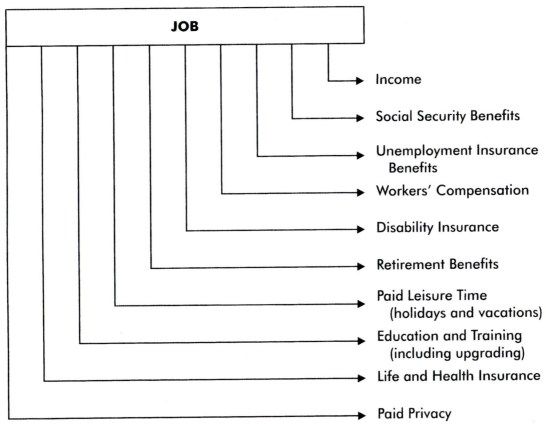

FIGURE 2–1. (Adapted from Daniel Kruger's position paper, "Economic Benefits Accruing Through the Job")

informed free choice related to their own self-concept and preferred lifestyle.

This crucial relationship of the job to adult life is closely related to the economic benefits of employment. In a position paper, "Economic Benefits Accruing Through the Job," Kruger outlined those shown in Figure 2–1.

He underscores his theme by saying:

> If a job and work is so important, it seems that a primary function of education must be to provide meaningful, significant experiences designed to equip the individual for work in which he will be successful, and which will be properly challenging to him in accordance with his aptitudes, interests, and total personality. For many students high school may be the only formal preparation for a life-time of work and living. Preparation for work can be, must be, and is, done within the schools.[3]

The job provides even more than economic benefits. The job is often the focus for the choice of friends, residence, leisure-time activities, and even socio-economic value systems. Being employed is crucial to how persons feel about themselves. The difference between feeling worthless and having a positive self-concept is more often than not closely related to the job. Getting the right job can make a difference between a life of success and happiness and a life of disappointment and failure. The right job is one that is appropriate to the individual. Not only does it provide income but it can also enhance or

[3]Daniel Kruger, "Economic Benefits Accruing Through the Job," a position paper, circa 1970.

otherwise influence the aesthetic, cultural, and ethical values of the individual.

Work is part of the economic process, but it is also more. Work is a social arrangement. It meets—or has the potential to meet—certain *non-economic* social and psychological needs, in addition to helping to produce goods, services, and income. Work is not merely a means to an end (money and goods); it is a worthy activity and an end in itself. Work—especially in an advanced, affluent society—can provide outlets for creative expression and self-development. It can contribute to building a satisfying identity and self-image for the individual worker. Because work is so important to the individual, the employment outlook is vital for young people and adults.

Occupational Awareness as a Primary Need

Career development education takes into account the crystallization of a career objective (goal). Even a tentative career objective will help promote relevance to "schooling"—the courses and experiences provided by the school. The dropout rate is staggering—3,468,000 high school students drop out each year. Frequently, these dropouts lack employable skills. The welfare and unemployment costs for students who have dropped out are estimated to be in the billions.

A person's career choices will be mere fantasies if provision has not been made for self-inventory and assimilation of relevant and realistic occupational information. These efforts must begin with career-technical educators working with guidance counselors from the elementary grades through adult education. For example, career-technical educators can offer their service by appearing before groups of students or adults to provide information about occupational choices. Career technical educators can be involved in curriculum building that integrates exploratory courses and experiences, especially in

grades 6 to 9. Career-technical educators from the occupational fields of agriculture, family and consumer sciences, business, marketing, industrial, and health occupations, and the like, can arrange team teaching situations in courses such as "Introduction to Career and Technical Education" and "Survey of Occupations" to expose students to alternative occupational choices. Resource persons from business and industry can be utilized, and on-the-job visitations can be made. The various U.S. Labor Department publications, such as the *Occupational Outlook Handbook* and the *Occupational Outlook Quarterly*, are good resources to consult.

The Cluster Concept of Employment Preparation

Students who have completed a CTE program planned around a cluster of occupations probably will be better prepared for technological change. Changes on the job more likely will be expected rather than feared. In the *cluster concept,* instructional programs are developed to provide the skills, knowledges, and understandings common to several related occupations. The content of the courses and the learning experiences include a set of core competencies common to several jobs in an occupational cluster.

Within such a program, the students are prepared to enter a specific job title (or several) within a cluster of occupations. Critics of the cluster concept say such instructional programs sacrifice in-depth competence in a job title. But, those who support the concept argue that graduates of CTE programs designed for a cluster of occupations retain an element of final choice because the program avoids the danger of closure if any one of the specific occupations within the cluster becomes obsolete or eliminated through technological advancement. Supporters also believe such programs prepare students to adapt more easily as job requirements change.

Other Factors Affecting Occupational Development

Many other factors in addition to labor market demands affect occupational choices and career development plans. A favorable job market is not the only reason students enroll in occupational programs. Factors such as changing educational demands, employment changes, fair employment practices, and occupational upgrading determine career choices. These factors are part of the *strategy* of designing an educational program.

CHANGING EDUCATIONAL DEMANDS

Many years ago, jobs were largely dependent on physical activities. Today, jobs are much more cognitive in nature. In other words, workers must depend more and more on their abilities to read instructions, to choose between alternative actions on the job, to make decisions based upon data provided by computers. Technology is changing rapidly, and the economic system is so sensitive that there are often surpluses of jobs in some occupations. And, there is the danger of overeducating individuals—it has become so popular for almost everyone to go to college that many individuals overlooked opportunities in semi-skilled and skilled jobs through secondary programs, post-secondary programs, and apprenticeships.

REASONS FOR EMPLOYMENT CHANGE

Some young people prepare for a first job as a stop-gap earning capacity while they get additional education to prepare themselves for a more advanced occupation, or even for very different employment. Sometimes they try out jobs in search of the one they really want. Having the opportunity to explore various jobs is a very important element of education.

FAIR EMPLOYMENT PRACTICES

In recent years, the employment outlook for minority groups has improved. Unfair employment practices have been discouraged, as illustrated by the "Equal Employment Opportunity Commission Guidelines on Employee Selection Procedures," set forth by the Equal Employment Opportunity Commission in August of 1970.[4]

These guidelines define discrimination in employment practices through employment testing procedures. They also are designed to serve as a workable set of standards for schools, employers, unions, and employment agencies in determining whether their selection procedures conform to the obligations contained in Title VII of the Civil Rights Act. Fair employment practices.

OCCUPATIONAL UPGRADING— A MULTI-ETHNIC APPROACH TO EMPLOYMENT

There is an increased awareness among educators and employers regarding their responsibilities for the training and employment of minority groups. Education must assume responsibility for fair, accurate, and balanced treatment of minority groups, not only through education and placement but also—just as important—in the development of realistic job goals. Education and training lay the groundwork for occupational upgrading.

The employment problems of minority groups are often complicated because individuals may lack the necessary employment skills and experiences. The situation is critical for minority youth, especially those living in inner-city areas and those in low-income rural areas. A large number of non-whites under the age of 25 who seek work cannot find it. The transition from school to work is very difficult for

[4]"Equal Employment Opportunity Commission Guidelines on Employee Selection Procedures," *Federal Register*, Vol. 35, No. 149, August 1, 1970, pp. 12333–12336.

many students in minority groups. These students need work experiences and the opportunity to develop employable skills.

THREE MAJOR OCCUPATIONAL COMPETENCIES ARE NEEDED

Prospective employees need to learn job application skills, knowledges, and attitudes in order to be employed by prospective employers. Employers ask that employees come to them with enough specific education and training to be employable, but with a broad enough background to be able to move horizontally in a cluster of related occupations as well as vertically for advancement or promotion. It is important for career and technical education students to recognize the necessity for developing a variety of *occupational competencies* desired by employers. These include *job intelligence competency, job adjustment competency,* and *career development competency.*

Job intelligence competency.—Job intelligence means the possession of skills, knowledges, and attitudes necessary to perform the duties involved in a given occupation. For example, the medical laboratory technician needs not only the usual general education but also the background in certain sciences that are concerned with bacteriological, biological, and chemical tests to provide data for use in diagnosis and treatment. Skill in using microscopes, micrometers, and other instruments is needed. Interest in learning about diseases and an inquiry into such matters are important.

Job adjustment competency.—Job adjustment refers to an employee's ability to develop successfully the interpersonal relationships needed in any new employment situation. These human relations include getting along with co-workers, getting along with supervisors, and knowing how to benefit from supervision.

Career development competency.—Career development includes a realization of the need for experience and further study in order to accomplish an identified career goal. An employee needs to learn to test personal aptitudes, interests, and competencies against the experiences on the job. The employee must then learn to modify the education, training, and goals established so that they remain realistic to his/her personal needs.

Success on the job and advancement toward a career goal are achieved as all three job competencies are developed. Cooperative occupational education programs maximize the opportunities for graduates to have developed all three competencies prior to full-time employment.

Broad Implications for Education

Predicting future labor force needs is no easy task. Some aspects of forecasting are more reliable than others. For example, there are currently severe labor shortages of (1) highly skilled manual workers—computer and office machine repairs, automotive body repairs, and general maintenance mechanics, and (2) low-status service and industry workers. These workers probably will continue to be in short supply because there will be fewer young people entering the labor market and because many do not want the low-end (status) jobs. In addition, the highly skilled jobs require highly specialized training and the retraining of adult workers for these jobs is not widely available. Therefore, forecasting future labor shortages in these areas is relatively simple.

Since the employment prospects for individual occupations differ from year to year, it is important to check the *Occupational Outlook Handbook* for each occupation in a current source. More detailed statistics on employment, replacement needs, and educational and training program completions are presented in *Occupational Projections and Training Data,*

Bureau of Labor Statistics Bulletin 2501. A Statistical and Research Supplement to the *Occupational Outlook Handbook*, Occupational Projection and Training Data, is published each year by the U.S. Department of Labor, through the Superintendent of Public Documents, Washington, D.C. The Web site for the Bureau of Labor Statistics is http:// stats.bls.gov/blshome.htm.

On the other hand, forecasting the number of automobile salespersons needed in the next five years is very difficult because their employment depends not only upon population growth, which affects both the supply of workers and the demand for automobiles, but also upon changes in buyer preferences; changes in the production of automobiles; changes in the level of over-all economic activity, employment, and personal income; and other developments outside the automobile industry that are very difficult, if not almost impossible, to foresee. To design effective labor force policies requires a strong pool of readily available human resource knowledge and information. The existing knowledge and information base is inadequate and does not accurately describe the "real" world of work. An improved knowledge–information base is equally important for forecasting labor needs, creating state and local government agency activities, and developing educational programs and policies.

SPECIFIC IMPLICATIONS FOR EDUCATION AND TRAINING

Of particular importance to those concerned with educating youth for employment is an understanding of the dynamic changes going on in the economy. New production methods, new products, and new lifestyles are continually causing changes in the kinds of jobs available to workers. This process of change calls for a broad foundation of training and education so that, if a shift in plans becomes necessary, a transition from one occupation or field of work to another may be supplemented by training for entry into the

initial job; otherwise, unemployment will be the immediate result.

Many of the fast-growing occupations require specialized technical training. It becomes obvious that a young worker's chances for a steady, well-paying job in many areas of the economy will be substantially less if the student does not possess employable skills. Today, most employers regard a high school diploma as the minimum "certificate of employability." For many fast-growing jobs— especially in the professional, sub-professional, and technical areas—more than a secondary school education is required.

Young people need specialized education to fill the many jobs as technicians, supervisors, and mid-management personnel. For example, with automated, computerized equipment being used on a wider scale in offices, banks, insurance companies, and government operations, the skill requirements for clerical and other office jobs have risen. The demand of employers for better trained personnel to operate complicated and expensive machinery is a reality. Advanced developments in machine design, new materials, and complex equipment have made increased skill and technical information necessary for those who design and produce products, sell and demonstrate them, and repair and service them. At the same time, the unskilled jobs that once absorbed many untrained people have narrowed. Young people who have acquired a skill and a good basic education have a better chance at interesting work, good wages, steady employment, and greater satisfaction with life in general.

WHAT IS AN EMERGING OCCUPATION?

Some labor market economists propose that occupational patterns do not change overnight; rather this takes about 12 years. An occupation that is sometimes labeled as being new, such as "geriatric social worker," is often an existing occupation applied in a specific setting. Although the terms "new occupation" and "changing occupation" are frequently used interchangeably, career-technical educators need to be aware of the difference, even if subtle, so as to plan educational programs responsive to labor market needs. A *new occupation* is one that has come into existence in a particular geographic area in the last 10 years, has employment levels large enough to measure, and requires tasks, skills, and duties not included in any currently existing occupation. On the other hand, a *changing occupation* is an existing occupation that has experienced a change in duties, skills, or tasks, significant enough to require training beyond a short demonstration, but not significant enough to reclassify into another existing occupation or to create a new occupation.

About 20 million new jobs (above the 1998 rate) will be created by 2008, compared to 3 million in the 1970's. Paramount in career and technical education program development and curriculum refinement are "changing" rather than "new" occupations. The greatest impact on educational program planning is the effect of technology and/or social changes on traditional occupations. For example, automotive technicians must be able to use computerized diagnostic equipment to perform traditional tasks. Social changes have provided the basis for occupations such as "child advocate" and "geriatric assistant." Restructuring duties within an established occupational area frequently creates new occupations—for example, "paralegal." Career-technical educators must be aware of changing occupations and be prepared to make accommodations in their curriculum responsive to labor market needs.

IMPLICATIONS FOR CAREER AND TECHNICAL EDUCATION

The implications of these *demographic,* economic, and technological *trends* are significant for CTE. Career and technical education programs can assist in lowering unemployment rates by training young people and older adults and by retraining the unemployed for skilled, service, and technical occupations. If national policy requires an increase in labor productivity, vocational education can assist in producing the desired increase.

No worker has any real assurance now that mastery of an occupation, when obtained, will last his/her lifetime. Although jobs may change, a worker who has mastered the skills of a trade or an occupation and who has kept abreast of new techniques and developments can reasonably expect to continue in the trade or occupation throughout his/her working life.

Pre-employment training for youth must provide a solid occupational foundation. In addition, the future employee should be aware of the continued need for self-development and self-improvement in order to be well-informed and up-to-date in the occupation. Since more and more workers will need a program of lifelong learning, continued educational opportunities will need to be provided so that workers will be able to cope with occupational change. Career-technical educators must train broadly for career patterns, for clusters of occupations, and for lifelong employment opportunities.

A Summary View

It seems appropriate to conclude this discussion on the development of human resources, as related to the interests of career-technical, by summarizing the following challenges and opportunities perceived by those in positions establishing human resource policy.

1. Women will continue to enter and re-enter the labor force in increasing numbers, especially as fewer young people enter the labor market. Women will be able to participate more fully in the economy if broader employment opportunities, retraining opportunities, part-time employment, child day-care centers, and flexible work schedules are available to them.

2. Although the total number of children born in the United States has risen, the birth rate of minority groups is significantly higher than for the general

population. Minorities will soon be the fastest growing group in the new labor force. Educators will need to help these individuals acquire the job skills and work attitudes needed for successful careers.

3. Aggressive actions must be taken to reduce the unemployment rate of individuals who are economically and/or academically disadvantaged, particularly those individuals who possess minimum skills, low income, and limited job experience. The unemployment rate of black adults has been double the national average for white adults in the past few decades.

4. Education, career and technical education in particular, must begin to assume some degree of responsibility in creating new jobs. Schools, primarily through what was then called vocational education programs, were previously assigned the task of preparing youth with employable skills—their mission was to supply the labor market demands. Now, CTE is being asked to assume responsibility in job development, thereby creating jobs.

5. Although the number of young adult workers entering the labor force will decrease in the next few decades, many young people will still experience some difficulty in the transition from school to work. Schools must infuse into their curricula more educational programs that use the work environment to provide experiences, thus bridging the gap between school and work. Young people who are economically and/or academically disadvantaged, who lack competence and skill, may never make an effective link to the world of work if they are not helped. Educational programs that will assist these individuals to become responsible citizens must be developed. Career-technical education can assist young adults to begin satisfying and productive careers by providing these individuals with opportunities and challenges in line with their talents and energy.

6. Prospects for the permanent dislocation of many workers are expected to increase. Since 1980, millions of people have lost their jobs through plant closings, relocations, or technological innovations.

Thus, the need exists to begin to develop an *"assured jobs" program.* An assured jobs program would provide work, on a voluntary basis, for those who could meet an income test—such as that required for JTPA assistance—and who had received unemployment compensation for more than 15 weeks.

7. Fewer teenagers will be joining the labor force—their employment prospects can be improved substantially, but special efforts will be needed to provide opportunities for the growing numbers of black youth.

8. Occupation and industry growth trends will require continued adjustments in career decisions—currently the need to adjust cannot be readily anticipated by students and workers, educators and employers, by using the information now available. But, a certain degree of forecasting, even if based on limited data, is essential.

9. Part-time work opportunities are reaching major proportions, and more part-time work will be offered by employers who are wanting to avoid paying for employee benefits, including vacations. Students, women with children, and older workers play a key role in the labor force in this area. It is predicted that the average work week will be reduced, caused by the shift towards part-time jobs, not by better working conditions.

10. By 2008, the number of people age 55 and over in the labor force is projected to rise slightly. But, because so many of these individuals will be living on limited retirement incomes, they will be interested in jobs that will supplement their income, and some will even assume second careers. Career and technical education will need to provide retraining programs to equip these older persons with the skills needed to re-enter the labor force.

All educators who make a contribution to occupational and career development education must continue existing occupational programs and must innovate new approaches that will assist in meeting these human resource challenges and opportunities.

Cooperative occupational education levels are one approach to this development of human resources.

Questions and Activities

1. Study the most recent issue of the *Occupational Outlook Handbook* and the appropriate issues of the *Occupational Outlook Quarterly* to gather information on five occupations for which your service specialty trains. Write a five-page paper summarizing significant information on changes in these occupations.

2. Secure a copy of any human resource report that shows occupational trends and opportunities statewide. Evaluate the material for completeness. Which, if any, of the material could be used directly with students in occupational information programs?

3. Visit your local city or county office of the state employment service. Determine what sources of human resource data are available to schools.

4. A professor of philosophy of education who looked closely at career and technical education charged, in effect, that career-technical educators had become slaves to the business–industrial establishment, producing only what it wanted. Do you think there is any truth to this charge? Explain your answer.

5. Do you believe information regarding labor force age changes is important to consider when career and technical education programs are being planned? Why or why not?

6. Discuss at least two factors in addition to labor market demands that affect the career choices students make.

7. Assume you are the director of career and technical education for a large urban school district (grades K–12). You have been asked to report to the school's board of education your perceptions about "what career-technical cooperative and work experience education should be like in this school district during the next 10 years." Write your report. It should be no longer than four pages.

8. You have been asked by the school's career counselor to draft a booklet that would be distributed to students about the importance of having a job. Using the information contained in this chapter, prepare an outline for the proposed booklet.

9. A parent has asked you, the cooperative coordinator, to meet with the school's task force responsible for attempting to address the concerns of students and parents regarding employment opportunities for high school graduates of CTE programs. What kind of information would you provide the task force?

10. Explain why you believe it is important for career and technical education programs to help students develop the three major occupational competencies.

For references pertinent to the subject matter of this chapter, see Reading Resources.

3

Public Policy Goals and Institutional Roles

Job supervisors discuss the importance of on-the-job training in career and technical education.

KEY CONCEPT The institution's role in career and technical education must reflect public policy, and the primary responsibility of career and technical education is to meet public needs.

GOALS

After successfully completing the study of this chapter, answering the questions, and carrying out the activities, the student should be able to:

- Trace the development of public policy expressed through federal vocational/career and technical legislation.

- Explain the importance of public policy in the planning, organization, and administration of local career and technical education programs.

- Identify and describe the key concepts of the career education thrust.

- Define the term "special needs populations" and then describe each of the major segments of the population typically included as special needs groups.

- Explain the reasons for the adult training and development movement in career and technical education.

KEY TERMS

area vocational and technical school

bilingual education

Carl D. Perkins Vocational Education Act of 1984

Carl D. Perkins Vocational and Applied Technology Education Act of 1990

Carl D. Perkins Vocational and Technical Education Act of 1998

Career Education Act

Comprehensive Employment and Training Act (CETA) of 1973

disabled

disadvantaged

Federal Revenue Act

George–Barden Act

George–Deen Act

Job Training Partnership Act (JTPA)

Manpower Development and Training Act (MDTA) of 1962

Morrill Act

National Defense Education Act

private industry council (PIC)

service delivery areas (SDA's)

sex equity

Smith–Hughes Act

special needs populations

state councils

state plans

tech-prep education

vocational education

Vocational Education Act of 1963

Vocational Education Amendments of 1968

Vocational Education Act of 1976

Vocational Rehabilitation Act

Vocational education (now called career and technical education) was created in response to a social need for a trained labor force and as a function and a responsibility of public education. Career and technical education has been, and must continue to be, even more responsive to the social, technological, and economic changes in our contemporary society. As our society changes, career and technical education must adjust accordingly. But, throughout these adjustments, its concern must be directed primarily to the needs of the people—it must be people-oriented.

Career and technical education is rooted in our society based on a free enterprise economy. In this system the demand for goods and services creates the jobs that people hold. From their jobs they gain the earning power to buy the goods and services that they individually require. Career and technical education must serve at least two needs in a changing labor market:

1. Develop individuals' skills so as to enhance their employability not only after they leave school but also throughout their lives.

2. Provide opportunities for individuals to improve their employment status and earnings and to help them adapt to a changing economic environment.[1]

In our society the name of the game is change. Human resource needs have changed radically in some CTE areas since (1) the advent of the computer—in particular, (2) the influx of knowledgeable, professionally trained managers into administration, and (3) the utilization of new processes brought by technology. Products and services must be produced, sold, and distributed. These, in turn, necessitate the adaptation and readaptation of workers, supervisors, and managers to new tools, new appliances, new

operations, and new methods. This adaptation can be made only as new skills and new technical knowledge can be transmitted to great numbers of producers and distributors in a wide variety of occupations, scattered over large geographic areas. Haphazard methods of "pick-up" training have proven inadequate in a modern industrial society. Individuals move to other areas. They react to changes in employment opportunities. They leave the labor market; then later, they re-enter it with outdated skills. All these factors point out the need for occupational preparation and at times retraining. Public policy and legislation recognize that an organized system of career and technical education results in utilizing human effort more effectively, in conserving natural resources and energy, and in increasing the wealth of the nation.

Legislation and Social Strategy

Public policy, as expressed in legislation, is one of the basic considerations in the strategy of state and local decision-making. Most secondary and post-secondary districts, including private schools, engage in program planning, which, while controlled locally, must be sensitive to public policy as expressed through federal and state legislation. This is particularly true with programs that provide the reimbursements to local districts for approved expenditures. There is in public policy the promise of financial reward for approved programs; there is also the approbation for offering what is considered by those who made the inputs to legislation to be sound and needed educational practice.

THE LOCAL ADMINISTRATORS AND PUBLIC POLICY

Those who plan, organize, and administer a program involving the work environment as part of the instructional plan need to understand legislation and public policy. First, the program, if it is to be reim-

[1]*Notes and Working Papers Concerning the Administration of Programs Authorized Under the Vocational Education Act of 1963*, Public Law 88–210, As Amended. Washington, D.C.: U.S. Government Printing Office, 1968.

bursed with state and/or federal funds, must carry out the intent of the law. Second, public policy often shifts and provides for new and innovative approaches; the local administrator will want to put these in perspective to the on-going programs. And, most important, the person who contacts the public—advisory committees, employers, and taxpayers—is often asked about the way in which the educational program is connected with state and federal funding. The director or coordinator must, and school board members as well as the faculty should, interpret legislation and public policy for the school's administrators.

THE SIGNIFICANCE OF FEDERAL SPENDING FOR VOCATIONAL/CAREER AND TECHNICAL EDUCATION

Authorities responsible for the allocation of tax monies must be convinced that local and state appropriations for career and technical education will be an efficient and effective way of expanding tax dollars. Therefore, it is imperative that teacher-coordinators understand existing career-technical legislation and the developments leading to passage of legislation.

Federal legislation has stimulated the spending of state and local funds for CTE. Most states allocate monies for CTE far in excess of what comes from the federal government. Local contributions for career and technical education are important to match federal and state monies.

Institutional Development Related To Legislative Intent

The establishment of a public school vocational education system at the beginning of the 20th century occurred as the result of the unprecedented expansion of industry, business, and agriculture between 1860 and 1910. The population increased from 31,433,321 in 1860 to 91,972,266 in 1910. Approximately 27,000,000 immigrants entered the United States during this time, and manufacturing establishments multiplied the number of wage earners on their payrolls seven times.[2] Preparation of these workers for useful employment necessitated a sound system of vocational education.

The first federal aid to develop institutions in vocational education was provided in 1862 by the **Morrill Act,** through which public lands could be used for the support of college-level education. Thus, land-grant colleges and universities sprang up.

Free public secondary education as we know it today did not exist as public policy until 1875. For several decades the secondary schools did not regard offering vocational education as a part of their role in meeting the demands of society. These conditions prompted the government to encourage industry to bring in great numbers of immigrants—laborers and skilled mechanics. The training schools and guild apprenticeships of Europe were the vocational education base for the United States. By 1906, the shortage of skilled industrial workers had become so critical that the progressive leaders of U.S. economic, political, and social life sought a solution to the problem. Public interest in the problem of providing adequate vocational industrial education was aroused in 1906 by the famous Douglas Report of Massachusetts.[3] This was the beginning of the movement that brought about the passage of the **Smith–Hughes Act** in 1917, permanently authorizing a federally promoted and supported national program of vocational education. The concept of federal sharing of state and local school costs was established.

The National Society for Vocational Education defined "vocational schools" as follows:

---

[2]Layton S. Hawkins, Charles A. Prosser, and John C. Wright, *Development of Vocational Education,* Chicago: American Technical Society, 1941, p. 63.

[3]A report of a study group appointed by the governor of Massachusetts, which revealed at that time the inadequacy of manual training to meet the needs of trained workers in industry.

Vocational schools include all agricultural, industrial, commercial, and household arts schools when their main purpose is to fit individuals for useful occupations and when they deal with pupils above 1 year of age and below college grade.[4]

THE CHANGING K–12 ORGANIZATION

The institutional structure has changed markedly in recent years, with an expanding range of types and sizes of institutions and government agencies sponsoring occupational and career development opportunities. What is remarkable is the pluralistic nature of this institution arrangement, with states and local communities deciding the arrangement best suited to their needs as they perceive them. The opportunity to offer a greater range of career and technical education from entry training to semi-professional training, daytime and nighttime, part time and full time, and to supplement this with exploratory experiences is enhanced by several developments.

At the K–12 level, the following developments have occurred. These are: (1) the middle school (usually grades 6–8), which has emerged as a school suited to the needs of an age group; (2) the consolidation of districts and urbanization of population, which has resulted in larger districts and larger high schools capable of more comprehensive programs; (3) innovations in modular and extended school-year scheduling; and (4) the shared-time and educational park concepts, wherein adjacent districts (or several schools within a district) combine to offer specialized CTE programs none could provide alone and provide alternative schools for students who are unable to adjust to the more traditional high school. There is also recognition of lifelong education and of the need to serve those whom the high school has not served well, such as individuals who are physically and/or mentally disabled and/or economically and/or academically disadvantaged.

Area vocational and technical education school.—The Carl D. Perkins Vocational and Technical Education Act of 1998 (Public Law 105-332) defined the *"area vocational and technical education school"* as:

(A) A specialized public secondary school used exclusively or principally for the provision of vocational and technical education to individuals who are available for study in preparation for entering the labor market;

(B) The department of a public secondary school exclusively or principally used for providing vocational and technical education in not fewer than five different occupational fields to individuals who are available for study in preparation for entering the labor market;

(C) A public or nonprofit technical institute or vocational and technical education school used exclusively or principally for the provision of vocational and technical education to individuals who have completed or left secondary school and who are available for study in preparation for entering the labor market, if the institution or school admits as regular students both individuals who have completed secondary school and individuals who have left secondary school; or

(D) The department or division of an institution of higher education, that operates under the policies of the eligible agency and that provides vocational and technical education in not fewer than five different occupational fields leading to immediate employment but not necessarily leading to a baccalaureate degree, if the department or division admits as regular students both individuals who have completed secondary school and individuals who have left secondary school.[5]

Community–junior colleges.—Most states have established community–junior colleges (13th and 14th years). These colleges usually offer the associate degree and employment-bound (terminal) programs.

---

[4]Hawkins, *et al.*, p. 68.

[5]*The Official Guide to the Perkins Act of 1998*, American Vocational Association, 1410 King Street, Alexandria, VA 22314, pp. 87–88.

Associate degree programs concentrate on semi-professional, technical, and mid-management preparation for employment in areas such as agribusiness; electronics; engineering (junior); secretarial administration; information processing, including both data and word; and marketing. Less-than-degree certificate programs (usually six months to one year) may be offered in occupational training programs. Some states have established separate area vocational and technical education schools (technical institutes) for career-technical training in grades 13 and 14, to complement the community–junior college system. The community–junior colleges are usually within commuting distance of potential students and provide low-cost post-secondary educational opportunities. The increased demand for large numbers of technically trained men and women has resulted in the expansion of CTE in the area vocational and technical education schools and in the community–junior colleges. Community–junior colleges through outreach programs and adult classes are providing CTE to many people who were previously ignored.

INDUSTRIAL TRAINING PROGRAMS

Business and industry often provide task training in organized programs of instruction for their employees. Apprenticeship training programs are joint efforts involving federal and state governments, unions, and management. Professional and trade associations through conferences, seminars, and workshops help their members upgrade skills and gain knowledge of advancements in the profession. Chapter 8 discusses in some depth adult industrial training and cooperative education.

PROGRAMS FOR
HUMAN RESOURCE DEVELOPMENT

The *Manpower Development and Training Act (MDTA) of 1962* authorized funds for training and retraining of unemployed and underemployed adults. This act was a milestone in providing training for those who were economically disadvantaged and who were not being served in regular vocational programs. Eligible trainees and potential job openings were identified by the state employment service. State vocational education departments contracted for the courses and experiences that matched the identified needs. If training was not available through public education agencies, contracts could be established with private institutions or agencies. Regional MDTA skills centers were established for daytime and evening training. A cooperative human resource planning system, which attempted to bring together state and regional planning efforts of all agencies involved in human resource and anti-poverty programs, was established.

Beginning in 1970, under the leadership of Sidney P. Marland, Jr., then the U.S. Commissioner of Education, career education became an important concept in the schools. Early efforts in career education were largely supported with vocational education dollars. In 1978, the *Career Education Act* was enacted by Congress. According to the comprehensive career development concept, the individual is viewed as progressing through various planned experiences, a series of dimensions that total a complete cycle (see Figure 3–1). These dimensions, to which work environment-based learning plans can relate, are:

1. **Career awareness**—Students study careers represented in the home, in school, and in the community, as well as characteristics of occupations— duties and responsibilities of occupations; educational and training requirements; advantages and disadvantages of occupations; employment outlook; data, people, and things; involvements; clusters of occupations and commonalities of characteristics and requirements; methods of clustering; and career ladders and lattices.

2. **Self-awareness**—Students create realistic self-images and self-awareness through the study of topics such as occupational interest, aptitude, attitude, and ability inventories; employer expectations; values

CAREER EDUCATION ELEMENTS

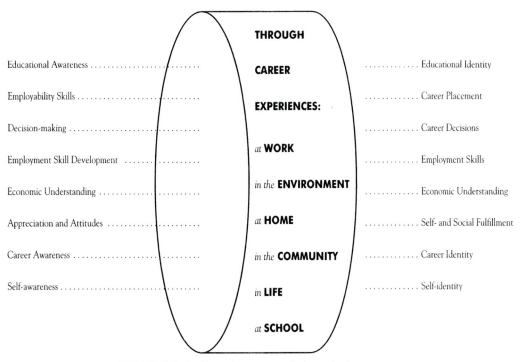

FIGURE 3–1. Development of the whole person.

and career choice; relation of self-information to education and training requirements of occupations and to job advancement avenues; work experience; and knowledge of individual differences and their importance in career planning.

3. **Decision-making**—Students examine bases for decision-making; methods of decision-making; activities for achieving decision-making skills; resources needed for the study of occupations; occupational hazards; occupations selected for in-depth study; education and training alternatives, career development processes and goals; and interviewing and reporting.

4. **Appreciations, attitudes, and self-fulfillment**—Students investigate the dignity of work; the interrelationships of the world of work and human needs; the rewards of work; and the hazards of work.

5. **Economic understanding**—Students consider the technological factors associated with eco-nomic development, along with production, distribution, services, and consumption in relation to jobs and economic structure; reasonable employment and wages; money management; and lifestyles associated with careers.

6. **Employment skill development**—Students avail themselves of work experience; simulated learning experiences; occupational adjustment; correlation of in-school education experiences with work experience; apprenticeship programs, cooperative education, industrial training and secondary occupational programs; and identification with machines, tools, materials, and processes.

7. **Educational awareness**—Students study the financing of education and training; the occupational and educational training opportunities and admission requirements; the relationship of occupational employment requirements to education and training programs; and study habits and skills.

8. **Employability skills**—Students examine legal and certification requirements for careers; job applications and interviews; pre-employment tests; employment agencies and sources; professional, union, industrial, and service organizations; motivation factors; work habits and personality factors related to career success; hobbies and avocations related to careers and home and community experience related to careers (part-time jobs, volunteer services, sports, family involvement).

Public Policy As Expressed Through Federal Vocational/ Career and Technical Legislation

The first federal participation in promoting education for agricultural and mechanical pursuits was expressed in the Morrill Act of 1862. But, the real beginning of public school vocational education through federal policy began with the passage of the Smith–Hughes Act of 1917. It and subsequent legislation are outlined briefly here as an expression of the types of ventures that public policy has promoted in public education.

SMITH–HUGHES ACT—1917

Beginning in 1917, vocational education received its first real impetus through the *Smith–Hughes Act*, which provided *in perpetuity* to the states an annual grant of approximately $7.2 million for the promotion of vocational education in agricultural, trade and industrial, and home economics education. Business, health, and marketing education were not included.

GEORGE–DEEN ACT—1936

Congress approved the *George–Deen Act* on June 8, 1936, as a continuing statute with no expiration date. Unlike the Smith–Hughes Act, it was an authorization, not a permanent act. The act had an annual authorization of $12 million for agriculture, home economics, and trade and industrial education. The money was divided equally among the three services, and allotments to the states were made on the basis of farm population for agriculture, rural population for home economics, and nonfarm population for trade and industrial education. The act was significant because marketing occupations were recognized for the first time, and an annual allotment of $1.2 million was authorized for them.

GEORGE–BARDEN ACT—1946

On August 1, 1946, Congress approved as a post–World War II measure the *George–Barden Act* (the Vocational Education Act of 1946), which authorized an appropriation of $28,850,000 annually for the further development of vocational education. This act replaced the George–Deen Act.

This act authorized $10 million for agricultural education, to be allocated among the states on the basis of farm population. Authority was given in the act for the expenditure of funds in support of two youth organizations in agriculture: the Future Farmers of America and the New Farmers of America. It authorized $8 million for home economics, the basis of allotment being the rural population of the state. It also authorized $8 million for trade and industrial education, to be allocated among the states on the basis of nonfarm population. As in the George–Deen Act, funds for marketing occupations were limited to support for part-time (cooperative) and evening courses for employed workers—no preparatory courses as in other fields were authorized.

NATIONAL DEFENSE EDUCATION ACT—1946

Title VIII of the *National Defense Education Act*, "Area Vocational Education Programs," authorized an appropriation of $15 million annually for four years to support programs limited to the training of highly skilled technicians in occupations recognized as necessary for the national defense. These provisions became Title III of the George–Barden Act.

Allotments were made to states according to each state's proportion of the total amount allocated under the George–Barden Act for agricultural, home economics, trade and industrial, and marketing occupations.

AREA REDEVELOPMENT ACT—1962

The *Area Redevelopment Act* was an emergency bill born out of a recession. It authorized $4.5 million annually to be used for vocational education until 1965. The legislation recognized the critical need for training that arises from unemployment and underemployment in economically distressed areas and, therefore, authorized vocational education retraining programs in certain geographic areas that had been designated as redevelopment areas by the Secretary of Commerce. The act further provided that the Secretary of Labor select and refer persons for training, while the Secretary of Health, Education, and Welfare contract with other public and private educational institutions if the required services were not available through state and local vocational education agencies. Funds appropriated under this act went to the state for use only in designated redevelopment areas; there were no state allotments and no requirements for matching of funds for training programs.

VOCATIONAL EDUCATION ACT OF 1963

Significant expansion and reorientation of vocational education were the goals of the *Vocational Education Act of 1963* (Public Law 88–210). The purposes of the act were varied. However, the major ones were to maintain, extend, and improve existing programs of vocational education and to provide part-time employment for youth who needed the earnings from such employment to continue their schooling on a full-time basis. The intent of the act was to insure that persons of all ages in all communities would have ready access to vocational training or retraining of high quality, suited to their personal needs, interests, and abilities. For the first time, vocational education was mandated to meet the needs of individual students and not just to meet the employment needs of industry.

The law stipulated that the funds be used in accordance with approved state plans for the following:

(1) Vocational education for persons attending high school;

(2) Vocational education for persons who have completed or left high school and who are available for full-time study in preparation for entering the labor market;

(3) Vocational education for persons (other than persons who are receiving training allowances under the Manpower Development and Training Act), who have already entered the labor market and who need training or retraining to achieve stability or advancement in employment;

(4) Vocational education for persons who have academic, socio- economic, or other handicaps that prevent them from succeeding in the regular vocational education program;

(5) Construction of area vocational education school facilities; and

(6) Ancillary services and activities to assure quality in all vocational education programs, such as teacher training and supervision, program evaluation, special demonstration and experimental programs, development of instructional materials, and state administration and leadership, including periodic evaluation of state and local vocational education programs and services in light of information regarding current and projected manpower needs and job opportunities.[6]

It is significant that this legislation did not stipulate funds for the various services; instead, it stipulated them for particular types and ages of persons.

---

[6]Excerpts taken from the *Congressional Record*, proceedings and debates of the Eighty-eighth Congress, First Session, Vol. 109, No. 205, December 13, 1963.

VOCATIONAL EDUCATION AMENDMENTS OF 1968

The purpose of Title I of the *Vocational Education Amendments of 1968* (Public Law 90–576) was to authorize federal grants to states to assist them in maintaining, extending, and improving existing programs of vocational education and to provide part-time employment for youth who needed the earnings from such employment to continue their vocational training. The act was divided into several parts, each related to a specific purpose. In general, the act's purposes were to:

(1) Provide the opportunity for vocational preparation in the high school;

(2) Provide vocational preparation for persons who have completed or left high school and who are available for study in preparation for entering the labor market;

(3) Provide vocational education for persons who have academic, socio-economic or other handicaps that prevent them from succeeding in the regular vocational education program;

(4) Provide vocational education for handicapped persons who because of their handicapping condition cannot succeed in the regular vocational education program without special educational assistance or who require a modified vocational education program;

(5) Support construction of area vocational education school facilities; and

(6) Provide vocational guidance counseling to aid persons in the selection of, and preparation for, employment in all vocational areas.[7]

To meet the special needs for new careers and occupations, the Vocational Education Amendments of 1968 provided financial support for research and training and for experimental demonstration programs. In addition, they provided support for exemplary programs and projects, residential vocational education, consumer and homemaking education, cooperative vocational education programs, and work-study programs for vocational education students.

VOCATIONAL EDUCATION ACT OF 1976

It appears that in instituting the *Vocational Education Act of 1976* (Public Law 94–482), Congress was not satisfied with what vocational education had done in the past in preparing individuals for the labor market. The Vocational Education Act of 1976 mandated vocational education to reflect the needs of the labor market more accurately and, in some cases, to become a partner with business, industry, and labor in attempting to change the work place to be more responsive to the needs of people. Vocational education was asked to assume a pro-active rather than a re-active role. The primary goals of the Vocational Education Act of 1976 were to help states in improving planning in the use of all resources available to them for vocational education and manpower training by involving a wide range of agencies and individuals concerned with education and training within the state in the development of the vocational education plans.[8] The major thrusts of the Vocational Education Act of 1976 were:

(1) to extend, improve, and, where necessary, maintain existing programs of vocational education;

(2) to develop new programs of vocational education;

(3) to develop and carry out such programs of vocational education within each State so as to overcome sex discrimination and sex stereotyping in vocational education programs (including programs of homemaking), and thereby furnish equal

---

[7]Excerpts from Grant Venn, *Vocational Education Amendments of 1968*, June 1969, pp. 1–12.

[8]*Vocational Education Act of 1976*, Title 11 of Public Law 94–482, Education Amendments of 1976, Part A—"State Vocational Education Programs," Sec. 101.

educational opportunities in vocational education to persons of both sexes; and

(4) to provide part-time employment for youths who need the earnings from such employment to continue their vocational training on a full-time basis,

(5) so that persons of all ages in all communities of the State, those in high school, those who have completed or discontinued their formal education and are preparing to enter the labor market, those who have already entered the labor market, but need to upgrade their skills or learn new ones, those with special educational handicaps, and those in postsecondary schools, will have ready access to vocational training or retraining which is of high quality, which is realistic in the light of actual or anticipated opportunities for gainful employment, and which is suited to their needs, interests, and ability to benefit from such training.[9]

The 1976 act identified several major areas of concern and established operational guidelines and rules and regulations for these areas. The primary concerns were: (1) program planning, including articulation with other agencies; (2) collection and dissemination of data about vocational education programs; (3) program evaluation, including accountability concerns, sex bias and sex stereotyping as they pertained to various program functions and components; (4) special populations, including those individuals who are economically and/or academically disadvantaged, physically and/or mentally disabled, unemployed, and/or limited–English-speaking; and (5) program improvement strategies.

CARL D. PERKINS VOCATIONAL EDUCATION ACT OF 1984

The *Carl D. Perkins Vocational Education Act of 1984* (Public Law 94-524), began with fiscal year 1985 and expired September 30, 1989. It amended

the Vocational Education Act of 1963 and replaced the amendments of 1968 and 1976. The act changed the emphasis of federal funding in vocational education from primarily expansion to program improvement and at-risk populations.

(1) Assist the States to expand, improve, modernize, and develop quality vocational education programs in order to meet the needs of the Nation's existing and future work force for marketable skills and to improve productivity and promote economic growth;

(2) Assure that individuals who are inadequately served under vocational education programs are assured access to quality vocational education programs, especially individuals who are disadvantaged, who are handicapped, men and women who are entering nontraditional occupations, adults who are in need of training and retraining, individuals who are single parents or homemakers, individuals with limited English proficiency, and individuals who are incarcerated in correctional institutions;

(3) Promote greater cooperation between public agencies and the private sector in preparing individuals for employment, in promoting the quality of vocational education in the States, and in making the vocational system more responsive to the labor market in the States;

(4) Improve the academic foundations of vocational students and to aid in the application of newer technologies (including the use of computers) in terms of employment or occupational goals;

(5) Provide vocational education services to train, retrain, and upgrade employed and unemployed workers in new skills for which there is a demand in that State or employment market;

(6) Assist the most economically depressed areas of a State to raise employment and occupational competencies of its citizens;

(7) Assist the State to utilize a full range of supportive services, special programs, and guidance counseling and placement to achieve the basic purposes of this Act;

[9]*Carl D. Perkins Vocational Education Act of 1984*, Conference Report, pp. 3 and 4.

(8) Improve the effectiveness of consumer and homemaking education and to reduce the limiting effects of sex-role stereotyping on occupations, job skills, levels of competency, and careers; and

(9) Authorize national programs designed to meet designated vocational education needs and to strengthen the vocational education research process.[10]

CARL D. PERKINS VOCATIONAL AND APPLIED TECHNOLOGY EDUCATION ACT OF 1990

In 1990, amendments to the Carl D. Perkins Vocational Education Act of 1984 became effective. This legislation was titled the *Carl D. Perkins Vocational and Applied Technology Education Act* (Public Law 101-392). This act condensed the nine stated objectives identified in the first Perkins Act to only one objective, emphasizing academic, economic, and occupational achievement. The act reads as follows:

It is the purpose of this Act to make the United States more competitive in the world economy by developing more fully the academic and occupational skill of all segments of the population. This purpose will principally be achieved through concentrating resources on improving educational programs leading to academic, occupational, training, and re-training skill competencies needed to work in a technologically advanced society.[11]

Federal funds were made available to help provide vocational and technical education programs and services to youth and adults. A majority of funds appropriated under this Perkins legislation was in the form of state grants supplied to state education agencies according to a formula based on states' age and per capita income populations.

The act established an interdepartmental task force on coordination of vocational education and related programs. This task force was to examine possible common objectives and ways to integrate research and development activities under the Adult Education Act, the Carl D. Perkins Vocational Education Act of 1984, the Job Training Partnership Act, the Vocational Rehabilitation Act of 1973, and the Wagner-Peyser Act. The task force was to submit a report on its findings to Congress every two years. Its recommendations were to be used in formulating legislation.

A new section of this legislation impacting education at the secondary and post-secondary levels dealt with *tech-prep education*. Planning and demonstration grants were to be provided to a consortia of local education institutions for the development and operation of tech-prep programs designed to establish four-year educational structures leading to a two-year associate degree or a two-year certificate.[12]

CARL D. PERKINS VOCATIONAL AND TECHNICAL EDUCATION ACT OF 1998

The new *Carl D. Perkins Vocational and Technical Education Act* (Public Law 105-332) was signed into law October 31, 1998. It established a new direction and vision for career and technical education to follow as it moves into the 21st century. Perkins 1998 encourages career and technical education reform, innovation, and improvement. The major purposes of the new Perkins legislation are to:

(1) further develop the academic, vocational and technical skills of vocational students through high standards;

(2) link secondary and postsecondary vocational programs;

(3) increase flexibility in the administration and use of federal funds;

---

[10]*Ibid.*, p. 4.

[11]Carl D. Perkins Vocational and Applied Technology Education Act Amendments of 1990. Public Law 101-392. Washington, DC.: U.S. Congress, 1990. (ERIC Document Reproduction Service No. ED 330 818), p. 7.

[12]*The AVA Guide to the Carl D. Perkins Vocational and Applied Technology Education Act of 1990*, American Vocational Association, 1410 King Street, Alexandria, VA 22314.

(4) disseminate national research about vocational and technical education;

(5) provide professional development and technical assistance to vocational educators.[13]

Tech-prep education was reauthorized in the new legislation. Perkins 1998 endorses tech-prep educational strategies using work-based learning and new technologies. It supports the development of partnerships with businesses, labor organizations, and institutions of higher education.

Many of the requirements and restrictions associated with the Perkins Acts of 1984 and 1990 have been removed to give states, school districts, and educational institutions greater flexibility in designing educational programs and experiences that meet the needs of their students and communities. The new Perkins Act seeks to develop high-quality career and technical education programs designed to work in conjunction with the Workforce Investment Act of 1998 and the Adult Education and Family Literacy Act. Career and technical education programs developed under these new provisions should promote student knowledge and attainment of challenging educational and occupational standards, offer students industry-related experiences and understanding, target the needs of special populations, involve parents and employers in educational processes, and insist on the application of advanced technologies. A number of core indicators specified in the law have been developed to guide state performance accountability systems.

(1) Student attainment of challenging State established academic, and vocational and technical, skill proficiencies;

(2) Student attainment of a secondary school diploma or its recognized equivalent, a proficiency credential in conjunction with a secondary school diploma, or postsecondary degree or credential;

(3) Placement in, retention, and completion of, postsecondary education or advanced training,

placement in military service, or placement or retention in employment;

(4) Student participation in and completion of vocational and technical education programs that lead to nontraditional training and employment.

States are required to use these core indicators in developing and establishing their plans and directions for career and technical education. States that exceed their agreed-upon performance levels will be awarded incentive grants. However, states that do not meet their agreed-upon performance levels may have their grants reduced.[14]

Legislation Reflecting Special Concerns

Other laws and government actions have influenced career and technical education. A discussion of some of these follows.

FEDERAL REVENUE ACT OF 1978

The costs of unemployment to society are high. There are the lost goods and services that those who are unemployed would have produced had they been working. The unemployment-related increase in crime, health problems, and family stress carry heavy costs for those affected and to the larger society. To encourage employers to provide employment and training to targeted populations who possess little if any job skills was the intent of the Federal Revenue Act of 1978.

The program provides dollar credits to employers who hire from seven specified groups. Included among the seven targeted groups are youth, ages 18 to 24, from low-income families; students, ages 16 to

---

[13]*The Official Guide to the Perkins Act of 1998*, p. 87.

[14]*Carl D. Perkins Vocational and Technical Education Act of 1998*, Office of Vocational and Adult Education, 400 Maryland Avenue, S.W., 4090 MES, Washington, DC 20202, www.ed.gov/offices/OVAE/VocEd/InfoBoard/legis.html.

18, in approved cooperative education programs; persons who are disabled, in vocational rehabilitation programs; and qualified summer youth employees. The credit to reduce federal income tax is the key provision of the Federal Revenue Act of 1978. The tax credit is relatively easy to claim. In essence, an eligible worker applying for a job under the program is given a voucher by the agency (in case of the cooperative education student, the agency would be the school) that certifies that the job seeker is a member of one of the targeted groups. If the employer decided to hire the individual, the voucher is endorsed by the employer and returned to the state employment service, which issues a tax credit certificate. If the employee was hired before January 1, 1986, the tax credit is equal to 50 percent of the first year's wages up to $6,000 (i.e., $3,000) and 25 percent (i.e., $1,500) in the second year. In the case of a qualified summer youth employee, the employer's credit is limited to 85 percent of the first $3,000 of wages paid to that employee for work done in any 90-day period between May 1 and September 15 of the same calendar year.

If the employee was hired after December 31, 1985, and before January 1, 1989, there are new rules that will or will not allow an employer the tax credit. The tax credit is now limited to 40 percent of the qualified first-year wages. There is no credit for any second-year wages. In addition, the tax credit is only allowable if the eligible employee works for more than 90 days or 120 hours (14 days or 20 hours in the case of a qualified summer youth). Employers cannot use the tax credit to reduce their tax liability below zero. Any unused tax credit can be carried back 3 tax years or forward 15 tax years.

COMPREHENSIVE EMPLOYMENT AND TRAINING ACT (CETA) OF 1973

In response to the high unemployment level, the increasing number of welfare recipients, the increasing number of economically disadvantaged rural and urban communities with hard core unemployed, the significantly large number of youth unable to find part-time or full-time employment, and the increasing number of minority group members who were unskilled and unemployed, Congress in 1973 enacted the *Comprehensive Employment and Training Act (CETA).* This legislation was intended to continue the goals of the Manpower Development and Training Act (MDTA) of 1962, plus expand the previous services MDTA made available. In addition, the funding patterns and coordination requirements among other federal and state agencies providing human resource training were more comprehensive than in the MDTA.

One of the unique features of CETA was its funding pattern. The act established the delivery concept of a prime sponsor. The occupational education, training, and other employment services programs were conducted in conjunction with local units of government known as CETA prime sponsors. Prime sponsors provided a variety of employment and training services by contracting with approved public and private education agencies.

When CETA became law in 1973, the concept of revenue sharing was popular—the federal government would give to both the states and their localities the tax revenues collected from their citizens and let those units of government make the essential decisions concerning the use of public funds. This view was popular among groups who were concerned about "big government" and strongly centralized federal control and among state and local politicians. Local planning was a major rationale for the CETA revenue sharing philosophy.

In October 1978, the Comprehensive Employment and Training Act Amendments of 1978 were signed into law. These CETA amendments authorized the extension of the activities and programs of the 1973 law and included provisions to improve the delivery, accountability, and coordination of the various employment and training programs and services. In addition, the 1978 amendments required program coordination and linkages with other human resource programs, including vocational education.

JOB TRAINING PARTNERSHIP ACTS OF 1982 AND 1992 (JTPA)

The 1982 *Job Training Partnership Act* (Public Law 97-300) replaced CETA, which expired September 30, 1982. The second *Job Training Partnership Act* (Public Law 102-367) was enacted in 1992. The major purpose of these acts was to establish programs to prepare youth and adults facing serious barriers to employment for participation in the labor force. Programs provided individuals with job training and other services designed to increase employment and earnings, increase educational and occupational skills, and decrease welfare dependency. The ultimate goal of these acts was to improve the quality of the workforce and enhance the productivity and competitiveness of the nation.

These statutes enlarged the role of state governments and private industry in federal job training programs, imposed performance standards, provided limited support services, and created new programs of retraining for displaced workers. Playing a leading role implementing the statutes was each state's governor, who designated local service delivery areas (SDA's), passed judgment on the plans drawn up for the local training programs, and drew up a plan for coordinating job training programs with other human services in the state. Each state was required to have a state job training coordinating council (SJTCC), appointed by the governor. Councils recommended SDA's. Every SDA had to have a private industry council (PIC). The legislation gave PIC's broad powers. PIC's were responsible for approving local job training plans and for establishing goals for accomplishing identified performance standards.

Participation in a local job training program was limited to individuals who were economically disadvantaged, except for 10 percent of the participants, who could be non-economically disadvantaged if they had encountered barriers to employment. An example of persons in this latter group would be displaced homemakers. JTPA required that vocational education be involved in the job training activities.

The acts identified a long list of activities for which funds could be used, including job counseling, job search assistance, remedial education in the basics, work experience, on-the-job training, vocational exploration, bilingual education, job development, customized industry training, supportive services, and pre-employment skills training for youths ages 14 and 15.

WORKFORCE INVESTMENT ACT OF 1998 (WIA)

The *Workforce Investment Act* (Public Law 105-220) was written to consolidate, coordinate, and improve employment, training, literacy, and vocational rehabilitation programs in the United States. It provides the framework for a unique national workforce preparation and employment system designed to meet both the needs of the nation's businesses and the needs of job seekers and those who want to further their careers. The WIA reforms federal job training programs and creates a new comprehensive workforce investment system. The reformed system is intended to be customer-focused, to help Americans access the tools they need to manage their careers through information and high-quality services, and to help U.S. companies find skilled workers. This new law replaces the JTPA and embodies seven key principles: (1) streamlining of services, (2) empowering of individuals, (3) universal access, (4) increased accountability, (5) strong role for local workforce investment boards and the private sector, (6) state and local flexibility, and (7) improved youth programs.[15]

The new system will be based on the "One-Stop" concept, wherein information about and access to a wide array of job training, education, and employment services is available for customers at a single location. Services provided will be streamlined through better integration of information. Programs and providers will co-locate, coordinate, and inte-

[15]*Workforce Investment Act of 1998*, U.S. Department of Labor, Employment and Training Administration, http://usworkforce.org/runningtext2.htm.

grate activities and information so that the system as a whole is coherent and accessible to individuals and businesses alike.

Provisions of the act promote individual responsibility and personal decision-making through the use of "Individual Training Accounts," which allow adult customers to "purchase" the training they determine best for them. This market-driven system will enable customers to get the skills and credentials they need to succeed in their local labor markets.

Individuals are empowered in several ways. First, eligible adults are given financial power to use Individual Training Accounts (ITA's) at qualified institutions. These ITA's supplement financial aid already available through other sources, or if no other financial aid is available, they may pay for all the costs of training. Second, individuals are empowered with greater levels of information and guidance through a system of consumer reports that provide key information on the performance outcomes of training and education providers. Third, individuals are empowered by the advice, guidance, and support available through the One-Stop system and the activities of One-Stop partners.

Universal access to the One-Stop system and to core employment-related services will be made possible for all persons. Information about job vacancies, career options, student financial aid, and relevant employment trends and instruction on how to conduct a job search, write a résumé, or interview with an employer is available to any job seeker in the United States or to anyone who wants to advance his or her career.

As individuals become empowered to choose the services they require, states, local areas, and providers of those services will become more accountable for meeting the participants' needs. The goal of the act is to increase employment, retention, and earnings of participants and, in doing so, improve the quality of the workforce to sustain economic growth, enhance productivity and competitiveness, and reduce welfare dependency. Consistent with this goal, the act identifies core indicators of performance that state and local entities managing the workforce

investment system must meet or suffer sanctions. However, state and local entities exceeding the performance levels can receive incentive funds. Training providers and their programs also have to demonstrate successful performance to remain eligible to receive funds under the act. And participants, with their ITA's, have the opportunity to make training choices based on program outcomes. To survive in the market, training providers must make accountability for performance and customer satisfaction a top priority.

The new act establishes a strong role for local workforce investment boards and the private sector. Local business-led boards, acting as "boards of directors," will focus on strategic planning, policy development, and oversight of the local workforce investment system. Business and labor have an immediate and direct stake in the quality of the workforce investment system. Their active involvement is critical to the provision of essential data on what skills are in demand, what jobs are available, what career fields are expanding, and how programs can be identified and developed that best meet local employer needs. Highly successful private industry councils under JTPA exhibit these characteristics now. Under WIA, this will become the norm.

States and localities have increased flexibility under the WIA. Significant authority is reserved for the state governor and chief elected officials to build on existing reforms in order to implement innovative and comprehensive workforce investment systems tailored to meet the particular needs of local and regional labor markets.

The act creates opportunities for improved youth programs. Youth programs will be linked closely to local labor market needs and to community youth programs and services, with strong connections between academic and occupational learning. Youth programs include activities that promote youth development and citizenship, such as leadership development through voluntary community service opportunities; adult mentoring and follow-up; and targeted opportunities for youth living in high-poverty areas.

CREATION OF A DEPARTMENT OF EDUCATION

On October 17, 1979, President Carter fulfilled a campaign promise by signing the bill creating a separate cabinet-level Department of Education. The presidential appointments include the secretary, the under secretary, and the assistant secretaries for elementary and secondary education, post-secondary education, vocational and adult education, special education and rehabilitation services, educational research and improvement, civil rights, and overseas dependents.

Legislation for Special Needs Populations

In the 1960's and 1970's, much of the legislation passed by Congress dealt with providing equal education for "all." Never before had there been such an emphasis on providing vocational education for all students, no matter what their race, sex, age, national origin, language, or economic level. Laws enacted during this period include the Civil Rights Act of 1964, the Economic Opportunity Act of 1965, the Elementary and Secondary Education Act of 1973, the Comprehensive Employment Training Act of 1973, the Education for All Handicapped Children Act of 1975, and the Vocational Education Act of 1976, which mandated changes designed to enable vocational education to serve better "all" people, including the special needs populations.

The Carl D. Perkins Vocational and Technical Education Act of 1998 defines "special populations" as "(a) individuals with disabilities; (b) individuals from economically disadvantaged families, including foster children; (c) individuals preparing for nontraditional training and development; (d) single parents, including single pregnant women; (e) displaced homemakers; and (f) individuals with other barriers to educational achievement, including individuals with limited English proficiency."[16]

VOCATIONAL/CAREER AND TECHNICAL EDUCATION FOR INDIVIDUALS WHO ARE DISABLED

In the past three decades, career and technical education, as vocational education is now called, has become more concerned with the role of serving persons with disabilities and has made progress in adapting and refining programs to prepare these "students at risk" vocationally. It is estimated that in the United States, more than 8 million children have disabilities. In 1996, children with disabilities receiving services in federally supported programs constituted 12 percent of all students enrolled in public schools (grades K–12), up from 8 percent in 1977. The number of students who participated in federal programs for children with disabilities increased 51 percent between 1977 and 1996, rising from 3.7 to 5.6 million. Between 1977 and 1996, children with specific learning disabilities as a percentage of total public school (grade K–12) enrollment rose from 2 to 6 percent, while children with speech or language impairments and those with mental retardation decreased by approximately 1 percentage point each. In 1995, 73 percent of public school children with disabilities were served in regular classrooms or resource rooms in regular school buildings, while 23 percent were served in separate classrooms in regular school buildings.[17]

Various laws (beginning with the Vocational Education Act of 1963) have charged career and technical education with the responsibility of assisting students who have disabilities to succeed in CTE programs. New federal and state laws have substantially altered how CTE must serve these persons. As a result of legislation, such as the Individuals with Dis-

---

[16]*The Official Guide to the Perkins Act of 1998*, pp. 90–91.

[17]*The Condition of Education 1998*, National Center for Education Statistics, 1990 K Street, NW, Washington, DC 20006, http://nces.ed.gov/pubs98/condition98/.

abilities Education Act (IDEA) of 1997 and the three Carl D. Perkins Acts, career and technical education must assess the skills of students who have disabilities, make CTE available as part of the total education plan, and provide career and technical education and counseling in a nondiscriminatory manner.

Two major legislative provisions impact career-technical cooperative education: (1) Children who have disabilities should be placed in the least restrictive environment—at the very least, this implies mainstreaming or, where possible, the inclusion of students with disabilities within CTE programs. (2) An individualized education program or plan (IEP) must be developed for every student with a disability. Teacher-coordinators need to work closely with special education personnel to build CTE programs that serve the individual needs of students with disabilities. No longer can the special needs department and the CTE department exist separately in the education of students who have disabilities.

For the last 30 years, career-technical educators have been urged to accept responsibility for helping individuals who need special assistance to succeed in vocational education programs. The Vocational Education Act of 1963 and that of 1976 charged vocational education with the responsibility of addressing the needs of individuals who are unable to succeed in regular vocational education programs. The 1984, 1990, and 1998 Carl D. Perkins Vocational Education Acts repeated the mandate and gave added attention to the requirements. The intention of Congress has been to provide special assistance to CTE students who are physically and/or mentally disabled and/or economically and/or academically disadvantaged.

VOCATIONAL/CAREER AND TECHNICAL EDUCATION FOR INDIVIDUALS WHO ARE DISADVANTAGED

At a time when serious concerns are being voiced about the staggering number of U.S. adults who are functionally illiterate, the number of U.S. workers who are structurally unemployed, the shocking increase in the number of high school students who drop out from school, the increase in the number of teenagers who are mothers, and the large number of youths, especially in minority groups, who are unemployed, CTE's commitment to improve career and technical programs for individuals who are *disadvantaged* has special meaning.

Problems of youth.—The costs to society of unemployment and delinquency among youth are a drain on societal resources that could be used for other purposes. Studies have clearly shown that youth who get a poor start in life are likely to develop lifelong patterns of unemployment and crime. Opportunities for those who do not finish high school will be increasingly limited, and workers who are not literate may not even be considered for most jobs. Those who do not complete high school and are employed are more likely to have low-paying jobs with little advancement potential, while workers in occupations requiring higher levels of education have higher incomes. In addition, many of the occupations projected to grow most rapidly between 1998 and 2008 are among those with higher earnings.

Few, if any, career-technical educators would disagree that schools must become more responsive to the employment needs of young people while they are in school and prepare them, upon their leaving school, to make an effective transition to the world of work. The task of preparing students, especially those who are disadvantaged and/or disabled, with the necessary job skills and work attitudes for success in the world of work is a major concern among educators who are responsive to the needs of young people in the critical transition from school to work and family formation.

Unemployment problems.—Unemployment is everyone's problem. It is economically dangerous to the nation and creates serious burdens on everyone. People who are unemployed deprive society of goods and services that, had they been employed, would have added to the national output. High unemployment brings with it many related social problems and pressures, including increases in crime, health problems, and family stress, all of which imply additional costs that must be met.

Getting and holding a job for many people is an unfulfilled dream. In December 1998, 6.2 million people in the United States were out of work, while the workforce numbered at 160.8 million. Since most unemployment data do not include the names of unemployed persons who are no longer registered with state employment agencies because they are no longer eligible to receive unemployment compensation, the actual number of people who are unemployed could be much greater than the 6.2 million reported. Historically, economic problems, including recessions, have unquestionably created serious unemployment problems.

The first federal legislation to promote language as an essential part of culture was Title VII, called the "Bilingual Education Act," a part of the Elementary and Secondary Education Act of 1965 (Public Law 89–10), which was enacted in 1968. It authorized funds for *bilingual programs* at the pre-school, elementary, and secondary levels. Concerns about those students with limited English or non-English proficiency prompted Congress, through the Vocational Education Act of 1976, to give special consideration to bilingual education for the first time. The Perkins Acts of 1984, 1990, and 1998 continued to authorize support for individuals with limited English proficiency.

Concerns among career-technical educators about those students with limited English or non-English proficiency have become more urgent. Minorities and immigrants will constitute a larger share of the U.S. population in 2008 than they do today. Substantial increases in the number of Hispanics, Asians, and Blacks are anticipated, reflecting immigration, and higher birth rates among Blacks and Hispanics. Substantial inflows of immigrants will continue to have significant implications for the labor force. Immigrants tend to be of working age but of different educational and occupational backgrounds than the U.S. population as a whole.

In addition, a substantial number of new faces arrive illegally. Estimates of the number of entering illegal immigrants vary widely. A study completed by the National Academy of Sciences estimated the total illegals in the United States to be no more than 2 to 4 million, while the Bureau of the Census estimates the number at between 3.5 and 6 million. These figures include anyone—from foreign students who deliberately overstay their student visas to the boat people who scramble onto U.S. soil in Florida.

A study completed by the National Center for Research in Vocational Education at The Ohio State University clearly shows that there are many barriers that keep limited-English- or non-English-speaking students from participating in and succeeding in CTE. Clearly, all career-technical educators, including coordinators of cooperative education plans, must meet the challenge by providing the leadership necessary to assure accessibility and success in CTE programs and plans for this special population.

Although many U.S. workers feel that opportunities for employment should be more plentiful, that the quality of work should be better, that work settings and structures should be more flexible and more in accord with new and varying lifestyles, and that workers should be more actively involved in decisions that affect their needs, "the great majority of Americans—be they young or older, affluent or poor, male or female—would prefer work to welfare."[18] Passage of the WIA in 1998 illustrates the U.S. government's commitment to providing opportunities for Americans who are unemployed and seemingly trapped in the grasps of welfare. Welfare-to-work grants have been established to help the hardest-to-employ and long-term welfare recipients gain education, work experience, and private-sector jobs.

[18]Reginald E. Petty, *Trends and Issues in Vocational Education Research and Development*, Columbus, Ohio: National Center for Research in Vocational Education, The Ohio State University, November 1978, p. 8.

Single parents and displaced homemakers.—The Carl D. Perkins Vocational Education Act of 1984 provided the largest set-aside of vocational education monies ever for single parents and homemakers. The act is an example of federal legislation responding to economic and social changes. In less than 15 years, the number of one-parent families in the United States doubled, and about 90 percent of these one-parent families are maintained by the mothers.

Older adults.—A new "special needs population is growing in our society—the graying population—senior citizens, retirees, and oldsters. Many oldsters are victims of age-ism discrimination. Age-ism takes several forms, the most significant being: (a) compulsory age for retirement, and (b) difficulty in reentering the labor force after leaving a job in one's later years."[19]

The "baby boom" generation—everyone born between 1946 and 1964—has shaken U.S. economics, politics, and culture. As a group, these individuals represent about one-third of the population in the United States and will make up more than half the labor force by 2008. The middle aging of these baby boomers (the first ones turned 50 in 1996) will no doubt bring about significant cultural changes. Some experts believe that more people will be affluent, more people will be conservative, and more people will focus on their home life.

The age distribution will shift toward relatively fewer children and teenagers and a growing proportion of middle-aged and older people into the 21st century. The decline in the proportion of teenagers reflects the lower birth rates that prevailed during the 1980's; the impending large increase in the middle-aged population reflects the aging of the "baby boom" generation born between 1946 and 1964; and the very rapid growth in the number of old people is attributable to high birth rates prior to the 1930's, together with improvements in medical technology that have allowed people to live longer.[20]

Adult Education Thrusts

INCREASING IMPORTANCE OF ADULT EDUCATION

As individuals live longer and as older adults make up a larger part of the population, the concept of lifelong education will become even more important to career-technical educators. The reasons for taking adult classes vary as widely as the needs of the persons enrolled: a homemaker wants to return to the labor market since the children have "left the nest"; an individual in one occupation is bored by it and wants to change; a person in a low-skill entry job wants to advance.

Enrollment increases in adult education will be a challenge to schools. Existing programs will need to be modified to meet the needs of this new clientele. In addition, new programs and courses will need to be developed to fulfill the lifelong educational needs of the public. Current educational policies regarding admission requirements, credit for experience, career counseling and placement, financial assistance, and transportation will need to be reexamined.

The urgent needs of our older adults have been recognized in Title II of the Workforce Investment Act of 1998. Title II, the Adult Education and Family Literacy Act (AEFLA), restructures and improves programs previously authorized by the Adult Education Act. AEFLA focuses on strengthening program quality by requiring states to give priority in awarding funds to local programs that are based on a solid foundation of research, address the diverse needs of adult learners, and utilize other effective practices and strategies. Factors that must be considered in awarding funds include whether a program provides

[19]Angelo C. Gilli, "Vocational Education for Retirees and Middle-Age Career Changers," *Vocational Education for Special Groups*, Sixth Yearbook, Arlington, Virginia: American Vocational Association, 1976, p. 86.

[20]*Occupational Handbook, 1994–95*, p. 12.

learning in real-life contexts, employs advance in technology, and is staffed by well-trained instructors, counselors, and administrators.

Many in the over-55 age group are searching for ways to stay active and productive. Whether still on the job or already retired, whether driven by inflation or just restless because of unwanted idleness, increasing numbers of older adults are wanting to get involved in adult education programs.

One writer believes that the U.S. economy is suffering a human capital crisis by not responding more swiftly to the needs of adult learners. The chief components of the human capital crisis are summarized under four "F's": Function, Fit, Flexibility, and Frustration.[21] The component "Function" refers to the crisis of functional illiteracy among 15 percent of U.S. adults. "Fit" pertains to the significant number of adults who have extensive knowledge and skills that no longer fit the technological changes in job requirements. The third "F," "Flexibility," describes the human capital crisis of about 36 percent of the adult population who find that their work is in transition or that they are anticipating transition in their work. The final "F," "Frustration," addresses perhaps the most dangerous element of the human capital crisis—the personal frustrations of today's workers.

Sex Equity Thrusts

Congress, in passing the 1976 Vocational Education Act, emphasized **sex equity**—the elimination of sex bias and sex stereotyping. The law required every state to set aside at least $50,000 of its federal vocational education grant to hire a person to work full time on sex equity. In addition, it specified the responsibilities of the sex equity coordinator. Decisions such as which department within the state department the sex equity coordinator would be located in, what the qualifications and the salary would be, how much authority the coordinator would have, and how much financial support the state would provide for sex equity above the federally required minimum were up to each state.

States varied widely in the ways they implemented the federal mandate. State-wide and regional meetings acquainting educators with the federal thrust to increase access for individuals to enter vocational education were held in many states. Some states developed and disseminated written materials and audio–visual aids to help local schools reach out and recruit nontraditional vocational education students.

The Vocational Education Act of 1976 required each state accepting funds under the act to submit in its five-year plan "policies and procedures which the state will follow so as to assure equal access to vocational education programs by both women and men."

The Carl D. Perkins Vocational and Applied Technology Education Act of 1990 called for states to reserve 10.5 percent of its allotments for single parents, displaced homemakers, single pregnant women, and programs that promote sex equity in career and technical education. A minimum of 3 percent had to be specifically designated toward sex equity programs.

The Carl D. Perkins Vocational and Technical Education Act of 1998 calls for state boards to spend between $60,000 and $150,000 of monies authorized for state leadership activities on services designed to prepare individuals for nontraditional employment. Nontraditional employment is defined as occupations or fields of work in which individuals from one gender make up less than 25 percent of the persons employed in those occupations or fields of work.[22]

Federal laws, including the Vocational Education Act of 1976 and the Perkins Acts of 1984, 1990, and 1998, challenged business, industry, labor, and education to develop policies, procedures, and practices promoting racial and sex equality. *Attitudes* must be changed before actions can be. However, changing

[21]Luis J. Perelman, *The Learning Enterprise: Adult Learning, Human Capital, and Economic Development*, Washington, D.C.: Council of State Planning Agencies, 1984, pp. 1–4.

[22]*The Official Guide to the Perkins Act of 1998*, p. 20.

attitudes is often a slow process because no change is possible without some discomfort. Thus, any change is always met with some form of resistance, with apathy being a very common form.

The Educational Reform Movement

It seems that the educational process is constantly being criticized for doing what it is doing poorly or for not doing what it should be doing at all. Career and technical education has certainly not been exempt from such criticism. Instead it has often been the recipient of criticism from other educators as well as from the general public.

One reflection of the public's attitude toward the nation's schools was the Reagan Administration–appointed National Commission on Excellence in Education—an 18-member panel—in August 1981.[23] Using presidential power to focus public opinion on education was enough to put schools on the front burner of discussion. The Commission's report, *A Nation at Risk: The Imperative for Educational Reform*, provoked alarm in each state's office of the governor, and a year later, all 50 states were implementing educational reform measures. Although the report's recommendations for reform were not necessarily new, the federal government's initiative was enough to spark national reaction.

National reports that followed the Commission's report tended to focus on the amount of education and seriously neglected the quality of learning and individual differences. Each of the reports stressed issues such as:

1. Lengthening the school day and the school year.

2. Increasing the number of units (credits) required for high school graduation by specifying additional units in English, mathematics, science, and social studies.

3. Raising the entrance requirements for state colleges and universities.

4. Restructuring the high school curriculum by prescribing five "new basics"—four years of English, three years of mathematics, three years of social studies, and one-half year of computer science, and by requiring college-bound students to have two years of a foreign language.

While it is difficult to find fault with such aspiring recommendations, career-technical educators expressed concern that the majority of students would be deprived of a meaningful education as states concentrated on the "pursuit of excellence." Career-technical educators in general questioned the wisdom of prescribing the new basics, with little or no regard for the aspirations, interests, and abilities of each student. In spite of such concerns, a significant number of states implemented a good number of the recommendations. Many career-technical educators believe that the increasing dropout rate is directly related to requiring students to enroll in additional new basics courses, thus reducing the number of electives students may schedule.

Public Policy Toward Work Education

Learning experiences in the world of work have long been an inherent component of career and technical education programs. Cooperative occupational education plans, characterized by blending related class instruction and on-the-job training, and coordinated by the school, have been, to some degree, supported by federal CTE legislation since the Smith–Hughes Act in 1917. Since the mid-sixties, coopera-

[23]National Commission on Excellence in Education, *A Nation at Risk: The Imperative for Educational Reform,* Washington, D.C.: U.S. Government Printing Office, April 1983.

tive education has expanded greatly. Part of this growth is due at least in part to the increased state and federal funding through career and technical education acts, higher education legislation, and various human resource acts. The growth of cooperative education can be related directly to the national belief that learning by doing has much merit—the world-of-work involvement has many educational values that cannot be otherwise learned. Most of us agree that work experiences build character and responsible citizen traits. Our nation has always believed that youth must learn to work so that they can develop into independent and productive adults, rather than adults who must depend upon others for their existence. Learning how to work is important not only to our economy but also to our political system, for individual participation in a democracy is largely dependent upon having the competence to participate in its economic life.

In January 1984, the National Commission on Secondary Vocational Education was formed, partially at least in response to the *Nation at Risk* report released the previous year, wherein the appropriateness of vocational education in secondary schools was almost entirely neglected. The National Commission on Secondary Vocational Education was charged with the task of examining the role and function of vocational education within the mission of the secondary schools. Its report, *The Unfinished Agenda*, strongly supports cooperative vocational education.

> Field-based learning is grossly underutilized. Therefore, we recommend that supervised, field-based learning experiences be made available to all secondary students. Cooperative education must be a "capstone" element in all vocational education programs.[24]

The National Commission on Secondary Vocational Education goes on to say in its report that ". . . cooperative vocational education, a form of field-based learning, has been one of the most successful aspects of vocational education.[25] The Commission found that all students benefit from field-based learning programs. These activities:

- Provide assistance to students in career decision making;
- Provide understanding of the reality of the workplace; allow for trial and error at a time when error is not devastating to a young person's work record;
- Include opportunities to acquire necessary work habits, develop job skills, and establish an employment record;
- Connect students and employers in mutually satisfactory arrangements; and
- Motivate students by providing earnings and potential for regular employment.[26]

Cooperative education and work experience education have grown rapidly at the post-secondary level as well. Indications are that as lifelong learning becomes a reality for more and more people, the rapid growth of cooperative education at the post-secondary level will continue.

School-to-Work Legislation

The School-to-Work Opportunities Act of 1994 (Public Law 103-239) marked a giant step toward the development of an educational system that matches the students' educational attainment and their corresponding academic and technical skills more closely to employment opportunities.

The major purpose of the legislation was to establish a national framework whereby all states created state-wide school-to-work opportunity systems that

---

[24]National Commission on Secondary Vocational Education, *The Unfinished Agenda: The Role of Vocational Education in the High School*, Columbus, Ohio: National Center for Research in Vocational Education, The Ohio State University, p. 28.

[25]*Ibid.*, p. 22.

[26]*Ibid.*

were part of comprehensive educational reform and integrated with systems developed under the Goals 2000: Educate America Act and the National Skills Standards Act of 1994. The School-to-Work Opportunities Act offered opportunities for all students to participate in a program of study that enabled them to earn portable credentials and prepare for their first jobs in high-skill, high-wage careers, while increasing their opportunities for further education. The cornerstone of this Act was to utilize the workplace as an active learning environment in collaboration with the educational system by making employers joint partners with educators in an effort to provide all students with high-quality work site–based learning experiences. Cooperative education is one component of this delivery system. The Web site address for school-to-work is www.stw.ed.gov.

Goals 2000: Educate America Act

The Goals 2000: Educate America Act created a National Skill Standards Board to certify voluntary performance standards for each industry or career field. The Department of Education issued grants to groups to develop skill standards for specific career fields. Career fields covered by the grants were heavy highway utility construction, chemical processing, advanced manufacturing, welding, human services, hazardous materials, photonics, agriscience/biotechnology, supermarket workers, and hazardous waste remediation workers.

Accountability As Reflected in Public Policy

Legislatures, as well as the general public, are demanding that education become more accountable. The 1976 Vocational Education Act prescribed for the first time an evaluation system for vocational

education. The law mandates that programs be assessed on one primary criterion: student placement and employer satisfaction.

The Carl D. Perkins Vocational and Technical Education Act of 1998 directs the Secretary of Education to conduct an "independent evaluation and assessment of vocational and technical education programs under this Act" and to appoint an independent advisory panel to advise the Secretary on the implementation of the assessment. Increased accountability is a major focus of the new act, and states are required to collect data and report on student achievement using specified core indicators. The act requires each state, in cooperation with the U.S. Department of Education, to set performance levels for CTE students in four categories:

(1) Student attainment of vocational, technical and academic skill proficiencies.

(2) Acquisition of secondary or postsecondary degrees or credentials.

(3) Placement and retention in postsecondary education or employment.

(4) Completion of vocational and technical programs that lead to nontraditional training and employment, meaning fields in which one gender accounts for less than a quarter of the participants.[27]

States must make continual progress not only toward meeting performance levels but also in making consistent improvement in performance levels of students enrolled in career and technical education programs. Progress of a state will be made public and will be compared to the progress of other states. Funding for career and technical education programs can be withheld from states failing to meet performance levels or not developing improvement plans. States exceeding established performance levels will be rewarded through additional funding.[28]

---

[27]*The Official Guide to the Perkins Act of 1998*, p. 12.
[28]*Ibid.*

A Summary View

An analysis of the various laws guiding career and technical education's policies and plans clearly reflects our nation's concerns with CTE and its relationships to the world of work. Career-technical educators are being told to provide greater flexibility in programs, to take a more active role in job planning and job development, and even to redefine CTE roles in the broader sense of education, training, and work.

Nowhere is the demand stronger for career and technical education to assume a more active role in economic development. Career-technical educators are being asked to develop meaningful work experiences for students so that they can begin to realize the emotional satisfaction and job enrichment opportunities employment can provide. In addition, career-technical educators are being challenged to assume leadership roles in job creation through preparing students to become entrepreneurs—to become employers. Many believe the private sector will not be able to provide sufficient jobs for these seeking work. New businesses and industries must be created to establish new employment opportunities. This is entrepreneurship. The problems are complex, but the need reflected in public policy is clear—career and technical education must accept the challenge.

The growth of CTE has been very rapid and often without unified direction. With the exception of the federal vocational education acts, which do exercise their controls over educational policy by "holding the purse strings," there has been no national framework of policies for CTE. The growth of vocational/career and technical education has usually been guided by a very general philosophy—to provide more work-relevant education to more students. In the years to come, career-technical educators will need to rethink the purpose of CTE. Is it simply education for work? Policies dealing with the needs of young and adult learners, advisory committees, and social and labor market demands will probably need to be reformulated. Many of the trends stated in the Carl D.

Perkins Vocational and Technical Education Act of 1998 will create the need for new policies. Issues such as providing CTE for economically disadvantaged individuals, retraining dislocated workers, providing CTE for the "graying" population, coordinating with WIA, providing training in correctional institutions, and implementing tech-prep programs are important issues that will have grass-roots input into policy development. State plans for career and technical education will spell out how each state perceives its priorities in allocating block grant funds for CTE.

Questions and Activities

1. Prepare a report on your community that describes the existing educational institutions (private and public) and their career preparation programs. Identify also any training programs conducted by government, or business, or labor or industry sponsorship. Which programs use some form of work exploration or experience as part of the plan of instruction?

2. Describe the major shift in emphasis of federal legislation. What seems to be the cause of this shift?

3. Obtain a copy of your state's five-year plan for CTE. What are your state's primary goals for CTE? Where and how is the use of the work environment covered as a plan of instruction?

4. Obtain data that describe the investment of your state and at least one local community in CTE. What sources of funds were used? For what programs were these funds spent? Prepare a report to the community (in the form of a speech) that shows what dollars were invested and who profited from this investment.

5. Some career-technical educators are not very pleased with the development of federal- and state-sponsored training programs, such as WIA, that are outside the public school programs. Why do you believe these programs were established? Whom are they intended to serve? Do they relate at all to the regular CTE program? Explain.

6. Prepare a defense for, or an objection to, the concept that career and technical education is really economic education and is needed to help each individual live in our present society.

7. Assume you have been asked to speak to a community group, such as the local chamber of commerce, on the topic of "How Government Is Involved in Career and Technical Education." Outline the points you would make in your participation.

8. Describe what you believe are the most significant differences between the Vocational Education Act of 1963 and the 1968 and 1976 amendments. Describe the major differences between the 1984, 1990, and 1998 Perkins Acts. Examine your own state plan to determine how career and technical education needs will be served in your state under current legislation.

9. Interview a vocational/career and technical director. Ask the director which of the following CTE needs most: Money? Better equipment? Community sup- port? Then ask the director how federal legislation can help with these needs.

10. Based on your reading, prepare working definitions for these terms:

   a. Disadvantaged

   b. Special needs populations

   c. Multicultural programs

   d. Disabled

   e. Human resource development programs

   f. Tech-prep

For references pertinent to the subject matter of this chapter, see Reading Resources.

The Structure of Curriculum Patterns—

How the Work Environment Is Used at Various Levels

A SYSTEMS APPROACH TO IMPROVING INSTRUCTION

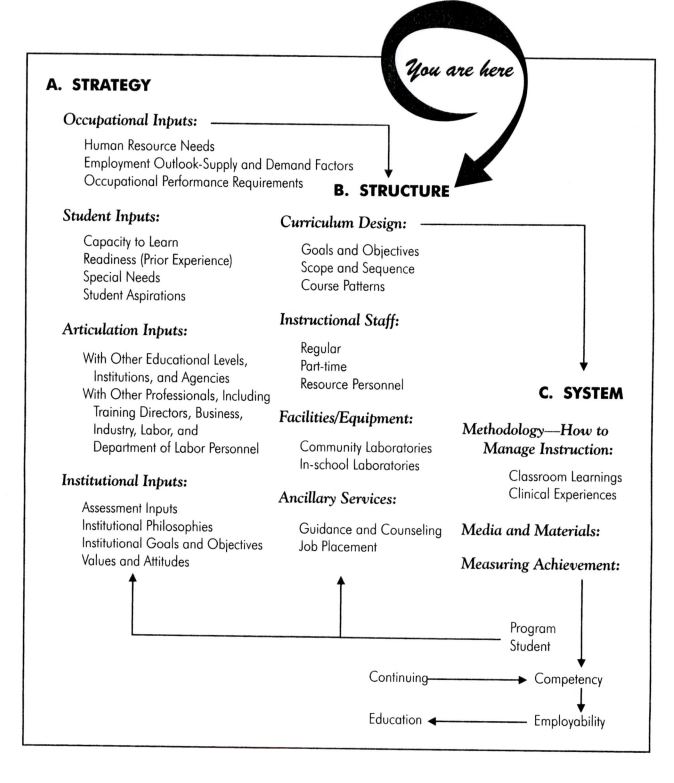

You are here

A. STRATEGY

Occupational Inputs:

Human Resource Needs
Employment Outlook-Supply and Demand Factors
Occupational Performance Requirements

Student Inputs:

Capacity to Learn
Readiness (Prior Experience)
Special Needs
Student Aspirations

Articulation Inputs:

With Other Educational Levels,
Institutions, and Agencies
With Other Professionals, Including
Training Directors, Business,
Industry, Labor, and
Department of Labor Personnel

Institutional Inputs:

Assessment Inputs
Institutional Philosophies
Institutional Goals and Objectives
Values and Attitudes

B. STRUCTURE

Curriculum Design:

Goals and Objectives
Scope and Sequence
Course Patterns

Instructional Staff:

Regular
Part-time
Resource Personnel

Facilities/Equipment:

Community Laboratories
In-school Laboratories

Ancillary Services:

Guidance and Counseling
Job Placement

C. SYSTEM

Methodology—How to Manage Instruction:

Classroom Learnings
Clinical Experiences

Media and Materials:

Measuring Achievement:

Program
Student

Continuing ⟶ Competency

Education ⟵ Employability

INTRODUCTION TO SECTION TWO

Recognizing that the "community can be a classroom," educational institutions have extended their curricula beyond the four walls of the school. To the contemporary educator, the job has become an occupational laboratory, which is controlled and supervised by the school to provide educational experiences that are not obtainable within the school itself. Consequently, the work environment is frequently used as a teaching–learning situation by junior and senior high schools, post-secondary institutions, and colleges and universities.

Regarding curriculum patterns, Section Two centers on the **structure** of decision-making, which is determined by the goals of the student, the educational institution, and society. Based upon the premise that the use of a work-based learning environment is a plan of providing career exploration and career and technical instruction, the decision-maker is asked to determine where and when the student should be involved in an occupational laboratory.

Section Two also provides models of curriculum plans, a description of which curriculum patterns are useful in helping students to obtain their goals, and a description of the role of the coordinator in directing community laboratory experiences in the post-secondary, collegiate, and adult training and development environments.

4

Cooperative Education Models

A cooperative education plan requires frequent consultation between the teacher-coordinator, the on-the-job supervisor, and the student-learner.

KEY CONCEPT: Cooperative education is a plan for achieving instructional goals within a particular occupational curriculum through a training station and in-school–related instruction.

GOALS

After successfully completing the study of this chapter, answering the questions, and carrying out the activities, the student should be able to:

- Explain how cooperative education is a plan of instruction.

- Discuss the place of the cooperative education plan in the curriculum.

- Identify and explain the characteristics of an effective cooperative education plan.

- Distinguish between Plans A, B, and C as cooperative education plans.

- Identify and explain specialized cooperative education plans.

KEY TERMS

combination plans

cooperative plan of career and technical instruction

coordination

method of instruction

preparatory course

program of instruction

sectional cooperative plan

specialized cooperative education plans

student-learner

teacher-coordinator

training station

training station sponsor

The essential differences between general work experience plans and cooperative occupational education plans were presented in Chapter 1. The major point was that although in both cases the work environment was a focus for learning, the purposes of the two plans were quite different; one was designed to develop general education outcomes as the primary objective, while the other was designed to develop career and technical education (occupational competence) outcomes as the primary objective. It is this fundamental distinction that undergirds the nature of the *cooperative plan of career and technical instruction,* which has frequently been used as a structural plan to provide education for job entry. The plan has been widely used in marketing education (ME) and in trade and industrial education (T & I) for many years at the secondary level in the public schools. In the past 25 years it has become popular in business education (BE) and in other specialized occupational areas, such as agricultural education (AG ED), health occupations (HO), and family and consumer sciences (FCS).

Applying the Plan

The general nature of the cooperative plan of career and technical instruction was introduced in Chapter 1. This chapter provides a more extensive treatment of its structure. In addition, cooperative plans are described briefly as they currently exist in the various occupational areas in CTE. It is important for the teacher-coordinator, counselor, and administrator to recognize the elements of the cooperative plan that are commonly applied in different occupational education areas. It is imperative to understand the variations in the everyday practice of program planning and operation that are made necessary by the differing instructional needs in each occupational area. These are further detailed in Chapters 17 through 22.

When secondary school cooperative CTE plans in the curricula of various occupations are analyzed, the common elements of program operation become apparent. This commonality of elements occurs because the cooperative plan of instruction is exactly that—a plan of instruction that is applied to different occupational curricula goals and curricula structures. One cannot conclude that everything is done in exactly the same way. The common elements of program organization and operation must be applied differently in each occupational area. The reason for this lies in several curriculum factors:

1. The nature of the skills, attitudes, and knowledges required in each occupational cluster job title and their relative degree of importance for performance.

2. The type of curriculum pattern of which the cooperative plan is part.

3. The degree to which learnings are developed primarily in school versus primarily on the job.

LOCATION OF THE INSTRUCTION

The cooperative plan represents a tripartite instructional situation in which the various psychomotor, cognitive, and affective (attitudinal) learnings make up the package of competencies required in any given occupation. Some are learned only in school in the related class and applied on the job. Some are learned solely on the job, while others are learned partially in school and partially on the job. In addition, the career objectives of the student mandate that certain competencies, such as how to train and how to supervise, be learned in school, even though they are not required at the student's training station job. Further, the cooperative plan also gives emphasis to discussion in school of problem situations encountered on the job, such as how to get along with other employees and how to benefit from supervision and constructive criticism.

COMPETENCIES TO BE ACQUIRED

The various combinations of skills, knowledges, and attitudes required in each occupation affect the way that cooperative instruction occurs. The occupational competencies vary not only in kind but also in the degree to which they are important. For example, retailing is an occupation in which human relations abilities are far more a "core" of occupational competencies than they are in many industrial occupations. It is also quite true that the occupations vary in the degree to which the skill can be developed in school and applied on the job, as opposed to being developed primarily on the job. For example, basic keyboarding and word processing skills must, and can, be learned in school under controlled conditions. On the other hand, in some health occupations, the equipment on which skill is to be developed is simply too expensive and specialized to be included in the laboratories of most schools. A somewhat similar situation exists in marketing education, where concepts and principles can be learned in school but where the job is needed as a "live" laboratory to provide real human relations situations for customers. However, in all occupations the job provides the student-learners with the realistic climate of co-workers and supervisors and the standards and conditions of actual employment.

THE PLACE OF THE COOPERATIVE PLAN IN A CURRICULUM

The cooperative plan, as has been explained briefly, is indeed that—a strategy, structure, and system of instruction that achieves predetermined goals and competencies in a given occupational field. As a plan of instruction, cooperative education can be the *capstone* (or senior and exit experience) of a curriculum consisting of two or more preparatory courses, or it can be *self-contained* as a complete curriculum in itself, in which case it provides all learnings without prior course work in the field.

In the first case, the faculty assess the career objectives of the students and decide that there are several courses needed to assist each student in meeting this goal. Beginning courses usually provide basic knowledges as well as help students develop skills foundational to an occupational cluster. The cooperative plan is used to "top-off" the student's training by providing for advanced development of skills, acquisition of advanced knowledges, application of these and prior learnings to a real occupational situation, and learning of new skills on the job. The plan affects the integration of prior and newly learned abilities into a complete set of behaviors that meet occupational requirements. The capstone approach thus integrates skill and knowledge competencies into final performance, while at the same time, providing the opportunity for each student to learn individual skills and knowledges not previously possessed.

In the second case, the cooperative plan of instruction is perceived as a complete curriculum. It does not rely on prior learnings by the student. Similarly to the capstone plan, the self-contained plan has competency-based performance goals, which are those of the marketplace. Dissimilarly, the self-contained plan does as much as possible for the student in "one shot"—it teaches beginning skills and knowledges and helps the student to integrate these into a final performance standard. The self-contained program has the great burden of taking on an untried student, whose goal may be less than clear, and moving that person to a final and finished performance level consistent with job-entry requirements.

It should be clear that the cooperative plan, when applied to these two approaches to curriculum-building, places very different stresses upon the planning and management of instruction in the related class and in supplementary courses in terms of the content to be taught and the sequence of instruction. Differences will occur in the job training station experiences needed because the student who is prepared by prior classroom and laboratory work needs more advanced rather than elementary experiences on the job.

This discussion of the nature of occupations and their learning requirements should demonstrate that while the elements of the cooperative plan of instruction appear similar at first glance, they are applied differently, depending upon the instructional area to be served.

Cooperative Education— A Program or a Plan?

Calling cooperative education a *"program of instruction"* rather than a "plan of instruction" has had a real fascination for local school personnel, their administrators, and state staff educational consultants (supervisors). Their rationale seems to be that the term "program" connotes an identifiable whole and can be used as a visible object of publicity. As an entity, it (the program) can "seek" resources in its behalf. It can be "sold" or merchandised to the local educational groups, to the students, to their parents, and to employer groups. This very real usefulness as an identifiable entity is what gives birth to some very real educational dangers. When called a "program," rather than a "plan of instruction," the entity has the disadvantages of:

1. Being difficult to change due to the vested interests of those who created it or operate it.

2. Developing in the minds of those who operate it a belief in its operational procedures and the effects of any change upon the publics who support it. In contrast, a plan of instruction exists primarily for the student.

3. Being seen by educational administrators asked to install it as a completely new program. New programs not only require a policy decision but also involve commitments for new funding expenditures.

4. Becoming, in at least some cases, a thing apart from the on-going instructional concerns of the faculty. This occurs especially where the cooperative pro-

gram is operated out of an administrative unit, such as a secondary school district office, and away from the mainstream of the high school or post-secondary classrooms, corridors, and teachers' lounges.

5. Emerging as the province of one person, the coordinator, who may begin to see the program in personal terms as being "my baby" and even "my program."

6. Showing at times the tendency to resist expansion lest control be lost with increasing numbers, with application to many learning areas within the school, or with the inclusion of undesirable students.

DIFFICULTIES IN EDUCATIONAL PLANNING

The viewing of cooperative education, or work experience education, or work-study, as a program rather than a plan of instruction has caused immense but hidden difficulties in educational planning and change. How much better it would be to see it for what it really is—an arrangement or a way of achieving learning outcomes. When thus perceived, the plan would be the one chosen if it truly served to bring about learnings needed by the student and if it truly served the goals sought by the teaching group using the plan. In the latter case, the faculty in health occupations education, as an example, would see the plan as their way of better achieving what it is they wish to achieve. Or, in another example, the counseling staff would see the plan as a means of facilitating the occupational exploration concerns with which they deal. Or, the human resource specialist would see the plan as a means of assisting a trainee who is economically disadvantaged to deal with and overcome problems of job adjustment after months or years of failure.

Lest this criticism of calling cooperative education a program rather than a plan seems too harsh, it must be remembered that dedication and zeal are to be praised. Most programs are very responsive to local needs. They exhibit adaptation to change. But, all too frequently, the program has become a "thing," and its institutionalism has retarded and even opposed change at the expense of those it was designed to serve—the students.

IS THE COOPERATIVE PLAN A METHOD?

From reading periodical articles and state handbooks and listening to speeches, a person may determine that the cooperative plan is defined as a *method of instruction.* The analogy is to liken the program to a teaching method, such as lecture, case problem, or project. But this analogy is entirely incorrect. The cooperative plan of instruction is a plan, a strategy of instruction. It uses those teaching methods, such as the lecture and the field trip, which are appropriate to the learning styles of the students and the outcomes sought. Further, any effective plan of instruction uses methods appropriate to the nature of the content of instruction, for example, the scientific method of teaching in the physical and natural sciences. The cooperative plan is not a method of instruction, although unlike most areas of instruction, it does use laboratory learning experiences in the community on a real job as *one* instructional method.

There can be no real doubt—the cooperative plan is not a program, nor is it a method. It is a way of arranging all kinds of learning experiences into a suitable pattern for bringing about the occupational learning experience of the students so that they achieve occupational competence.

Characteristics of the Cooperative Plan

The cooperative plan is exactly that—a plan for achieving instructional goals within particular occupational curriculum. The successful plan in operation exhibits the following characteristics:

1. **Instruction in school**—Two kinds of instruction are necessary: (a) basic related instruction—those concepts, knowledges, skills, understandings, and attitudes needed by all the students as a core of their occupational preparation—and (b) specific related instruction—those concepts, skills, and attitudes needed by each student-learner to handle the duties and responsibilities at the *training station* and to prepare for advancement toward the chosen career objective.

2. **Selected training station**—A cooperating business, industry, or government agency is selected according to criteria that measure its ability to provide the opportunities for those supervised educational experiences that prepare the student for his/her intended career objective.

3. **Student-learner with a career objective**—A *student-learner* is a person who has enrolled in the cooperative education plan for the expressed purpose of preparing for an occupation or an area of occupations. This person is considered to be a *student* in the secondary school and a *learner* in an occupation in the supervised business laboratory experiences.

4. **Preparatory courses**—*Preparatory courses* stem from a plan of instruction that includes a planned series of courses to develop an individual's occupational information and those skills, concepts, and attitudes needed before placement as a cooperative student-learner.

5. **Step-by-step training plan**—This is a written plan indicating what is to be learned by a particular student-learner and where it is to be taught—the classroom or the training station. The plan is derived from an analysis of the tasks, duties, and responsibilities of the student-learner in the part-time occupation and in the future career position. It is developed jointly by the teacher-coordinator and the training station sponsor, with assistance from the student.

6. **Adequate on-the-job supervision**—This involves the appointment of a *training station sponsor* (an experienced employee, supervisor, or manager) who is directly responsible for the occupational learn-

ing experiences of the student-learner on the job as shown in the step-by-step training plan.

7. **Qualified teacher-coordinator**—The *teacher-coordinator* is a person employed by the educational agency to direct the cooperative education plan of instruction. The teacher-coordinator possesses the technical education, professional education, and occupational experience necessary for success as a career and technical education teacher. The teacher-coordinator teaches the daily CTE classes at the school and coordinates the employment learning experiences with the school learning experiences of each student-learner. Sometimes the teacher-coordinator of the secondary education program handles adult education and training in the community.

8. **Adequate coordination time**—There must be adequate time, depending upon the number of student-learners and the geographical distribution of the training stations, for the coordinator to carry out efficiently the organized program of activities that unite the training station experiences with the classroom learning experiences of each student-learner into one meaningful plan. *Coordination* may be thought of as all those activities outside the classroom performed by the teacher-coordinator to organize, administer, operate, and/or improve the cooperative education plan.

9. **Suitable classroom facilities and instructional materials**—Facilities and materials adequate for teaching career-technical knowledges, skills, and attitudes are necessary for occupational preparation. As a minimum, the classroom should accommodate not only group instruction around movable tables and chairs but also small group or committee work with reference materials available in the room and private coordinator-student conference activities. In some curricula it must also provide for skill development and laboratory practice. A special budget should be available for instructional materials, supplies, equipment, and maintenance.

10. **Well-defined school policies regarding the plan**—A local plan of administration and operation regarding student-learner recruitment and accep-tance of the plan, class scheduling, school credit for both the classroom phase and the on-the-job phase of the plan, minimum essentials for training stations, responsibilities and duties of the teacher-coordinator, and so forth, should be prepared.

11. **Well-organized records**—A records system with physical facilities to accommodate it should be implemented so as to maintain current records on students and training stations as well as follow-up records on graduates. Library references, text materials, and individual instructional materials should be cataloged and systematized.

12. **Use of an advisory committee**—A committee representing students, schools, parents, business, and industry should be appointed to serve in an advisory capacity to the administrator and the teacher-coordinator.

13. **Organization of a student organization**—This organization provides the opportunity for leadership training in marketing activities.

Common Secondary Curriculum Patterns

The state plan of career and technical education in almost every state usually allows local boards of education the option of using any one of three patterns for the cooperative plan. These patterns were originally known as Plans A, B, and C[1]. Plan A is a two-year plan, typically at the eleventh- and twelfth-grade levels, which consists of related instruction for one period daily for both years, accompanied by a minimum of 15 hours per week of occupational experience during both years. Plan B is a one-year plan at

[1]Some state plans revised under provisions of the Vocational Education Act of 1963 have dropped the designation of Plans A, B, and C. However, they are used in this text partly for historical reference, but primarily because they are a useful classification system. Some states also allow modification of these basic plans where local needs demand it, as in modular schedules and programs for individuals who are economically and/or academically disadvantaged.

the twelfth-grade level and consists of a double period daily of related instruction plus the 15 hours minimum per week of occupational experience. Plan C is also a one-year plan at the twelfth-grade level. Like Plan B, it calls for one year (450 hours) of occupational experience. However, the related instruction is reduced to one period daily, but a preparatory curriculum is required, which is described as two units of subjects that form the common core of the particular occupational curriculum. Plans differ in the exact requirements for Plan C. In some states the preparatory subjects must be taught by the cooperative teacher-coordinator; in other states a related subjects teacher may be employed. In some states the preparatory units are one-year courses, while in others they are one-semester courses or quarter courses.

All three plans have merit according to their adherents.

1. Plan A is superior because it provides twice as much on-the-job learning over a two-year period. It also helps prevent students from dropping out by allowing them to earn and learn earlier in their high school careers. Plan A was widely used in many states in earlier years. It is much less used now because of increasing general education requirements and a higher number of credits required for graduation by many high schools and by many states.

2. Plan B fits better in contemporary high school curricula and is useful for students who choose CTE late in their high school careers. It also provides for intensive and in-depth instruction during the double period of the senior year.

3. Plan C is very popular in some states, particularly in those in which high schools operate on a six-period day with two required general education subjects (allowing only one period for CTE instruction). Plan C also is believed by some educators to provide the advantage of the preparatory class being a "screening" device, giving the student salable skills, knowledges, and attitudes *before* the student is placed in a training station.

There is little evidence to support the merits of one plan over another. Each appears empirically to be useful in certain local schools.

Specialized Cooperative Plans

Section Four is devoted to *specialized cooperative education plans* in agricultural occupations (Chapter 17), business occupations (Chapter 18), health occupations (Chapter 19), family and consumer sciences occupations (Chapter 20), marketing occupations (Chapter 21), and trade and industrial occupations (Chapter 22). As mentioned in the Preface, students interested in a specialized area should study the appropriate chapter in depth but familiarize themselves with the other specialized area plans as well. This familiarity will enable them to operate intelligently in a community having more than one plan in operation.

Plans Involving More Than One Occupational Area

Some school systems have developed cooperative plans that involve special occupational areas. For convenience, these plans may be referred to as "combination plans." They have evolved from several needs. Small high schools (defined as those having less than 300 to 500 enrollment in grades 9 through 12) have adopted combination plans because the number of students wishing training in a given occupational area, such as the trades, is too small to support a full cooperative plan in that specialized occupational area. In addition, small towns often do not provide enough training stations in a given occupational area, and high-cost school laboratories are not feasible. This situation of too few students or too few training stations also occurs occasionally in larger schools in suburban areas without much business or industry or with heavy enrollment in the college-bound program. Some larger schools also inaugurate cooperative education with a combination plan and then split the plan into separate occupational plans as enrollments rise. This latter course of action is not recommended in most cases because there is a tendency to continue the plan rather than to split it, on the grounds that "it works, and besides, additional coordinators would put too much strain on the budget." On the other hand, it must be recognized that in smaller communities, particularly in rural areas, the combination plan is the answer to career and technical education—without it there probably would not be any CTE training in areas where on-the-job learning is needed.

Types of combination plans.—The term *"combination plans,"* as used in this chapter, refers to cooperative plans that involve more than one broad occupational grouping. The problem of semantics is a difficult one in the field of cooperative education and causes great confusion when educators in different states use either the same term to describe different plans or different terms to describe the same plans.[2] Agreement among career-technical educators on terminology is sorely needed.

There are basically two different kinds of combination plans. These are:

1. **A true "diversified occupations plan"**—In this plan, one person is the teacher-coordinator for all cooperative students who represent a broad spectrum of occupational areas—business, marketing, and trade and industrial, as well as health and family and consumer sciences. In some programs, agribusiness students have also been included. In such programs the typical pattern of instruction is one of two:

a. The teacher-coordinator has all students together in class, using group instruction on topics common to all occupations (general related instruction) and individual instruction (specific related instruction) applicable only to each student in his/her occupation.

b. The teacher-coordinator has all students together in one class, as in "a," but in addition "farms out" each student to another class, in which the instruction is related to the student's occupational goal.[3]

2. **A sectional cooperative plan**[4]—This is a plan in which directly related instruction is given separately to students in each occupational area, but in which one teacher teaches both classes and coordinates both groups. The most common combinations are business and marketing, although a few marketing and industrial combinations are found. The problem of any combination plan is to find a teacher who is prepared in both areas in terms of technical content, professional methods, and occupational experience.

There are many problems in operating combination plans. Combinations tend to be less effective instructionally than separate plans, particularly when a coordinator trained in one field must teach and coordinate in other fields. The means of providing directly related instruction for all students presents a paradox. If the student-learners are kept together in one class, then their specialized instructional needs must be met by individual instruction—a relatively inefficient method, considering the instructional materials that must be found and purchased and the

---

[2]For example, in Ohio, the term *"diversified occupations plan"* refers to a plan of various industrial occupations; whereas, in Michigan and Florida, the term refers to a plan involving office, marketing, and trade and industrial occupations. On the other hand, Minnesota uses the term similarly to Michigan but calls the Ohio plan a "miscellaneous trades" cooperative plan.

[3]Some schools in some states operate plans in which coordinators do not teach their students. Instead, the students are farmed out to other classes for all instruction. Such plans are not true cooperative plans because the instruction is not directly related to the job. Rather, they meet the definition of "work-study" plans, as described in Chapter 1. Unfortunately, the quality of many such plans affects the reputation of true cooperative plans.

[4]The term *"sectional cooperative program"* was used in the 1949 edition of the policy bulletin of the U.S. Office of Education, Vocational Division, but was dropped following the passage of the Vocational Education Act of 1963.

reliance that must be placed on the student's intellectual maturity to cope with individual learning and on a teacher who may not be competent in the student's field. On the other hand, if student-learners are farmed out to other classes, they have a teacher who is not the coordinator, who cannot really know their job situations and their needs and problems, and who is likely to teach primarily for the non-cooperative students in the class.

The best solution to the problem is either (1) the sectional cooperative plan when a teacher-coordinator can be hired who is competent in both areas or (2) a diversified plan which has two periods of instruction, one taught by the teacher-coordinator for all student-learners and the other in which students are farmed out to appropriate related subjects teachers. Such plans are further strengthened by requiring prerequisite preparatory courses in the student-learners' occupational interest areas.

In summary, it must be recognized that in a small high school, particularly in a rural area, a combination cooperative plan of some type is better than no career and technical education at all. On the other hand, school administrators and CTE leaders must guard against the installation of combination plans in schools where it is possible to gain the effective specialization in separate occupational clusters.

Combination plans using the team approach.—Career-technical educators are well aware that the changing nature of the occupational world has created the need for workers who have competencies in two or more occupational areas. For example, the food sales and service trades encompass knowledges and skills from both the marketing education area and the family and consumer sciences area. Some occupations involve both agricultural abilities and marketing abilities. Other occupations involve varying combinations of competencies derived from the areas of business education, marketing education, trade and industrial education, agricultural education, and family and consumer sciences education. There is no doubt that such occupational combinations exist,

---

CASE-IN-POINT

Specializing a Cooperative Plan

Allen Lloyd, teacher-coordinator of diversified cooperative education in Center Point, has 15 student-learners with training station placement and career objectives in health occupations in related classroom instruction. He has also determined that there are several more students interested in health occupations for the new school year.

Therefore, with the advice of the general advisory committee and the permission of his local director of career and technical education and the superintendent of schools, he decided to remove these health students from the diversified plan group and set up a specialized health occupations plan. This action, ideally, would necessitate the hiring of a teacher-coordinator certified in health occupations.

He proceeded to set up an advisory committee for health occupations and to survey the business community to confirm that there are adequate training station possibilities and is now helping the administration recruit and hire a certified teacher-coordinator.

After the new coordinator has been hired, Allen and the director of CTE will help get the health occupations plan started on the right basis.

---

although much research is needed to determine the competency "mix" of such occupations. There is a need in career and technical education for developmental programs that attack the problem of how to devise cooperative occupational education plans that will serve these "occupational mixes." One answer may lie in the diversified occupations plan described earlier in this chapter. The plan of instruction might consist of three hours a week of group instruction related to common problems of all workers (taught by the teacher-coordinator), with student-learners being farmed out to related classes in their career fields for five periods per week; and this may be supplemented by a two- or three-hour block of instruction one afternoon a week in the individual occupational specialty. An alternative solution might be a double-period block in instruction, which is team-taught by specialists in various occupational areas. Exact answers are not known; creative experimentation is needed. In many cases, this problem has been obviated by the creation of shared-time area career-technical schools.

Questions and Activities

1. Select from "Characteristics of the Cooperative Plan" those that are the absolute essentials—the areas critical to success of the cooperative plan in a particular occupational curriculum. Defend your choices.

2. Describe specific examples of why and how the elements of the cooperative plan differ between two occupational curricula. Which elements, if any, remain essentially the same in practice?

3. Check the terminology used in your own state plan and manuals of operation for the various cooperative occupational education programs. Pay particular attention to your own interest area of cooperative education. How does your state plan or local plan compare with the suggestions in this chapter?

4. Be prepared to discuss the program approach to education in any one of the CTE areas (agricultural, marketing, family and consumer sciences, trade and industrial, business, and technology education). Does the series of courses and experiences available to the career-oriented student end at the twelfth-grade level? Explain your answer.

5. Using a rational argument and a specific example, defend or oppose the approach of using the cooperative plan for less than one year.

6. Should the cooperative plan as described in this chapter accommodate young people who are dropouts or potential dropouts? College-bound? Defend your stand.

7. Check the current sources of information on work opportunities in your special field of interest. (a) Determine if there have been any appreciable changes in the types of positions for which young people should train. (b) Describe the career ladders.

8. Prepare a brief paper summarizing cooperative education enrollments in your state. Describe trends and, if possible, analyze data showing the sizes of schools in which cooperative plans are found.

For references pertinent to the subject matter of this chapter, see Reading Resources.

5

Coordinators and
Their Roles

The teacher-coordinator must correlate classroom instruction with on-the-job training.

KEY CONCEPT: Successful cooperative education plans are built by competent teacher-coordinators.

GOALS

After successfully completing the study of this chapter, answering the questions, and carrying out the activities, the student should be able to:

- Tell other interested parties about the many unique roles of the teacher-coordinator.

- Identify and explain the characteristics possessed by successful teacher-coordinators.

- Identify and explain the competencies needed by teacher-coordinators.

- Demonstrate how to administer the cooperative plan.

- Discuss the reasons why teacher-coordinators join and serve professional organizations.

KEY TERMS

Association for Career and Technical Education
annual contract
annual report
Cooperative Education Association

counselor-coordinator
directed occupational experience
learning manager
monthly enrollment report
professionalism

In final measure, the quality of educational programs does not depend upon space and equipment and instructional materials but upon the competence of the personnel involved. This is perhaps more evident in cooperative occupational education plans than in most other school endeavors because teacher-coordinators are highly visible figures. Their teachings and other actions are scrutinized by many publics

Students are highly responsive to them not only as classroom teachers but also as persons who are a part of their out-of-school lives. Administrators see them as representatives in the community, who, at times, see more people than they do. Counselors are aware of coordinators' close relationships with students at critical times in their lives. Other teachers know that coordinators are daily aware of what is happening in the occupational life of the community. This visibility of teacher-coordinators leads to an inevitable comparison of performance by many publics, according to their standards. It is compounded by the cooperative plan (or any work experience plan) being a shared-time program in which the institution and the community join hands in a common endeavor. This sharing of the inputs of time and money resources, as well as the responsibility for a student's learning, results in a unique situation in education. There is pressure on the coordinators for management abilities that will assure the success of the plan. The teacher-coordinator is in the spotlight, just as is the marching band director or the football coach.

The Basic Role

It has been said by those who are not well-informed about the cooperative plan of occupational education that a coordinator "is a guy (or gal) who teaches a couple of hours and then goes downtown to drink coffee!" All too often this statement is evidence that coordinators have failed to make administrators and teachers understand what their full range of responsibilities consists of and that what constitutes getting action in a world of work may be quite different from what the in-school person believes it to be.

Coordinators must be recognized as being unique educational professionals, who, at one time or another, are:

1. **Teachers,** who must be among the best, especially at individual instruction.

2. **Student organization advisors,** who assist students in developing employment skills and leadership abilities.

3. **Public relations persons,** who may come in contact with more taxpayers and voters than the superintendent does.

4. **Counselors,** who deal firsthand with educational, social, occupational, and personal problems.

5. **Artisans,** who know a trade and the language of the trade and have their employers' respect for it.

6. **Administrators,** who keep reports and records and arrange the schedules of others.

7. **Evaluators,** who engage not only in classroom measurement but also in the measurement of students on the job and of the contributions of the plan.

8. **Planners, organizers, and managers,** who supervise an instructional system.

Teacher, counselor, administrator, student organization advisor, and public relations person—a successful teacher-coordinator is all of these. Although no teacher-coordinator of career and technical education (CTE) is perfect, a person considering entering the field of cooperative education should be cognizant of the qualifications of the ideal model, just as students should perceive the qualifications of the model worker.

Types of Coordinators and Their Roles

As described in Chapter 1, there are many and diverse educational plans that use the work environment to achieve learning goals. These plans are so varied in goals and operational procedures that there are widely differing roles for the person who directs them. Some of these positions are as follows, though not necessarily in order of importance:

1. Director of a general work experience plan or a work-study plan.

2. Work exploration *counselor-coordinator.*

3. Coordinator of professional internships or practicums (at the baccalaureate or graduate level).

4. Teacher-coordinator of a specialized occupational training plan at the secondary, post-secondary, or adult level.

5. Coordinator of all directed occupational experiences, regardless of the field, in a small school or a shared-time area career-technical school.

Of these types, only the roles of the teacher-coordinator, the coordinator of the occupational education plan, and the director of the work exploration plan are discussed here because they deal directly with the process of occupational and career development programs at the secondary, post-secondary, and adult levels.

A person designated as coordinator may be involved in three major activities: (1) coordinating on-the-job learnings related to a career goal, (2) teaching the class in school, which is related to job needs, and (3) career guidance and counseling. There will be many times that a coordinator must assume the role of a teacher . . . or a counselor . . . or a combination of the functions involved in these positions. The job description of the primary role of the coordinator will have a direct effect on the kind of individual needed because of the type of job to be done and the relationship between the school and the community.

THE TEACHER-COORDINATOR OF THE COOPERATIVE OCCUPATIONAL EDUCATION PLAN

The CTE teacher-coordinator has two main responsibilities: (1) being the instructor of the occupational orientation and related skills training laboratory class in which all trainees are enrolled and (2) being the coordinator of the job laboratory experiences for the occupation.

Responsibility includes the training and cooperative work experience activities for only one group of trainees. For example, if there are 20 trainees in the related instruction, coordination covers the cooperative experiences for those 20 trainees only. An obvious advantage to this kind of plan is the constant and meaningful communication with regard to the learnings at school and on the job between the school and the training sponsor at the training station. Another prime advantage is the opportunity for the coordinator to develop instruction that is responsive to the needs of the trainees.

The teacher-coordinator has, from contacts in the community, an intimate knowledge of industry and its requirements. This not only assures the training needed to meet employment standards but also paves the way for job placement opportunities later. The teacher-coordinator, who is a competent trades worker, technician, and businessperson, earns the respect of colleagues at the school as well as that of associates in the business or industrial world. If this basic approach of occupational competence is used, the school will employ one teacher-coordinator in each occupational area being taught. Therefore, eight teacher-coordinators would be needed if eight occupational areas were to be offered. Further, the teacher-coordinator should be a staff member in only one building and not be asked to shuttle between several schools in the district. Shuttling frequently causes the teacher-coordinator to be considered an outsider by the building staff.

With a storehouse of contacts in business and industry, the teacher-coordinator can be one of the most effective occupational resources the school has. As a "floating" teacher, the part-time coordinator can teach classes on topics such as employability skills and personal attire and can invite individuals from industry to speak to these classes, because of an acquaintanceship with their contributions to the occupational courses. As a resource person, the coordinator can provide the counselor with information about the personal and educational problems a trainee faces.

THE COORDINATOR OF THE COOPERATIVE OCCUPATIONAL EDUCATION PLAN

The main responsibility of the coordinator who does not teach the related class is to place all trainees in proper training stations and to provide the necessary coordination between (1) the employer and the teacher of the related class and (2) the employer and the counselor. Coordination of communication within the school (administrators, instructors, counselors) is an essential part of the coordinator's responsibilities. In some schools there is but one coordinator handling trainees in all fields in what is commonly called the diversified occupations plan. Typically, in these cases, the coordinator does not teach the related class, and trainees are usually farmed out to an occupational course in each trainee's career interest area. When the coordinator is not assigned the basic responsibilities of teaching, that person is able to devote full time to the coordination functions, *i.e.,* contacts in the community, communications with others concerning the plan, and placement of trainees. On the other hand, the coordinator suffers the disadvantage of not directly teaching students and must attempt to get another teacher who understands the student's needs and who will change the sequence of instruction to meet on-the-job needs.

THE COUNSELOR-COORDINATOR

The two main responsibilities of the counselor-coordinator are: (1) counseling individual trainees in a general work experience or work-study plan and (2) directing on-the-job experiences for these students. These and other guidance- and counseling-related duties can account for approximately 50 percent of a counselor-coordinator's total responsibility. The counselor-coordinator is also capable of providing personal and work adjustment counseling needed by the trainee because of his/her ability to relate to those involved with the work experience plan and to understand the needs of individual students. There is an immediate advantage in using this plan: the counselor-coordinator has a direct and positive influence on the trainees, particularly those who are classified as learning-disadvantaged. However, the counselor-coordinator's lack of occupational knowledge in regard to job requirements and opportunities may be a handicap. In addition, there is usually no class or regular group instruction time when the trainees can relate to the person who knows them and their job situation or when they can profit from relating to other students with similar problems.

The Coordinator's Image

Coordinators, whether of work experience or occupational career education, must sell **themselves,** the **school plan,** and the **individual trainee.**

Themselves.—Coordinators must believe that employers will feel it is worth their time to listen. This will be demonstrated if the coordinator has an obvious enthusiasm and belief in the cooperative plan. As one coordinator said, "The employer is always interested in talking about the business, and you will need to be familiar with the product and procedures, as well as the elementary technical language of the field. The employer will give you time if you are willing to give of your time. The employer will appreciate your having taken the time to listen to problems and to discuss the program and the trainees."

The school plan.—Coordinators will have to sell their school's plan of instruction to many employers who have never participated in an endeavor as a part of a school's curriculum, as well as to employers who dislike anything that "hints" of federal money. Although many employers are interested in helping young people, they are primarily interested in what the plan will cost in time and money; thus, the coordinator should always talk in terms of what the plan can do for a particular business, inform the employer of others in the industry who are participating in the plan, and explain why this plan will work in comparison to other government training programs with which the industry may have been associated. Coordinators will have to sell the plan to teachers and counselors. By asking their advice and help on individual trainees' placement and by involving them in the process, coordinators may help them to see the value of the plan.

The individual trainee.—It is much easier to sell an employer on providing a training station for a specific person than on a plan that is impersonal. Employers should be told in advance about the trainees' strengths and weaknesses. As programs respond to the mandates for the mainstreaming or inclusion of students who are economically disadvantaged, who are members of minorities, and who are physically and/or mentally disabled, it is especially important that the employers be informed about the trainee and agree to be a partner in the student's instructional program. Basic to this communication between the coordinator and the employer is the philosophy of developing the station to fit the trainee, rather than developing the trainee to fit the station.

given job in a cooperative occupational education plan. In no way can the administrator interpret the assignment of one coordinator as being equal to that of another, either within the buildings of the district or in relation to others nearby.

VARIABLES AFFECTING THE TEACHER-COORDINATOR'S LOAD

The cooperative occupational education plan of providing learning experiences is individualistic to a particular situation, and the time arrangements made for the teacher-coordinator to carry out the assignment must be evaluated in terms of the variables of the situation. The task list in the following section reflects one state's viewpoint about the tasks performed by teacher-coordinators and the variables affecting the coordinator's load.

THE TEACHER-COORDINATOR'S CONTRACT

The teacher-coordinator's contract should include the following elements:

1. **Employment period**—extended year and/or extended day (after school hours or nights and weekends).

2. **Salary**—paid according to scale as teacher with credit for years of occupational experience. (The authors do not support paying the teacher-coordinator on the supervisory scale in a school system.)

3. **Benefits**—paid for local mileage and supported by travel for professional meetings and conferences.

The Job of the Teacher-Coordinator

The administrator who is understanding of load factors for the staff will recognize that there is no magic formula for calculating the time needed to do a

A State Viewpoint About Teacher-Coordinator Tasks

Each state that has ground rules for cooperative occupational education plan operation has a viewpoint about what teacher-coordinators are expected to do. The following is an example of a task list derived from a bulletin in a state that has been involved in cooperative occupational education for many years.[1]

Guiding and Selecting Students

Describing the plan to students.
Working with guidance personnel.
Providing occupational information.
Counseling students about entering the plan.
Gathering information on students.
Helping enrollees with career planning.

Placing Students in Training Jobs

Enlisting participation of cooperating employers.
Selecting suitable training station for each student.
Orienting employers, training supervisors, and co-workers.
Preparing students for job interviews.
Placing students on the job.

Assisting Students in Adjusting to Their Work Environment

Helping students on their jobs.
Dealing with job problems.
Planning personal development with training supervisors and students.
Evaluating job progress.

Improving Training Done on the Job

Establishing responsibilities on the job.

Developing training plans.
Consulting and assisting training supervisors.
Maintaining training emphasis.

Correlating Classroom Instruction with On-the-Job Training

Determining needed instruction.
Assembling instructional materials.
Preparing for instruction.
Teaching classes.
Directing individual projects and study.
Obtaining assistance from other teachers.
Advising training supervisors concerning applications of classroom instruction to be made on the job.
Evaluating learning outcomes.

Assisting Students in Making Personal Adjustments

Aiding students in correcting poor personal habits.
Counseling students with personal and socio-economic problems.
Assisting students with educational problems.
Resolving behavior problems.

Directing Career-Technical Student Organization

Advising student organization.
Guiding students in organization of activities.
Participating in student organization activities.

Providing Services to Graduates and Adults

Providing guidance and placement service for graduates.
Participating in the planning and operation of adult training and development programs.

---

[1]*An Articulated Guide for Cooperative Career Education,* Bull. No. 34–571, Springfield, Illinois: Division of Vocational and Technical Education, State Board of Vocational and Technical Education, State Board of Vocational Education and Rehabilitation, pp. 7–10, 99–101.

Characteristics of Successful Teacher-Coordinators

Coordinators are often asked to be "all things to all people." Successful coordinators possess a combination of characteristics and abilities that enable them to meet the objectives of the cooperative occupational education plan: to train trainees for a given occupation and to assist them in keeping the jobs for which they are trained.

Coordinators who have the basic competencies (discussed later in this chapter) should stress their own professional and personal developmental programs to attain the following characteristics:

1. Knowledge about the occupations in the geographic area and about their skill requirements.

2. An understanding and a feeling for all antipoverty and work training programs in the area, as well as for CTE training programs in schools, colleges, and industries.

3. A commitment to the equality of individuals and to the elimination of discriminatory practices.

4. Sensitivity to the needs and feelings of people—not just trainees—understanding the success, failure, or frustration an individual might feel.

5. Public relations skills—to be aggressive and outgoing, persuasive and communicative as salespersons of the plan, to be able to speak clearly (even if not eloquently) before an audience, and to be self-confident and determined.

6. Self-realization through career education of a continuing nature.

7. The ability to administer the program effectively—to influence and guide without appearing too direct and to handle reports, calls, and appointments efficiently.

8. A personal appearance that is in accord with the accepted standards in industry.

9. A positive self-concept—able to deal with and overcome rejections.

10. The ability to work with little or no supervision.

11. The understanding required to handle the criticism of trainees without becoming defensive.

12. The ability to turn individuals with an obvious lack of knowledge into informed citizens supportive of the special needs of certain groups.

PHYSICAL STAMINA AND VITALITY

The life of a teacher-coordinator is dynamic and challenging, but demanding of physical and mental stamina. Just as important, it should be remembered that coordination is not an "8 to 4" job and that there are some long days because students are employed at nights (even on the late nights) and on weekends. Energy over long periods is a most important asset. The successful coordinator is not a "telephone coordinator," for there is no substitute for personal contact between the coordinator and the various publics—job supervisors, parents, school authorities, other faculty members, students, professional association members, and others. Activities call for foot work, local travel, and out-of-city travel. Good physical health, mental alertness, and emotional stability are "musts."

What does it take, then, to be a successful coordinator? Among other characteristics—understanding, sincerity, sensitivity, empathy, vitality, and, finally a venturesome spirit that responds to the challenge of everyday problems.

Competencies Needed by Teacher-Coordinators

KNOWLEDGE ABOUT THE WORLD OF WORK

Teacher-coordinators teach technology to technicians, public relations to receptionists, selling to salespersons, and the principles of management to future managers. Teacher-coordinators are more

> ### CASE-IN-POINT
> **The Coordinator Who Turned into an Errand Runner**
>
> John Jones, the coordinator at Metropolitan East High School, was a congenial person. His congeniality and ability to get along with others was one of the reasons he was hired for the position.
>
> As John became familiar with his duties as a coordinator, he soon realized that his business contacts in the surrounding area were very important—to him and to the district. Perhaps his mistake was in mentioning to the principal and other staff members, in an off-hand manner, how impressive and meaningful these contacts were becoming.
>
> The first time he was asked to hand-deliver a purchase order for auto parts to the Smith Auto Company and then wait for a few "important" items to bring back—he didn't mind. Nor did he mind the second or third time. When he was in the office one morning, he even volunteered to pick up some supplies for the librarian at the district office "because I'm going that way anyway." It didn't take long for word to get around that John "would be glad to do it on his way." Eventually, he realized that out of his day, two or three hours were being used for other people's errands. He began to realize that he was being "used" and, even worse, the trainees were being cheated.
>
> Luckily for John, he realized early that it was becoming a problem. All he had to do to correct the problem was to be more selective of the kinds of errands he took care of. Luckily for the program, he learned not to be an errand runner.

than classroom teachers. They are combination teachers and personnel managers who deal with students, parents, other teachers, administrators, industry supervisors, trades workers, union leaders, managers, and owners (entrepreneurs). Since they deal with those in the community who are influential, they must possess mature judgment. It is for this reason that certification requirements include recent, varied occupational experiences relevant to the field. Prospective coordinators need real-world business or industrial experiences. Preferably, the experience should be in the kinds of jobs and firms in which their students will be working. If possible, they should have the responsibility of supervising or training workers on the job, thereby knowing firsthand the problem of maintaining work standards. Coordinators must administer the cooperative plan in a manner that encourages students to assume responsibilities in the classroom and on the job. Essential is the knack of stimulating students to maximum accomplishment. The ability to think in terms of the roadway and to talk in the language of the trade comes primarily from occupational experience or from close association with those in the trade. Teachers with occupational experience make classroom instruction more meaningful by bringing their experiences into the classroom. Thus, teacher-coordinators have credibility by "having been there."

The *length* of the occupational experience gained by a teacher-coordinator is not as important as the *quality* of the experience. For example, supervised occupational experience gained through an internship, a cooperative plan, or an apprenticeship program of 6 to 12 months and of relatively recent occurrence might very well be better than several years of repetitive work experience.[2]

It is comparatively easy for prospective teachers (or in-service teachers) to obtain part-time or full-time jobs in business or industry to gain valuable basic experience. However, it is often difficult to

---

[2]The state plan in an individual state will indicate occupational experience requirements in general. Certification requirements will translate these into specifics for CTE areas.

obtain currently both the college training necessary to qualify as teachers and the practical experience needed as coordinators. Many individuals do accomplish both during their college careers by working in business full time during the summer and part time for credit during the academic year or on their graduate programs. This alternative is the college-level cooperative education or supervised work-study plan. Coordinators should also acquire additional occupational experience after they begin coordinating in order to keep abreast of change and to expand their areas of competencies.

DIRECTED OCCUPATIONAL EXPERIENCE

An alternative to developing and measuring occupational experience and competence by clock hours has gained wide acceptance. This alternative is the creation of **directed occupational experience** courses within CTE teacher preparation programs. These courses are designed to include two complementary components: actual occupational experience concurrent with a seminar designed to supplement that experience. The seminar serves to develop and reinforce professional competence required of CTE teachers. An added benefit of directed occupational experiences is the opportunity for preservice teachers to integrate theory and practice, causing changes in their behavior and attitudes toward work and education. Within the realm of professional career and technical education, a curriculum that supplants or supplements actual clock-hour occupational experience can also be effective in determining occupational competency.

PROVEN CLASSROOM TEACHERS AND LEARNING MANAGERS

Teacher-coordinators, as learning managers, must:

1. Be excellent classroom teachers with a demonstrated ability to work with students.

2. Be particularly adept at project-type teaching.

3. Be able to plan assignments for individual students.

4. Be flexible and willing to teach *what* is needed *when* it is needed. This flexibility should prevail both in and out of the classroom—constantly challenging students and job supervisors to their best efforts.

5. Have a genuine desire to assist young people and adults to develop into competent employers, supervisors, crafts workers, and managers.

6. Be prompt, courteous, ambitious, respectful, and sympathetic—if they are to expect the same qualities in those they teach.

7. Above all, be able to help their students to develop as human beings in order to realize their ultimate potential, regardless of what it is.

No matter how successful the teachers' prior experiences have been, effectiveness in cooperative classrooms depends upon their depth of understanding of the learning process and their knowledge of how career-technical learning is implemented by proper procedures and techniques in the classroom and on the job. Teacher-coordinators should be able to project into situations confronting their students. If they understand and can empathize with their students, are courteous and considerate, and know their subject matter well, and how to put it across to their students, they will not have to resort to an overt display of authority.

Most student-learners have the desire to learn. Many have had unsatisfactory results in the past and view themselves as failures. They want and need success. They want to feel good about themselves. The cooperative plan, with its immediacy of feedback from the job, to the classroom, to the student-learner, and to the teacher-coordinator, provides motivation for the students to learn. Feedback further reinforces learning experiences. The effective teacher-coordinator harnesses the student-learner's job enthusiasm to learning experiences in the classroom.

Educational Preparation

Teacher-coordinators must usually fulfill certain academic requirements, including graduation from an approved four-year college, with a major in the technical content area associated with their specialized field in CTE.[3] College preparation may be at the undergraduate level or at the graduate level for those who have a non-teaching degree and includes professional and technical education as well as related general education. Many states are now requiring that prospective teacher-coordinators pass competency exams covering professional and technical education before a teaching license can be issued.

Professional education.—The professional education requirements usually are similar for all the cooperative career and technical education areas. For initial certification, most states require professional preparation of at least 6 to 15 semester hours in courses covering principles of career and technical education, organization and administration of cooperative programs, coordination techniques, instructional methods, development of instructional materials, educational psychology, special populations, and career-technical guidance. Student teaching in a cooperative plan is desirable, as is professional course preparation in the specialized occupational field to be taught. Some states require additional professional training for continued or permanent certification.

Technical education.—Technical course work requirements do, or should, include both breadth and depth in the subject matter areas in which the coordinator will teach. The teacher-coordinator should have a major of the kind deemed appropriate by accrediting associations for a secondary teacher. For example, the

following technical education requirements, ranging from 18 to 27 semester hours, might prevail.

Marketing Areas	Business Areas	Technical/ Industrial Areas
Marketing	Economics	Electronics
Management	Management	Machine tools
Economics	Accounting and finance	Hydraulics
Accounting and finance	Information management	Metals
Retailing	Desktop publishing	Woods
Advertising and sales promotion	Keyboarding	Plastics
Sales	Office management	Drafting
	Word processing	

Undergirding related preparation.—Teacher-coordinators need not only sound general education but also related course work in areas such as sociology, human resources, economics, psychology, speech, and counseling theory and practice. They may especially wish to consider the need for courses relating to members of minority groups and teaching of individuals who are disadvantaged and/or disabled.

To reiterate, teacher-coordinators, because they are teachers, should meet at least the minimum standards in content preparation required by accrediting associations of secondary teachers in that field, as well as have the preparation to function in the coordination role. Some states operating area schools do not hold to this standard.

Ability to Coordinate

Teacher-coordinators must understand the variety and complexity of their out-of-the-classroom responsibilities, collectively called "coordination."

Can any teacher coordinate?—If an individual has the basic characteristics for being an effective cooperative education teacher-coordinator, enjoys making contacts with adults in the business community, and is willing to meet certification requirements

[3]Prospective teacher-coordinators should check the certification standards of their own state board of CTE because state requirements vary considerably.

in the state, then the question "But, can I coordinate?" is the main question remaining. This question, expressed outwardly and feared inwardly, is the biggest hurdle that has to be faced when a person is starting a career as a teacher-coordinator. Although a successful career in CTE can lead to above-average personal and financial rewards and the fulfillment of goals, one must clear any hurdle resulting from this question.

It is logical and natural to ask "Can I coordinate?" And, the question deserves an honest answer. If one can answer it affirmatively, then that individual is well on the way to becoming a good coordinator.

Anyone considering becoming a teacher-coordinator must be willing to leave the four walls of the classroom and the confines of the school building to reach the invigorating environment of business and industry. Successful coordinators would say, then, "If people will work at it systematically, they can coordinate." Given the qualifications of successful teachers and the means of meeting certification requirements, the key characteristics common to the most effective coordinators are systematic work, persistence, and empathy for others. But, what is meant by work in coordination endeavors? Simply stated, coordinators are working when they are using coordination time wisely to help student-learners, to develop training stations, and to develop the plan. All these efforts put specifically into coordination. Systematic planning of coordination time and the willingness to work at coordination take the fear out of the task. Coordination activities have many characteristics of direct selling in that both take systematic planning and work and direct contact with people.

Simplifying Program Administration

The development of useful records and reports is recognized by effective coordinators as a vital activity. It should not be considered as wasted effort. Effective coordinators recognize early the need for records that reflect a true image of their plan, allowing them to evaluate, to back up their requests, and

A teacher-coordinator explains the cooperative plan to a school administrator. Maximum support of the program depends upon complete understanding by the school administrator.

to demonstrate their effectiveness. Successful coordinators recognize and understand the administration's needs to justify every aspect of a curriculum and to give fiscal evidence to the local board, to the state department of public instruction and to any other agency that funds the plan. Public relations built upon facts about the local plan will be more effective than unsupported opinion.

Administrators need to justify every part of the curriculum in terms of its contribution to the educational program of the school in proportion to the budget allocated to its support. It should be recognized also that the administrator is fiscally responsible to the local board, to the state department of public instruction, and to any other agency that funds the program.

Coordinators typically make reports to two groups: the state department of public instruction and their local school administration.

STATE REPORTS

If some coordinators feel that state reports are bothersome, they probably do not know that it is the responsibility of state government to justify, by evidence, the expenditures made of public funds. They may not perceive the need of state staff members for the information that is necessary to fulfill their role as consultants, to provide leadership in program development, and to engage in research. In most states the state department of public instruction requires only the following basic reports, although some states require additional reports, such as those on wages and hours, individual training plans, and names of employers.

1. **Annual contract** (a contract for reimbursement for career and technical education)—The *annual contract* is filed by the district administrators at the beginning of each school year. It usually provides for the following types of information.

    a. Name of each coordinator, field of coordination, and instructional schedule.

    b. Daily schedules of coordinators and related class teachers.

    c. Total salary and other expenses, such as planned instructional materials, supplies purchased, and travel of coordinators and related class teachers.

    d. Number of students enrolled.

This contract is sent to the state board of CTE, where it is reviewed by the state staff members who use it as a basic both for reviewing the program and for planning the funds necessary for the total state program. The approved contract is returned to the school where it becomes the basis for subsequent reimbursement.

2. **Monthly enrollment report**—Some states ask for a monthly enrollment report, aiding the staff in determining the amount of training being accomplished. This report normally includes total wages earned, taxes paid, and hours worked.

3. **Annual report and request for reimbursement**—On the basis of the contract approved by the state board, the local superintendent requests reimbursement at the end of the fiscal year by submitting data on actual expenditure of funds.

LOCAL REPORTS

The coordinator and the administrator should decide which local reports will best provide for evaluation of the plan and for its further development. The following local reports might be considered:

1. Annual descriptive and statistical report.
2. Records of achievement of individual trainees.
3. Weekly wage-and-hour reports.
4. Records on training stations.
5. Step-by-step training plans and training profiles.

Since reports 2 through 5 are discussed elsewhere in this text, only the first one, the annual descriptive and statistical report, is discussed here. An annual report on the plan will be valuable to the following: superintendent, principal, director of CTE, head counselor, department head, members of the advisory committee, state staff, and teacher-educators. The topics to be covered might include the following:

1. A brief description of the plan, including the purpose, number of student-learners in comparison with previous years, wages, and hours, list of training stations and their locations, and types of related instruction given.

2. A review of the coordinator's activities, including the number of hours spent on various activities, such as number of visitations, number of contacts to establish new training stations, professional activities, professional course work taken or practical experience gained, adult education duties, and student organization activities.

Upgrading Professional Competence

THE PROFESSIONAL FUTURE FOR TEACHER-COORDINATORS

Cooperative education has been taking on new forms under vocational education, or what is now CTE, since 1963. The National Commission on Secondary Vocational Education's report. *The Unfinished Agenda: The Role of Vocational Education in the High School*, gave a new thrust to cooperative education. The Commission found that even though " . . . cooperative vocational education . . . [was] one of the most successful aspects of vocational education . . . ,"[4] it involved too few students. The report urges school districts to expand cooperative education offerings, and it encourages business, industry,

labor, and public agencies to provide more training stations so that more students could profit from on-the-job experiences. Among the Commission's recommendations was that "cooperative education . . . be a 'capstone' element in all vocational education programs."[5]

But even though many public agencies recognize the values of cooperative education, coordinators frequently find themselves justifying the cooperative plan and their own activities. Those who choose to become teacher-coordinators must be professional in dealing with their peers in the teaching profession, with administrators, with the student-learners, and with businesspersons in the community.

During the period of adjustment to, and justification of, new activities fostered by expanded opportunities for cooperative education in the years ahead, teacher-coordinators must seek out, criticize, and oppose those factors, individuals, and activities that hamper the growth of sound cooperative education plans. Career-technical educators will need to cooperate with other career-technical educators and to support professional organizations of their choice that work for the good of their area of career and technical education in particular and the good of CTE in general.

For successful CTE coordinators, there is a very promising professional future, because of the competencies and the experiences brought to the cooperative plan and learned from working with the cooperative plan. Teacher-coordinators who have several years of experience have demonstrated the ability not only to plan and teach successfully but also to administer, to devise public relations schemes, to work with many segments of the school hierarchy, and to relate to the business community. In many of these activities, there have been opportunities for experiences not available to the regular teaching faculty, as well as opportunities to demonstrate leadership.

Successful teacher-coordinators are many times offered other professional opportunities. Some of

---

[4]National Commission on Secondary Vocational Education, *The Unfinished Agenda: The Role of Vocational Education in the High School*, Columbus, Ohio: National Center for Research in Vocational Education, The Ohio State University, 1984, p. 22.

[5]*Ibid.*, p.28.

these are as director of career and technical education for the district, training director for a firm or a trade association, teacher-educator, and specialist in a project of a government agency, including appointment to the state department staff. Teacher-coordinators should recognize these opportunities and capitalize upon them with additional training at the graduate level. They can contribute much to CTE, whether in business or government, if they plan their own career development and acquire the necessary advanced education.

BUILDING FOR PROFESSIONAL ADVANCEMENT

Following their initial certification and acquisition of teaching experience, teacher-coordinators should reassess their career plans and determine the steps that are necessary for professional advancement. Competent coordinators have many opportunities open to them because, to a large degree, the qualities that make them successful coordinators are those that make them valuable for more responsible positions. They have demonstrated depth in subject matter, the ability to work harmoniously with many groups, the ability to teach youth and adults, drive and personality, emotional stability, physical stamina, administrative skills, and flexibility to meet changing situations.

All of these factors contribute to *professionalism,* a necessary trait for professional advancement.

Teacher-coordinators should prepare for advancement by completing at least the requirements for a master's degree, and probably one year of study beyond. They should seek preparation in professional courses within their specialized teaching fields and within the broad area of CTE. They should supplement their initial preparation in general professional education with further course work in areas such as guidance, learning theory, curriculum, and administration. They should also secure advanced preparation in technical content areas and get more occupational experience.

Coordinators who seek advancement will also recognize that while advanced study is necessary, real growth will occur from experience and involvement. Therefore, they will seek out areas of service on various committees and projects within their schools, in civic and trade groups within the community, and in activities within their professional organizations. Coordinators who are concerned about growth recognize that increasing competence comes from seeking out and engaging in new experiences. Professional service, as well as service in the community, makes coordinators known, further enhancing their opportunities for advancement.

SERVING PROFESSIONAL ORGANIZATIONS

Teacher-coordinators have a stake in the activities and projects carried out by professional organizations. Membership and the seeking of active roles in local, state, and national education associations, particularly those dealing with career-technical education, are important considerations.

As pointed out in the *Vocational Education Journal,*[6] professional CTE organizations may set up programs of action, including goals such as quality, programs, public relations, cooperative relationship with other organizations, research and development, guidance and counseling, membership, legislation, organization and operation, and international relations. For coordinators, the involvement in professional associations as officers, as committee members, or as participants in professional meetings means opportunities to extend their horizons of understanding as career-technical educators. It means opportunities to meet leaders in the field. It means contributing personal and financial support.

Individuals who are concerned with professional improvement should consider joining various career-technical and cooperative education organizations. The following are some examples:

[6]R. Kirby Barrick, "So You Want to Be an AVA Leader," *Vocational Education Journal,* American Vocational Association, Vol. 61, No. B, April 1986, pp. 34–37.

General Career and Technical Education and Training Programs

Association for Career and Technical Education (formerly American Vocational Association)

This is the largest national education association dedicated to the advancement of education that prepares youth and adults for careers. Services include *Techniques,* a professional journal; the monthly newsletter, *Update*; insurance; a placement office; an annual convention; and professional relationship services with other associations and with the federal government. Address: 1410 King Street, Alexandria, VA 22314. Phone: (800) 826-9972. Email: acte@acteonline.org. Web site: www.acteonline.org.

State CTE Associations

Dues and services vary according to state, but there is usually a newsletter, as well as an annual convention and sometimes regional or state conferences. Each state department of education has the correct address of the state CTE association.

American Society for Training and Development

Founded in 1944, ASTD is a professional association focusing on workplace learning and performance issues. Services include various publications, research, and reports. Address: 1640 King Street, Box 1443, Alexandria, VA 22313-2043. Phone: (703) 683-8100 or (800) 628-2783. Fax: (703) 683-1523. Web site: www.astd.org.

Cooperative Education Association

Post-secondary and college or university cooperative education personnel should be particularly interested in membership. Members receive a quarterly journal and other services. There is an annual convention. Address: 8640 Guilford Road, Suite 215, Columbia, MD 21046. Phone: (410) 290-3666. Fax: (410) 290-7084. Web site: www.ceainc.org.

A few of the many organizations associated with specialized occupational fields are listed below.

Organizations Relating to Specialized Occupational Areas

American Association of Family and Consumer Sciences (AAFCS)

This is the only national organization representing family and consumer sciences professionals across practice areas and content specializations. Address: 1555 King Street, Alexandria, VA 22314. Phone: (703) 706-4600. Fax: (703) 706-4663. Email: info@aafcs.org. Web site: www.aafcs.org.

Marketing Education Association (MEA)

This national organization of education and business people is committed to the career development of youth and adults in the areas of marketing, management, and entrepreneurship. Address: 3300 Washtenaw Avenue, Suite 220, Ann Arbor, MI 48104-4200. Phone: (734) 971-6690. Fax: (734) 677-2407. Email: mea@nationalmea.org. Web site: www.nationalmea.org.

National Association of Agricultural Educators (NAAE)

This national agricultural education organization is dedicated to the professional interest and growth of agriculture teachers as well as to the recruiting and preparation of students to teach agriculture. Address: 1410 King Street, Suite 400, Alexandria, VA 22314. Phone: (703) 838-5885 or (800) 772-0939 (Ext. 4367). Fax: (703) 838-5888. Email: naae@teamaged.org. Web site: www.naae.org.

National Association of Health Occupations Teachers (NAHOT)

This national organization is devoted to the needs of all secondary and post-secondary health occupations teachers. Address: 1410 King Street, Alexandria, VA 22314. Phone: (800) 826-9972. Web site: www.acteonline.org/about/div-hoe-affil.html.

National Association for Trade and Industrial Education (NATIE)

This national association of trade and industrial educators is committed to trade and industrial education. Address: P.O. Box 1665, Leesburg, VA 20177. Phone: (703) 777-1740. Web site: http://skillsusa.org/NATIE.

National Business Education Association (NBEA)

This is the nation's largest professional organization devoted exclusively to serving individuals and groups engaged in instruction, administration, research, and dissemination of information for and about business. Address: 1914 Association Drive, Reston, VA 20191-1596. Phone: (703) 860-8300. Fax: (703) 620-4483. Email: nbea@nbea.org. Web site: www.nbea.org.

International Technology Education Association (ITEA)

This professional organization of technology teachers is dedicated to promoting technological literacy for all by supporting the teaching of technology and promoting professionalism. Address: 1914 Association Drive, Suite 201, Reston, VA 20191-1539. Phone: (703) 860-2100. Fax: (703) 860-0353. Email: itea@iris.org. Web site: www.iteawww.org.

In addition, coordinators will want to associate in the local community with professional and trade associations and other non-educational groups that relate to their occupational fields and to community efforts. Not only do coordinators make valuable contacts in such groups, but they also find ways to contribute to an improved community. Coordinators often find in these memberships the social–personal activities and relationships they desire. Examples of these to be considered are:

For all coordinators—Kiwanis, Rotary, Lions, etc.; Chamber of Commerce; Business and Professional Women's Club; Toastmasters; Junior Chamber of Commerce; local chapters of associations such as personnel directors and training directors; model cities; and community action groups.

For those in specialized occupational fields—groups such as local chapters of the Sales and Marketing Executives; Administrative Management Society; Data Processing Management Association; National Secretaries' Association. (See the local chamber of commerce for such groups existing in the community, and ask employers for suggestions.)

Questions and Activities

1. Describe the major differences in the roles of the three types of coordinators described in the text. What variations in competencies would be sought for each type?

2. After reviewing the competencies described as needed by a teacher-coordinator, which ones would you add? Use a specific community and curriculum on which to base your decision.

3. Describe two situations that use the cooperative occupational education plan: one a new plan, the other, one of quality and of several years' longevity. Describe the differences in the tasks of the coordinator in each case and indicate for each which competencies would be more or less important.

4. Review the tasks of teacher-coordinators listed in this chapter under the section "A State Viewpoint of Teacher-Coordinator Tasks." Put the tasks in order of importance and defend the priority sequence. Add other tasks you believe are needed.

5. Given only the choice of employing, for an on-going cooperative plan, a graduate trained as a teacher of occupational subjects and coordination or a person of substantial experience in the occupation with a degree and limited professional preparation, which would you hire? Prepare a statement of recommendation to the administration, defending your choice.

6. If a coordinator is believed by some people in the school and in the community to be a "guy (or gal) who goes downtown to drink coffee," what steps could the coordinator take to allay this impression and create a better image?

7. It is said that in some schools that classroom teachers see the CTE coordinator as competing for students and in other ways becoming a threat. In what ways might this be true? If so, what can the coordinator do about it?

8. Make up a list for the school administration of the convincing arguments for the coordinator to be employed for additional weeks beyond the school year and additional time beyond the customary school day.

9. Should coordination visitation time differ because of the location of the training stations? Consider the situation where an ME coordinator has most trainees placed in a shopping mall approximately 2 miles from the school. The ICE coordinator, on the other hand, has 20 students in 14 different occupations placed in three counties, with driving time up to 35 minutes.

Should each coordinator have the same amount of time? Why or why not?

For references pertinent to the subject matter of this chapter, see Reading Resources.

6

Initiating the Plan

A teacher-coordinator reports to her school's administrator about the contacts she has made in setting up training stations in the community.

KEY CONCEPT: Planning cannot be done at the last minute—it involves an organized sequence of activities over a period of months.

GOALS

After successfully completing the study of this chapter, answering the questions, and carrying out the activities, the student should be able to:

- List the initial steps that are used in structuring the cooperative plan.

- Distinguish between a steering committee and an advisory committee.

- Conduct a community survey.

- Prepare a calendar of events that the teacher-coordinator can use in developing the cooperative plan.

KEY TERMS

advisory committee

calendar of events

community survey

steering committee

The success of any venture, whether in education, government, business, or private life, rests with the initial planning done to establish the strategy of the venture and to provide the time and resources necessary for supportive service. Short-range and long-range planning, the organizing and marshalling of resources, the planning and staffing, the commitment to public relations, and the management of learning experiences are just as important to the educational venture as they are to a newly formulated business enterprise, a community swimming club, a sporting league, or a local fund-raising campaign. Some of the steps in initiating a venture may be skipped, but for each basic one omitted there is a price to pay both in terms of effectiveness and quality and in terms of human and material resource losses.

Initial Steps in Structuring the Cooperative Plan

The success of the cooperative education plan is dependent in many ways upon careful long-range planning and proper attention to organizational details. A school that makes last-minute decisions in the spring or summer to start a plan the following fall faces many difficulties. Problems in recruiting a qualified teacher-coordinator, in selecting and scheduling student-learners, and in selecting and developing training stations on a sound educational basis cannot be solved quickly. The long-range success of the plan will be jeopardized by hasty actions. When a well thought-out plan is properly executed, fewer problems arise if the people involved understand their roles.

Why should emphasis be given to the process of initiating a plan? There are several reasons. First, some districts do not yet have any cooperative plan, or, at best, they have a loosely supervised released-time work program. Second, some districts have an on-going cooperative plan in one area, such as business education, but they want to start up others, for

example, in health and in trade and industrial. Third, some districts have an on-going plan in an area such as marketing education and want to start specialized programs, such as wholesaling and petroleum marketing. Regardless of the situation, the calendar of events described later in this chapter can be used to initiate a plan or to renovate an existing one.

Initiating a Plan

The idea of using the cooperative occupational education plan in a school or adding the plan to another department in a school may originate with an administrator, with a department head or teacher, with a high school principal, or with a business or an industrial group or association within the community. Whatever the initial source may be, the chief school administrator or an assigned deputy is the person who must be consulted regarding the possibility of using this instructional strategy. This person is the one who is concerned with developing a quality addition to the curriculum and with maintaining minimum standards to qualify for career and technical education (CTE) funds for reimbursable programs. However, the actual responsibility for organizational activities is delegated—usually to the prospective teacher-coordinator, who has been hired in advance, or sometimes to the head of the appropriate department, another teacher in the department, or the administrator of CTE.

INVOLVING KEY INDIVIDUALS IN THE SCHOOL AND THE COMMUNITY

Planning and organizing a cooperative plan should be a team effort in involving key figures in the school and the community. School personnel involved are: (1) the administrators, including the superintendent, the director of the curriculum, the principal, and the director of CTE or the department head; (2) counselors; and (3) other teachers in the department. From outside school, help should be sought from: (1) the local state employment service office; (2) local leaders in business, industry, and labor; (3) representatives from the state department of CTE; and (4) a teacher-education institution. Naturally, in any given community, the type and size of the school and the kind of cooperative plan to be established will dictate the individuals to be involved in the development of the plan. The coordinator will be the key figure in the organizational effort, for it is the coordinator who has been assigned the responsibility of solving problems.

USING A STEERING COMMITTEE

The logical next step is to seek the consultative advice of interested publics within the school and the local community. An appropriate vehicle is a *steering committee.* This committee is usually temporary in nature and is designed to give an advisory "yes" or "no" reply to the question "Should we start a cooperative occupational education plan in our school system?" This reply is sought after proper and adequate orientation to the goals and objectives of the plan. A given school system may be thinking about starting more than one type of cooperative plan. If so, the steering committee might consider all of them and their interrelationships when it is advising whether to start one or all of them. A steering committee should be appointed by the superintendent of schools in a letter of invitation. One might consider as feasible in most communities a membership of 18 to 25 persons selected from the following:

From education: District office
Director of career and technical education
High school administration
Guidance
Department heads in occupational areas
Student council
Professional education association (local)

From business: Owners, managers, personnel directors, from businesses and government agencies
Professional and trade associations
Chamber of Commerce
State employment service

From labor: Labor unions and councils
High school graduates working in business, industry, or government

From the news media: Newspaper, radio, television

From parents: PTA and/or groups such as the League of Women Voters

From the community: Community action groups and social/civic clubs

The steering committee's first meeting may be planned around an orientation of the members to the objectives of a cooperative occupational education plan and how it operates. Resource persons from the state board of career and technical education or from a teacher-education institution may be called in. Visual aids my be shown. Committee members should then be instructed to take this information to the organizations they represent, gather reactions to the plan, and be prepared to come back to a second steering committee meeting to assist in deciding "yes" or "no" about organizing a plan in the local schools. Working committees may be appointed to gather information through surveys and research of various types before the second meeting.

SURVEYING THE STUDENT BODY AND THE BUSINESS COMMUNITY

In deciding whether to start a plan, secure the following types of information through surveys of the student body and the business community:

1. Need for trained, career-minded persons in marketing occupations, trade and industrial occupations, business occupations, health occupations, fam-

ily and consumer sciences occupations, and others in the community.

2. Opportunities for part-time training placements in the business community.

3. Any changing pattern of business enterprises in the community that would affect employment opportunities.

4. Career-technical interests of high school students that could be met by cooperative occupational education plans.

5. Post-secondary educational interests of high school students that could supplement, or be supplemented by, cooperative education offerings.

6. Physical facilities for cooperative education.

7. Locations of recent graduating class members and the occupations they entered.

Letters with return post-card forms are frequently used to gather information. Personal contact interviews are most effective if personnel can be assigned to these duties. A double post-card mailout form is not as effective as the personal interview but often will provide leads that can be followed up personally by the coordinator. A sample *community survey* form is shown as Figure 6–1.

USING AN ADVISORY COMMITTEE

The steering committee is *temporary*. It should be dissolved after the decision has been made as to whether to start a cooperative occupational education plan. At that time an *advisory committee* should be appointed, if the decision regarding the plan has been affirmative. The duty of the committee is strictly advisory; educational policy remains under the control of the superintendent of schools. The advisory committee should remain a sounding board for advice on operating procedures. The committee may be asked for advice on public information programs and may assist in these activities. It may suggest sources of training stations and sources and types of instructional materials, assist with student organi-

zation events, provide resource personnel for related classroom instruction, recommend minimum standards for students, assist with the employer-employee appreciation events, and identify adult education needs.

Certain members of the steering committee may be asked to serve on the advisory committee. An advisory group of 7 to 12 persons selected from the following representation is suggested.[1]

Education:	The administration Guidance personnel Teacher-coordinator
Business:	Chamber of commerce Civic or professional clubs Training stations—employers
Labor:	Organized labor
Parents:	PTA, or similar representation
Students:	A capable student-learner, perhaps the president of the student organization

Staggered terms for advisory committee members will provide for membership of both experienced and inexperienced persons in the future. A separate advisory committee for each cooperative occupational education area is recommended to take advantage of specialized interests and talents. It is possible for a single advisory committee to serve all cooperative plans, but a single committee cannot devote its attention to the specialized problems of the individual plans as effectively as can separate committees.

The committee should be chaired by a non-school person, with the coordinator acting as secretary to keep accurate minutes for future reference. Formal meetings should be called only at times when planned agendas justify such meetings. Businesspersons, like teachers, are busy people. Their time should be used efficiently. Usually about three formal meetings a year will suffice, supplemented by the

---

[1]The size of the advisory committee may vary depending upon what is considered an effective size for the job to be done.

COMMUNITY INTERVIEW SURVEY FORM

I. Name of business: _____

II. Address: _____

III. Phone: _____ IV. Type of business: _____

V. Approximate number of employees: _____

VI. Approximate number of office employees: _____

VII. Approximate number of sales employees: _____

VIII. Approximate number of trade or crafts employees: _____

IX. This business has the following resources available for school use:

    A. **Advisory committee:** A committee of interested citizens to advise the school on various phases of a cooperative occupational education plan. The following persons would be willing to serve on such an advisory committee if invited to do so.

Name	Position	When Available
_____	_____	_____
_____	_____	_____

    Remarks: _____

    _____

    B. **Training station possibilities:** On-the-job experiences supervised jointly by the cooperating employer and the coordinator of the cooperative occupation education plan.

      1. Does this business believe that supervised work experience is valuable training for employees? Yes _____ No _____

      2. Would this business be able to provide work experience opportunities for cooperative occupational education students? Yes _____ No _____ If no, would the business have training stations available in the future? Yes _____ No _____ If yes, about when? _____

      3. If affirmative, how many part-time training positions would be available?

        a. Office positions: _____

        b. Marketing positions:

          (1) Sales positions: _____

          (2) Non-sales positions: _____

        c. Trade and industrial positions:

          (1) Technicians: _____

          (2) Repairers: _____

          (3) Crafts workers: _____

FIGURE 6–1

CASE-IN-POINT

In our CTE classes we learn of the importance of advisory committees. Unfortunately, the importance of these lectures is often lost. Many programs have "paper committees," or a list of names of people who rarely if ever meet. To illustrate the importance of an advisory committee, let's examine the case of one new coordinator and how a successful advisory committee helped launch his program and career.

Stewart W. Husted, one of the co-authors of this text, took his first job as a marketing education teacher-coordinator in Decatur, Georgia, a suburb of Atlanta. Stewart's responsibility was to start a new program at Towers High School. During the summer months before school started, Husted recruited 32 students from a list given to him by a guidance counselor and his department chairperson. In addition, he put together an advisory committee of seven business, school, and community leaders.

On the committee was the general manager of a small local radio station. The general manager had the idea of teaching the students about radio advertising and personal selling. He volunteered one hour of air time each Saturday morning and dedicated the programming to a teen show. The marketing education students, working with a station DJ, then produced their own show called "The Morning After."

The entire DECA chapter became involved. The focal point was the selling of ads and then producing them for the show. Students were given the opportunity not only to sell ads and develop copy but also to do live on-the-air ads. As an additional motivator, the general manager donated the income earned to the Towers DECA chapter.

The success of the advisory committee member's ideas was phenomenal. The DECA chapter developed its idea into a project that was entered in the Sales and Marketing Executive International Creative Marketing Event. Working with an account executive from J. Walter Thompson, an ad agency, the students conducted marketing research and expanded their project to include publicity. The project was featured on several Atlanta TV stations and in one national publication. The students' reward, besides a world of new knowledge, was the second place in the state event and a fourth place in the state Chapter-of-the-Year Event. Furthermore, the publicity generated created tremendous student interest. The following year over 120 students applied for the marketing education program.

coordinator's seeking counsel from individual members informally during the school year. Figure 6–3 is an example of a letter of invitation to an employer that offers a general explanation of the committee's purpose and time commitment.

HIRING A TEACHER-COORDINATOR

The *teacher-coordinator* is hired by the administration soon after the steering committee has given the go-ahead for organizing a cooperative occupational education plan. The person who is hired proceeds to carry out the activities outlined in the calendar of events. Coordination time is provided as part of the contract time, but it must fit the conditions of the teacher's contract in the district. An experienced teacher who considers qualifying as a teacher-coordinator might well investigate the demands of program activities on physical stamina and emotional stability and general health. The administration needs assurance that the person selected can measure up to these demands. Businesspersons expect enthusiasm and leadership from the coordinator. Because of contacts with so many publics (school staff, businesspersons, students, and parents), the teacher-coordinator must have the abilities, background, and aptitudes to deal successfully with all of them.[2]

Sources of qualified teacher-coordinators.—There are three main sources of qualified teacher-coordinators: (1) graduates of teacher-education institutions who have prepared themselves during their undergraduate programs as coordinators; (2) experienced teachers with depth in occupational subject matter content and with practical business or trade experience background, who, in graduate school, have prepared for coordination through professional and technical courses in their specialized fields; and (3) individuals from business or industry who take course work for conditional certification and

---

[2]See Chapter 5 for an extensive discussion of the roles and the competencies required of teacher-coordinators.

WILL YOUR PRESENT SCHOOL DISTRICT OVER-ALL OCCUPATIONAL ADVISORY COMMITTEE BE USEFUL FOR YOUR COOPERATIVE PLAN?

		Yes	No*
1.	Do the businesses represented on the committee help to provide training stations, through their businesses or friends?	☐	☐
2.	Are the members genuinely interested in the welfare of the trainee and in a quality plan?	☐	☐
3.	Is the committee willing to meet regularly, perhaps bi-monthly, to asses the successes and failures of the plan?	☐	☐
4.	Are the members willing, if the need arises, to be resource speakers for a group of trainees—to "level" with them about the realities of the business world, or to provide someone to help develop this linkage for those who are being prepared for co-op?	☐	☐
5.	Are the members willing to sell the plan through the various professional agencies to which they belong?	☐	☐
6.	Do the members come in close contact with the occupational group your trainees represent?	☐	☐
7.	Do the members belong to the same trade groups and business/social associations?	☐	☐

NOTE: the above checklist can be very helpful in determining whether your current advisory committee is useful for your cooperative plan.

---

*If one or more answers to these questions are "No," your present occupational advisory committee should be organized to relate closely to the cooperative plan and the occupation it represents.

FIGURE 6–2

Business Education Department
Lincoln High School
East Parkway Drive
Cambridge City, Indiana 47327
Phone 317–478–3261, Ext. 27

September 1, ——

Mr. David Clements
Golay and Company, Inc.
1345 West Church Street
Cambridge City, Indiana 47327

Dear Mr. Clements:

You have been recommended as a possible candidate to serve on the advisory committee for the Lincoln High School Business Education Department.

This advisory committee meets an average of three times each year for two hours in the evening. The committee membership is listed on the enclosed sheet. Also enclosed is a Handbook for Members, published by the Indiana State Advisory Council on Career and Technical Education. This booklet helps explain the purposes of an advisory committee and the responsibilities of an advisory committee member.

If you have any questions after reading through the enclosed materials, please call me. I will try to answer any questions you may have. Our first meeting is scheduled for Tuesday, September 28, at 7 p.m. in Room 9 at Lincoln High School.

I hope you will agree to serve on this advisory committee. I feel that you will be a great asset to the committee and that you have much to offer to the business education department.

Yours truly,

Larry L. Shinn

Larry L. Shinn
Department Chair

SL
Enclosures

FIGURE 6–3. (Courtesy, Larry Shinn, *Advisory Committees: A Guide for Organization and Use.* Cincinnati: South-Western Publishing Co., 1983, p. 27)

who agree to undertake additional professional and technical training for certification.

A Calendar of Events— Stages in the Development of the Plan

The planning and organizational elements described in the following pages are often best acquired by the use of an outline of activities to be accomplished. This outline is called a *"calendar of events."* It will serve as a guide for school officials who have been delegated the responsibility of getting the cooperative plan under way. The outline shown here may be adjusted to fit the needs of any local community; it is a general guide to planning.[3] A glance at the calendar of events indicates the prudence of long-range planning for a cooperative occupational education plan.[4] Many school systems are wisely employing their teacher-coordinators one year in advance of the start of cooperative classes. A typical situation would find the teacher-coordinator with a half-time load for organizational duties for the proposed cooperative education plan and a half-time teaching load. In addition, assigning the new teacher-coordinator to public relations responsibilities during the first year, such as soliciting advertising space for the yearbook, provides acquaintanceship with the business community. Further, the school administrators can determine how well the person adjusts to, and is accepted by, the community and the school. For clarity, the steps are shown on a 24-month schedule, but, of course, this could be shortened into a

school year if sufficient resources are available. For this example, a high school district is used.

September–January

Interest in the plan is expressed by the principal, a department head and/or a teacher, by a business or an industrial group within the community, or by the superintendent. Regardless of who originates the idea, it is the superintendent or a designated deputy who must be consulted regarding the possibility of starting a cooperative occupational education plan.

The superintendent informs the department of education, division of career and technical education, of interest in cooperative education and may desire to request consulting service from the division of career and technical education.

A steering committee from business, education, labor, news media, and parents is appointed by the administration via a written invitation. This committee, temporary in nature, has for its purpose to render a decision as to whether or not there should be a cooperative occupational education plan started in the school system.

February

The first meeting of the steering committee should accomplish the purpose of orienting the group to the objectives of cooperative education and the way in which this type of plan operates.

A working committee may be appointed from the steering committee to gather information through surveys and research for presentation at the second meeting, at which time a final decision is made as to the desirability of cooperative education in the school system. The purpose of these surveys and research should be to obtain answers to the following suggested questions.

1. Is there a need for trained, career-minded persons in various occupations to be enrolled by the proposed cooperative education plan?

---

[3]Adapted from *A Handbook for Michigan Teacher-Coordinators of Cooperative Occupational Education Programs,* East Lansing: Michigan State University, Research and Development Program in Vocational–Technical Education.

[4]In some cases the school officials may hire a coordinator to do the entire organizational job or may delegate it to a teacher or a department head who has expressed interest in becoming a coordinator.

2. Are there opportunities for part-time training placement in the community?

3. Are there apparent changing patterns either in the growth or in the decline of business in the community that would affect employment opportunities?

4. Do the students have career-technical interests that could be met by cooperative occupational education plans?

5. Do the students have post-secondary interests that could supplement, or be supplemented by, cooperative education?

6. What physical facilities and equipment are available or needed in the school to provide cooperative education?

7. What types of occupations have recent graduating class members been able to enter?

8. Are parents and businesspersons supportive?

March

The second meeting of the steering committee should be an action meeting where the results of the various surveys and research would be presented and analyzed. The committee would then be expected to act upon the findings and assist in making a decision as to "yes" or "no" about organizing a cooperative occupational education plan in the school system. After this decision is made, the steering committee is dissolved.

April

If the steering committee's decision is affirmative, action is taken by the administration to select a CTE-certified teacher-coordinator and to designate organization responsibilities to school administration until the person is hired and on the job. The administration can usually rely on the following sources to obtain a qualified teacher-coordinator.

1. Graduates of teacher-education institutions who have taken undergraduate programs to prepare themselves as coordinators in the specialized areas.

2. Experienced classroom teachers, with depth of subject matter content and with practical industrial and/or business work experience background.

3. Individuals from business and industry who take course work for conditional certification and who undertake additional profession and technical training for permanent certification.

With more teachers studying cooperative education in preservice or graduate classes, the administration may find a member of the existing teaching staff who has the proper background to qualify for the provisional position of teacher-coordinator.

May or June

The new teacher-coordinator is hired to organize the plan during the coming school year.

September

The advisory committee is appointed by the administration. Duties of this committee should be strictly advisory; all policy remains under the control of the administration. The advisory committee should act as a sounding board for advice on operating procedures. Certain members of the steering committee may be asked to serve on the advisory committee. It is suggested that the advisory committee consist of 7 to 12 persons selected from education, business, labor, parents, and students.

Having staggered terms for the advisory committee members will provide for the future membership of both experienced and inexperienced persons. In a large school system, a separate advisory committee for each cooperative occupational education plan might be advantageous; however, in a small system a single advisory committee might serve all cooperative plans very effectively. The committee should be chaired by a non-school person, with the teacher-coordinator acting as secretary. Formal meetings should be called only when there is a need and a planned agenda justifies such meetings. Formal meetings can be supplemented by the coordinator's seeking counsel from individual members informally.

October

The first advisory committee meeting should be for the purpose of orientation. This would be a meeting where the goals and objectives of the plan and the purpose of the committee could be discussed. In order for the committee to be an effective part of the plan, each member of the committee should have a definite part in planning the over-all operation of the plan. Some of the activities of the advisory committee could include:

1. Suggesting training stations.

2. Suggesting and helping to secure classroom materials and equipment.

3. Acquainting the coordinator with influential businesspersons and labor leaders in the community and making arrangements for the coordinator to talk to various groups.

4. Making recommendations as to minimum employment standards for student-learners and reviewing the course content or courses to see whether they are occupationally sound and in keeping with occupational needs of the community.

5. Helping to identify adult education needs in the community.

6. Explaining the cooperative plan to their friends and their co-workers.

7. Working with the school in developing evaluation procedures for the cooperative plan.

8. Assisting in organizing a follow-up program of cooperative education graduates.

A public relations program should be planned for the school and the community. Plans should be made to inform the public of the advantages, principles, and goals of the plan in order to get the support of the entire teaching staff, employers of the students, workers in the participating firms, and other members of the community.

November–February

The teacher-coordinator appears before civic clubs, professional clubs, and school assemblies; writes newspaper releases for local and school newspapers; plans radio and TV appearances; and consults with guidance counselors, principals, labor leaders, and management leaders. The coordinator should always be careful to follow proper channels of communication.

The teacher-coordinator establishes and secures administrative approval of policies dealing with:

1. Coordination duties and time allotments—when, how much classroom teaching, responsibilities for adult education, etc.

2. Criteria for student selection.

3. Amount of credit for student participation in the plan.

4. Number of hours a student-learner should work.

5. Pay scale that should be expected or recommended.

6. Provision for extended contract for teacher-coordinator.

March–June

The teacher-coordinator counsels and interviews prospective student-learners. These prospective student-learners may have been obtained from lists procured as a result of:

1. An assembly program presented by the teacher-coordinator.

2. Information and application forms filled out by interested students through counselors or other teachers.

3. A talk made by the teacher-coordinator in a homeroom or in regular classes.

4. Newspaper articles read by interested students.

5. Faculty and counselor recommendations.

After all the information on the prospective student-learners has been gathered, the teacher-coordinator should have a final personal interview with each student-learner, explaining the various facets of the cooperative occupational education plan and what role the student-learner is expected to perform.

Even though the teacher-coordinator should utilize all the resources available to help in the selection of student-learners, the final decision of a particular student's participation in the plan should be the teacher- coordinator's.

The teacher-coordinator visits the parents of prospective student-learners at their homes or asks the parents to visit the school. Through this type of visitation the coordinator is able:

1. To better understand the student and to utilize this understanding in the development of the student's program.

2. To interpret the program to the parents and to enlist their aid in making the program a success for their child.

Application forms for reimbursement of instructional costs and equipment should be made to the state department of education, division of career and technical education, in June.

July

The teacher-coordinator selects and establishes business training stations on the basis of their educational value to cooperative occupational students. The coordinator should consider the following items in making the selection of training stations.

1. Opportunity for a variety of experiences.
2. Working conditions.
3. Reputation of business establishment.
4. Facilities and equipment.
5. Wages.
6. Accessibility of station to student-learner.
7. Type of supervision available.

The teacher-coordinator assists in structuring class schedules for student-learners. It is advisable, in scheduling student-learners, to arrange a dual class schedule. With the assistance of the guidance counselor and the teacher-coordinator, each cooperative student-learner prepares a schedule of classes accommodating the required academic subjects, the related instruction, and the supervised

laboratory work experience. Then, a second schedule of classes is prepared, omitting the cooperative occupational education plan. This system makes it possible to accommodate students who, for one reason or another, are not placed in a training station two or three weeks after school has started.

The teacher-coordinator arranges for related classroom facilities. This might involve providing for both a room and adequate equipment to allow desired related instruction to student-learners. The teacher-coordinator also should make plans to provide for a coordinator's office and a phone.

The teacher-coordinator requisitions textbooks, reference books, and supplies. Materials used should be current. There should be provision for a supplementary library of books, magazines, and business literature.

The teacher-coordinator should assess adult education and training needs, if any, for the coming year.

August

The teacher-coordinator:

1. Establishes additional training stations.
2. Checks established training stations to see if any changes have taken place.
3. Starts student-learner placements.

September

The teacher-coordinator:

1. Completes student-learner placements.
2. Starts related classroom instruction.
3. Continues public relations activities.
4. Initiates the student organization program. This is a very vital part of a total learning experience. People need to have friends, and they need to identify with a group whose members have common interests. Participating in this type of group activity provides for the development and exercise of student leadership.

The teacher-coordinator inaugurating a new plan will find it advisable to undertake systematically the steps outlined in this chapter. The established coordinator can use this list profitably to evaluate the present plan. The experienced coordinator taking over an established plan should use the list to review the plan and to look for aspects needing to be overhauled or redirected. Local situations always make some steps more important, relegate others to later disposition, and omit others entirely.

Once the plan has been approved and initiated, the focus shifts to the everyday activities of the coordination process.

Questions and Activities

1. Describe a community and the institutional plan that will use on-the-job experience as a learning laboratory. Plan a calendar of events that will meet this need to start a plan.

2. Assume the role of coordinator of a new plan. Describe the plan and the community. Then write your beliefs about what steps are needed and why. Make this a very personal letter to yourself.

3. Describe the reimbursement policies of your state for those instructional programs using the cooperative plan. Refer to Chapter 1 and indicate whether or not your state department of education will provide funds for plans using the work environment.

4. Using the community survey form included in the chapter, design a form that you believe would be appropriate for your community.

5. If an institution decides to open a new cooperative occupational education plan without the year's planning described in this chapter, what would be its attack? Formulate a plan or a calendar of events for a "crash" program.

For references pertinent to the subject matter of this chapter, see Reading Resources.

7

Coordinator Responsibilities at the Secondary Level

An important duty of a teacher-coordinator in a new plan is the recruitment of students.

KEY CONCEPT: *"Coordination"* is the process of building and maintaining harmonious relationships in the school, with students, with employers, and with other individuals.

GOALS

After successfully completing the study of this chapter, answering the questions, and carrying out the activities, the student should be able to:

- Define the coordination process.

- Explain the process whereby student-learners are interviewed, counseled, and admitted to the cooperative plan.

- Promote the cooperative plan of instruction.

- Understand the reimbursement policies of local and state education agencies.

KEY TERMS

news release

optional scheduling technique

public relations

reimbursement rate

In Chapter 6, the process of initiating a cooperative occupational education plan within the career and technical education (CTE) curriculum was described. Now that the new offering has been approved, the teacher-coordinator begins to implement the plan as a part of an instructional program. The responsibilities are mainly the teacher-coordinator's. It can be a "lonely" job because the plan (strategy) of instruction is quite different from that used in other instructional areas of the school.

The teacher-coordinator is a different "breed of cat"; the role is quite different from that of most school personnel—the hours are different, the instruction extends beyond the four walls of the school, and the teacher-coordinator is intimately involved on a daily basis with the community in a learning partnership.

What Is Coordination?

Coordination is a *process* but the term "coordination" is defined differently by different people. A simple definition would be "the process of building and maintaining harmonious relationships between all groups involved in the cooperative plan, to the end that the student-learner receives the very best preparation for a chosen occupation."[1] By this definition, coordination would include all activities of the teacher-coordinator, including instruction within the classroom. On the other hand, some individuals use a narrow definition of "coordination," referring only to those activities that occur outside the school when visits are made to the training stations.

According to the Beaumont definition of "coordination," the major components of the process are:

1. Refining the student's career objective and determining needed learning experiences—the *training plan.*

2. Developing the appropriate training station for the student.

3. Making arrangements with the training station for the placement and entering into a *training agreement* between the school and the training agency.

4. Orienting the training station sponsor.

5. Making evaluative visitations to the training station to determine if appropriate learning experiences are being provided.

6. Carrying out needed community public relations activities.

7. Relating training station experiences to in-school laboratory learning experiences.

8. Relating to the student's home as a partner in the learning process.

9. Achieving terminal job placement after training or arranging for additional or continuing education.

ARRANGING THE TEACHER-COORDINATOR'S SCHEDULE

How much time during the school week and year does the teacher-coordinator of a cooperative occupational education plan need to do an effective job? The answer of course lies in both the description of the range of responsibilities and the personal effectiveness in getting the job done. The question of assigning load time is a difficult one and is often complicated by teacher contracts in educational institutions that set forth definitions of teaching time. And, in the contemporary scene of public education, the tendency of educational administrators and boards to define faculty productivity in terms of student credit-hours (SCH) presents a most difficult factor in load calculations, since a lecture of one hour for 30 or 300 represents a definable number of student clock-hours of instruction, while the coordination visit of an hour

---

[1]This definition is attributed to John Beaumont while he was Chief of the Distributive Education Branch, U.S. Office of Education. The statement was made at a national conference, circa 1960.

in behalf of one student represents but one clock-hour of instruction.

Regardless of these present problems, the administrator who understands the cooperative occupational education plan will recognize the need of the coordinator for one-half hour per student per week for coordination activities (see Figure 7–1). Teaching, counseling, and other assigned duties are to be added on. In other words—30 students equals 15 hours of coordination time each week if each student-learner is to be accorded the professional practice needed. This rule-of-thumb is the result of experience, and most local administrators and state supervisors who have carefully analyzed coordination activities would agree with it. Adequate coordination time for the teacher-coordinator continues to be a vital factor for developing and maintaining quality educational experience for students under the plan.

The school administrator must understand that coordination time does not fit the customary definition of the school day or year. Coordination activities involve not only school personnel but also business and industrial leaders. The employer works evenings, Saturdays, and even Sundays. The coordinator meets these people on their schedule, not the school's. An empathetic coordinator knows that in some firms or agencies, lunch is not the time for rapport. At other times it may be just before 5 o'clock in the afternoon. In some places employers do not wish to be bothered in the early morning when the operation is getting under way or during the closing hours. Some firms now operate on an almost round-the-clock schedule, and the coordinator must be there when the training sponsor or supervisor is present. This may mean the night shift, or even the weekend. The point to be made is that the coordinator must be available when the job sponsor is available. Businesspersons operate at hours that are not necessarily attuned to the traditional school-day schedule. If teacher-coordinators are to be the school's liaison with the community, they must operate by the community's schedule; a wise administrator understands this in determining both load and work schedule.

Teacher-Coordination Activities Before Classroom Instruction Begins

Major planning and organizational activities of the teacher-coordinator before the classroom and job training begin are:

1. Disseminating information to school personnel, guidance counselors, staff, students, and parents—building the image.

2. Making promotional contacts in the community for training stations.

3. Counseling, interviewing, and selecting prospective student-learners.

4. Assisting in arranging class schedules, including related instruction periods.

5. Arranging for related classroom facilities, including furniture and fixtures.

6. Selecting and requisitioning textbooks, reference books, and supplies.

CREATING IMAGE AND INTEREST WITHIN THE SCHOOL

Through initiative and ingenuity, the teacher-coordinator must obtain school staff and student body understanding of the cooperative occupational education plan. Information sheets and application blanks are left in each counselor's office to be made available to interested students. The educational aspects of the training plan are stressed, indicating that dependability, employability, and the ability to benefit from the training are primary requisites for prospective student-learners. In plans where other courses are prerequisites, the counselor should understand the prerequisite curriculum. Short talks before home room groups or other classes, articles in the school newspaper, bulletin board displays, corridor displays, and assembly programs have been effec-

GETTING A QUALITY JOB DONE: CRITERIA FOR JUDGING COORDINATOR – TRAINEE RATIO

The suggested coordination time is at least *one-half hour per week* per student-learner, all other things being equal.

More Time	Less Time	Variable
☐	☐	Scope of coordinator's responsibilities _____ Social adjustment status of trainees _____ Teaching responsibilities _____ Counseling responsibilities _____ Correlation of related classroom instruction with on-the-job training _____ Administration duties, including preparation of required reports and records
☐	☐	Number of trainees in plan and special supportive services they require
☐	☐	Size of institution and number of people involved
☐	☐	Number of different occupational training plans coordinated
☐	☐	Geographic area (rural–urban) and travel required
☐	☐	Degree and diversity of disadvantagement in trainees
☐	☐	Degree to which the company takes over the training
☐	☐	Complexity of the occupation involved
☐	☐	Size of the specific occupational plan coordinated
☐	☐	Number of trainees actually placed in a given firm
☐	☐	Number of former trainees requiring follow-up services
☐	☐	Number of weeks in student-learner's cooperative experiences before a new training station must be developed
☐	☐	Difficulty of finding stations because of type of student-learners, type of local employers, condition of the labor market locally
☐	☐	Amount of job competition from high school and post-secondary cooperative plans
☐	☐	Newly installed cooperative plan requiring development time

FIGURE 7–1

tive means of disseminating information and creating an image. In some schools it is possible administratively to call student assemblies of certain classes, perhaps the junior class, to present a short explanation of each CTE program. At such an assembly the use of a form similar to Figure 7–2 aids in getting a list of possible interested students.

Research indicates that after the first year the students are the best salespersons for quality education. The students and the career-technical student orga-

---

CAREER AND TECHNICAL EDUCATION INTEREST BLANK

Name _____ Home room or student center _____

Date of birth _____ Present age _____

I am interested in learning how the following career and technical education plan would fit into my school plans and would like further information.

_____ Marketing education   (one-year business course, on-the-job and classroom training — seniors only)

_____ Health occupations   (one-year course in health-related occupations, on-the-job and classroom training — seniors only)

_____ Family and consumer sciences occupations   (one-year course in family and consumer sciences occupations, on-the-job and classroom training — seniors only)

_____ Trade and industrial occupations   (two-year trades course, on-the-job and classroom training — seniors only)

_____ Business occupations   (one-year business course, on-the-job and classroom training — seniors only)

_____ Agricultural occupations   (two-year course, in class, open to juniors)

My present class schedule in school is:

1st Period _____ in Room _____

2nd Period _____ in Room _____

3rd Period _____ in Room _____

4th Period _____ in Room _____

5th Period _____ in Room _____

6th Period _____ in Room _____

FIGURE 7–2

nization (CTSO) chapter should have discussed with the coordinator how they might best answer questions about their experiences. They can be encouraged to put up displays of their work. Awards they win in club competition should be given school-wide publicity. Their occupational certificates can be presented at the annual school awards ceremony, just as those for academic honors. Another image builder is the classroom, for the laboratory-like facility and its display will tell the story of the effective and relevant instruction that takes place there.

Students can also tell their story by producing a brief multimedia presentation of their experiences or by producing a descriptive brochure in cooperation with the journalism class. Image is built by a continuing series of communication devices with the school publics.

INTERVIEWING, COUNSELING, AND ADMITTING STUDENT-LEARNERS

The first task of the coordinator before admitting is recruiting. At times, this term, "recruiting" seems to be one that some coordinators are reluctant to accept; yet, they have only to look around their school to find the band director recruiting, the coaches recruiting, and many academic teachers recruiting. The term means in each case that educators are trying to find those students who can profit from what their instruction has to offer. As in all selling situations, prospects must be made aware of their needs and be convinced that the plan meets those needs. The career-technical educator must be aware of, and react to, the fact that the offerings may not be well known or may be in conflict with the many values and pressures that the student feels from peer groups and parents.

If recruiting is successful, the coordinator begins the process of interviewing, counseling, and admitting by remembering and practicing the dictum that CTE instruction is "for those who need it, want it, and can profit from it." Guidance counselors and other teachers may assist in refining the list and in

adding to it prospective student-learners through their student interviews at programming time. The finest working relationship within a school results when all likely prospects are referred to the teacher-coordinator for final interview and for the final decision on a particular student's participation in the plan.[2] To refine the list of interested students, the teacher-coordinator may check school records, check with other teachers, and then conduct personal interviews of those students who remain on the list after matters of appropriate age, sincerity of purpose, and possibility of class scheduling have been determined.

Counseling of prospective student-learners begins with reviewing the information on applications from prospective student-learners, supplemented by information gathered, with the student's permission, from the school records relative to aptitude, interest, and intelligence test scores. If adequate test information is not available, appropriate tests may be administered. Some teacher-coordinators work with the local state employment service, using a series of tests appropriate to the particular occupation.

During the interview with a student, the teacher-coordinator attempts to make certain the student understands the difference between the work experience (approach) plan and the cooperative occupational education plan (an educational plan preparing students for careers). Although all the students will not have had adequate information at hand to decide on careers or will not be able to decide on specific career objectives until after a preparatory course or brief on-the-job experience, the career approach seems most useful. Otherwise, a cooperative plan may become filled with job seekers who only wish to earn extra money or to get out of school or are "dumped in" because no one else wishes them in the classroom.

---

[2]Suggested minimum standards for student-learners in trade and industrial occupations are set forth in Chapter 22; for student-learners in business occupations, in Chapter 18; for student-learners in marketing education, in Chapter 21; and for student-learners in other cooperative plans, in Chapters 17, 19, and 20.

The list of prospects should be narrowed to student-learners who are employable.[3] The number of student-learners should conform to reasonable limits determined by school policy, physical facilities and equipment, and availability of training stations. Most programs have 15 to 30 students; 22 is the real maximum for individualizing instructions. The exact number depends upon the amount of coordination time the coordinator has, the number of other courses taught, and the class size in a district wherein teachers have a negotiated contract.

SETTING UP INTERVIEWS OF PROSPECTIVE STUDENT-LEARNERS BY PROSPECTIVE EMPLOYERS

Final selection of student-learners as employees is really made by the participating employers, not by the coordinator. But, the coordinator must decide whether more students are to be sent for an interview with each employer and for each training station available and, if so, which student or students are to be sent. Several methods of placement are possible. These are:

1. All interested student-learners are allowed to apply at a specific training station.

2. One student is selected and then others are successively chosen if the first one is not hired.

3. Several students are sent for an interview.

4. The students are allowed to find their own jobs, subject to the coordinator's approval.

---

[3]The word "acceptable" is used here to mean meeting the criteria of the coordinator and the advisory committee. There is some evidence in research studies that in prior years only students with above-average grade averages were accepted. In fact, coordinators were obviously accepting only students with an above "C" average. What is more important than grades is the capacity and motivation to profit from the instruction. There is benefit to the student if the cooperative plan meets the individual needs and if the employer finds the person employable—there is no merit in achieving for the cooperative plan by only those who are superior. Emphasis on the mainstreaming or inclusion of students who are economically and/or academically disadvantaged challenges the coordinator to seek out a wide range of training stations to accommodate a wide range of student capabilities.

If the career objectives of the student-learners have been well defined by the teacher-coordinator, the first method allows for the most natural competitive employment situation and leaves the final choice with the employer. In the second case, the coordinator is almost choosing a particular student-learner for a specific training station, even though the employer has the option of not selecting the student who comes for the interview. In the third case, where several are sent for an interview without much effort having been made to match career objectives to opportunities offered in established training stations, one is opening up the possibility of many adjustments and perhaps training station changes early in the school year. The weakest method is the last one, where students are allowed to find their own jobs. This is usually an indication that the coordinator has made little or no attempt to establish training stations for the purpose of matching students to training opportunities. Programs attempting to operate in this manner are doomed to difficulties and failure and often become, in reality, not cooperative plans, but just plain work experience plans.

All students should be trained in employability skills, specifically in interview techniques, during a preschool orientation period, during the first week of school, during the prerequisite course, or during a spring orientation program. Instruction may be given to the class as a whole or to each student individually. Interview training should include the following information: when and where to report for the interview, how to develop a written statement of qualifications for the position, how to answer the kinds of questions that employers usually ask, how to fill out written applications, and what to do following the interview. Role-playing practice interviews should be undertaken.

The teacher-coordinator should arrange and schedule the employment interviews, completing follow-ups after the interviews. Student-learners should go to interviews and should be assigned to training stations before school opens or as soon as possible after it opens. The process of attracting, interviewing, counseling, and admitting students to the cooperative plan should include several components. As a practical manner, Figures 7–3 and 7–4 illustrate two activities that will assist in getting the job done— completing the process. Asking each student who has indicated an interest in the program to complete a personal profile can be very helpful to the coordinator in better knowing the student. Letters of recommendation from other teachers, counselors, and former employers can be valuable in providing additional information about the student's strengths and weaknesses. The steps used in the process need to be an integrated sequence of activities.

ARRANGING CLASSROOM SCHEDULES AND RELATED CLASS INSTRUCTION

The use of the computer in school scheduling of offerings has been helpful in providing greater flexibility and response to student choice. But, even with this aid, coordinators should be cognizant of the need to make direct input into the prescheduling process where departmental offerings are proposed and students' needs reviewed. Coordinators should work closely with the scheduling officer (registrar) to be sure sections of courses are available to the trainees and to stress the trainees' scheduling limitations because of the timing of on-the-job laboratory hours and requirements for related instruction and supplementary courses. Coordinators should also be sure that their own teaching schedules allow them prime time for out-of-school coordination activities. These requirements should be established early with the administration as a matter of agreement regarding the coordinator's responsibilities.

The registrar should be asked to consider reserving sections of required courses, such as government and history, for the cooperative students. This is particularly important in small schools with limited offerings and teacher assignment flexibility or where teacher seniority is a vital factor in scheduling. The related class deserves special scheduling attention, since it is the primary source of instruction for the student's career development and because it requires priority for assignment of specialized instructional space. Scheduling the related class the last period(s) before the training station hours facilitates two quality conditions: (1) students do not have to waste travel time returning to the school and (2) trainees may be released from related instruction in order to profit from extra laboratory experiences, particularly during periods of peak business activity or when a special training session is available within the firm.

Registrars should be aware of the value of what is called the *optional scheduling technique* for the cooperative student-learner. A second schedule of classes is prepared and followed; the cooperative plan related courses are omitted in favor of other courses. This optional plan is especially valuable in situations where (1) the training station placements are scarce and trainees may not be placed until well after classes begin and (2) there is a question of a student's employability, and several interviews and after-school try-outs with prospective training stations may be needed. The optional scheduling system makes it possible to leave in the alternate academic program those student-learners who have not been placed in training stations by the opening of classes in the fall. As soon as a suitable training station placement has been arranged, a student switches to the cooperative class schedule without problem. Those who have not been appropriately placed within a reasonable time after school starts (perhaps three weeks) continue in the alternate academic schedule of classes. It is difficult to transfer a student from a cooperative class to another class after school starts, but seldom will a teacher object to giving up a student from class within a reasonable length of time after classes begin.

OBTAINING AN APPROPRIATE FACILITY

The in-school instruction phase of the cooperative plan is a laboratory-type situation, with emphasis on individualized instruction, individual study, and participatory methods. As such, the instruction requires facilities appropriate not only to this form of

APPLICATION FOR COOPERATIVE EDUCATION

Name _____
                      Last   First   Middle

Social Security Number _____

Grade _____ Date of Birth _____ Age _____

Sex _____ Home Phone _____

Home Address _____

Father's Name _____ Place of Employment _____

Mother's Name _____ Place of Employment _____

School Record:

Total Credits Earned _____

Circle the letter indicating overall grade average: A   B   C   D   F

List subjects failed if any: _____

How many days have you been absent this year? _____ Last year? _____

What school organizations do you belong to? (List offices held, honors, and awards received)

_____

_____

List hobbies and interests: _____

Would any of these conflict with a work schedule? _____

Former Work Experience:

Firm _____ Job Title _____

How long were you employed there? _____

Why did you leave? _____

Future Plans:

What do you plan to do after graduation? _____

What would you like to be doing 10 years from now? (Career Interest) _____

_____

Briefly describe why you are interested in this program. _____

_____

If employed, do you have your own means of transportation? _____

Have you discussed this work program with your parents? _____

What type of work do you desire:

1st choice _____ 2nd choice _____

3rd choice _____

I hereby certify that all the above information is factually correct and if I enroll in the program, I agree to follow all policies governing the program here at Jefferson Forest High School.

Date _____ Signature _____

FIGURE 7–3

PERSONAL RECOMMENDATION FOR
COOPERATIVE EDUCATION

_____(Name of student)_____ has applied for enrollment in the _____ cooperative plan. Student-learners receive classroom instruction _____ hours a day and are placed in training stations (businesses) to develop technical skills and to obtain valuable supervised work experience. Students enrolled in the _____ plan need to have a positive attitude toward work and school if they are to profit from the on-the-job training and are to make progress toward their particular career objectives. Based on your contact with _____(Name of student)_____, please rate him/her by checking the appropriate box to the right of each quality. Please leave the form in my mailbox or mail it to me at _____. Thank you for your assistance.

_____
Teacher-Coordinator

	Superior	Above Average	Average	Below Average	Unsatis-factory
1. **Dependability:** Is able to work with little supervision; is prompt, truthful, sincere, consistent; follows instructions					
2. **Personality traits:** Is courteous, considerate, respectful, mannerly					
3. **Leadership:** Is assertive, forceful, imaginative, resourceful, able to inspire others to act; exercises good judgment					
4. **Industriousness:** Is persistent; has good habits of work; makes wise use of time					
5. **Mental alertness:** Is attentive, interested, observant, eager to learn; has good memory					

FIGURE 7–4

PERSONAL RECOMMENDATION FOR
COOPERATIVE EDUCATION (Continued)

	Superior	Above Average	Average	Below Average	Unsatis-factory
6. **Thoroughness:** Is accurate, careful; completes work					
7. **Personal appearance and grooming:** Is clean; has neat and orderly appearance; is poised					
8. **Ability to get along with others:** Is tactful, friendly, cooperative; has sense of humor					
9. **Social habits:** Has positive attitude; exercises self-control; is honest; is not argumentative or boisterous					

Employability: If you were an employer or a job supervisor, would you want this student working for you? Explain. _____

_____

_____

Are you willing to have this student represent the school on the job in a local business? Explain.

_____

_____

_____

Comments: _____

_____

_____

_____

FIGURE 7–4 (Continued)

instruction but also to the function of the coordinator as a counselor and part-time administrator. The classroom and a private office adjacent are requisites to quality performance of these functions.

The exact specification of the laboratory is not possible here, since it depends upon the instructional demands of the occupational area involved. For example, a business occupations program demands a simulated office environment and considerable space for office equipment. Likewise, the marketing curriculum demands space and equipment appropriate to the marketing functions, the cooperative industrial education curriculum needs a related classroom and shop space, and so on. What is required is adequate table space for group discussion, filing and shelving for the host of reference materials needed for individual study, media for individualized instruction, filing space for the student study materials and records, private telephone service, and counseling areas offering privacy.

Creating specialized space of the type suggested may seem beyond the means of many schools. However, the administration should understand that this space can be used ideally for other laboratory classes and is well suited to evening adult instruction and human resource contract courses after school.

SELECTING INSTRUCTIONAL MATERIALS AND MEDIA

Because the cooperative plan at the secondary level involves employed youth who put into practice the theory of the classroom and study to meet their individual objectives, the instructional materials and media become most important. An overhead projector is a "must" to assist with group discussion and learnings, individual study media such as carrels must be accessible. A substantial supply of reference materials is needed to facilitate individual study and small group projects. Because the instruction is correlated with contemporary business and industrial practice, the materials must be up-to-date and must consist not only of textbooks but also of technical manuals,

handbooks, trade and professional magazines, and business literature. Students will need a project manual of some type as well as one or more sets of texts basic to the field.

The coordinator should consider the acquisitions program over a five-year period and get administrative approval for that time. With this plan used as a priority, a budget of $100 to $150 per student will enable a program to begin if the annual budget thereafter for five years averages from $75 to $100. Coordinators should consistently solicit from employers their technical literature, which can sometimes be classified as tax-free donations. Cooperative students should be encouraged to use their student organization as a source of funds for providing supplementary materials and media.

The Initial Promotion and Community Contact Campaign

The cooperative plan of instruction which uses the community as an instructional laboratory is still a relatively new idea in some communities. An initial promotion and publicity campaign of high impact and high intensity is vital. The idea must be sold quickly, and to the right publics. The coordinator who is skilled in **public relations** knows that an effective campaign requires having "a nose for news," an ability to recognize the opinion leaders in the community, a talent for knowing whose values to seek to supplement, and the tenacity for carrying through ideas by gaining access to the best media.

Before classroom instruction starts, there are several major promotional avenues that need to be utilized. The first is informal selling through everyday living.

"GETTING AROUND"

Educators who have something to sell to the community are prone to emulate business and to rely heavily on public relations efforts through the printed or visual media or through addresses to large groups. Certainly these mass efforts are economical in terms of time, energy, and cost. But they are, as the media describe them, "shotgun" approaches and are suited mainly to disseminating the over-all concept of what the school is trying to do.

The real task of teacher-coordinators is "getting around"—calling upon people and transmitting the message in a face-to-face and one-to-one situation. Coordinators do this not only through scheduled appointments but also through their personal life activities—when they grocery shop, they talk to the store manager; when they get their teeth cleaned, they sell the dentist; when they fill the car with gasoline, they talk about the school's new plan with the station manager; when they socialize, they sell the parents and the businesspersons, who also may be their friends. This informal approach may seem beyond a coordinator's responsibilities and the contractual work day, but successful coordinators believe in the plan's product, in what the school is doing for students, and in the benefits to the community. To support these beliefs, they use the informal communications network that exists in every community.

ADVISORY COMMITTEE MEETINGS

The teacher-coordinator plans the agenda for advisory committee meetings with the committee chairperson. Successful meetings are streamlined to the extent that the business is limited to one hour. The mailing of the agenda to committee members in advance of the meeting expedites accomplishment at sessions. It is advisable to select a time when businesspersons can free themselves from obligations most readily. Short noon luncheon meeting and "early-bird" breakfasts are popular so that busy days can be avoided.

TALKS WITH COMMUNITY GROUPS

Scheduled appearances before major civic groups and those professional groups directly related to the

coordinator's program provide opportunities to create understanding of a new program during its organization stages. Presentations ranging in length from a 5-minute announcement to a 20-minute slide presentation or videotape are effective. Visuals in the form of flannelboard presentations, posters, chalkboard and overhead projector presentations, and role-playing skits are received well.

PRESS, RADIO, AND TV

Every opportunity utilized for telling the story of cooperative education assists in reaching the many segments of the public served by such a plan. Newspapers are receptive to carefully planned, well-written articles. The writer of a **news release** is well advised to structure the article around answering the five main questions: Who? What? When? Where? Why? An acceptable form is illustrated in Figure 7–5. Copy sent to an editor is usually triple spaced to facilitate easy editing and to provide room for additions and/or corrections. Radio and television stations provide program time free of charge for certain civic presentations. Panel discussions, role-playing skits, demonstrations, and interviews based on cooperative education qualify for this free time. National Education Week, Careers in Retailing Week, Administrative Professionals Week, and Career and Technical Education Week are examples of appropriate times, although other times during the school year may be used.

Members of the press are cooperative if a few fundamental guidelines are followed by the person contacting them. Following are several basic suggestions for those who deal with local media.[4]

1. Have only one person from a particular phase of cooperative education contact the news media.

2. Quickly establish personal contact with the right persons at each newspaper and each radio and television station in the area.

---

[4]Updated from a publicity handbook published years ago by Sperry Hutchison Company.

NEWS RELEASE

Contact: _____   For Immediate Release

         _____

         _____

_____, _____ Coordinator at
      (Name)   (Type of Program)

_____ High School, participated in the 30th annual

Coordinator's Conference at _____, August 14–16.
_____, _____ (Refer to your special activity at
      (Name)

conference — spot on program, committee, discussion, etc.) _____

More than 100 coordinators of cooperative occupational education plans representing all parts of the state attended the conference, held under the sponsorship of the state board of vocational education in cooperation with the Career and Technical Education Department of the University's College of Education.

At one of the sessions, it was emphasized that coordinators and guidance counselors need to work together in planning training for young people. Realistic, up-to-date information regarding career opportunities in the fields is essential in assisting high school students to make career choices.

One of the features of the conference was a visit to the Jewel Grocery Company warehouse. Modern business techniques and procedures and warehousing phases of mass food distribution were observed by the group. Coordinators visited data processing installations on the university campus.

The marketing education plan, in cooperation with retail and wholesale businesses, places advanced high school students in occupations for a minimum of 15 hours per week of supervised on-the-job training that is coordinated with classroom study.

The business occupations plan operates much like marketing education, with job placements being made in businesses in which employers cooperate with the school by providing supervised training in preparation for secretarial and general office careers.

Robert F. Kozelka, Chief Consultant, Business and Marketing Education, and Bernard M. Ohm and Patricia Rath, Supervisors, represented the state board of career and technical education. Mr. Kozelka indicated that demand for qualified students in both programs continued at a high level.

During the ____–____ school year, _____ students participated in _____ _____'s coordinated in-school, on-the-job training plan. Business
   (Your Name)
persons interested in assisting the high school by providing training stations for advanced students seeking careers in business may obtain information from _____ (Teacher's Name)
at the high school.

FIGURE 7–5

3. Write everything down so details are not forgotten.

4. Meet every deadline promptly.

5. Keyboard all news releases, and send clear printed copies.

6. Check dates, names, and places before copy is sent in.

7. Give credit where credit is due.

8. Avoid using the pressure of friendship or business connections to get a story released.

9. Appreciate the time and space given cooperative education by the media.

10. Invite members of the press as guests to special functions. Don't ask them to buy tickets.

COORDINATION ACTIVITIES AND APPRENTICESHIP

A person who administers and supervises the related instruction for apprenticeship programs in a local community is involved with activities similar to those carried out in a cooperative education plan. For example, following is a list of those activities:

1. Meet with the joint apprenticeship committees of the particular trades to determine the related instruction needs of the apprentices.

2. Plan appropriate related instruction topics.

3. Secure appropriate instructional materials.

4. Arrange for classroom facilities and instructional equipment.

5. Recruit and hire instructors, some from the school system, some from the trades.

6. Prepare a sequential list of on-the-job experiences with the joint apprenticeship committee to be provided to the apprentices.

7. Provide teacher education (how to teach) for trade instructors who are inexperienced as teachers.

8. Prepare progress reports and evaluations on each student in cooperation with classroom instructors and job supervisors.

9. Maintain attendance records and submit financial reports required by the state board of CTE if state and federal funds are involved.

10. Help plan recognition ceremonies for apprentices as they become journeypersons.

Understanding Reimbursement Policies

Most state departments of career and technical education reimburse the local districts at the end of the fiscal year on the basis of the teacher's salary (that part of the contract time of the teacher-coordinator devoted to related classroom instruction and coordination), the coordinator's approved travel expenses, and, in some cases, certain types of equipment. For example, let us assume a situation in which the state department of CTE maintains a 40 percent *reimbursement rate*. At Central High School the marketing education teacher-coordinator has eight years' experience and the following schedule: one period teaching general business, two periods for the marketing education related class, and three periods for coordination. His salary is $40,000. Reimbursement is calculated as follows:

1. Time for CTE is five periods out of six. Thus five-sixths of his salary is eligible for reimbursement.

2. $40,000 × five-sixths is $33,000. To this eligible amount, the state reimbursement rate of 40 percent is applied.

3. $33,000 × 40 percent is $13,200, which is the amount the school district will get back from the state.[5]

---

[5]In this example, travel costs and other items were not included, in order to simplify the calculation. Some states reimburse on a unit basis regardless of costs. In other states, reimbursement is on a formula basis that takes into account items such as the per capita income of the district and the number of student contact hours.

■■■ CASE-IN-POINT ■■■

Brian Roberts is the agricultural teacher-coordinator for his local high school. While in college, Brian took a journalism course in news writing. When Brian began his job, he soon started applying theory to practice.

Using the techniques learned in class, Brian got into the habit of preparing news releases for every newsworthy event relating to his program. Of course, the FFA chapter he advised was often the focal point of a release. Other ideas for releases came from advisory committee meetings, employer—employee social functions, curriculum changes, student success stories, Career and Technical Education Week, and many others. It was not unusual for Brian to write 20 to 35 news releases per year.

One tip Brian learned early was never to give himself credit or play up the teacher-coordinator role. If the program is successful, that fact speaks for itself. In Brian's case, the publicity he generated for his program paid off. The local Optimist Club chose his FFA chapter as the Student Organization of the Year for his community. The Optimist Club gave a special banquet for his students and their parents and presented the chapter with a plaque and a $250 check to defray costs to the FFA state conference.

State reimbursement policies vary considerably among states, and some use what is known as project funding. In this system some local programs are called "special projects" because they meet a designated need and their costs are reimbursed as high as 75 to 100 percent of the local cost. In other states a minimum foundation payment per student or per program is paid. The coordinator should find out from the state department the reimbursement policies of CTE since the school's administration will expect the coordinator to be informed. And the community and employers will often ask questions about the use of federal and state funds which the coordinator is expected to answer.

State funds may be used in some cases for other expenses, such as travel of the coordinator in the local area, reimbursement for professional conferences and meetings approved by the state, classroom instructional materials, and classroom equipment. Since state policies differ, the coordinator should check carefully and often because rules change and fund availability differs from year to year.

Questions and Activities

1. Describe a cooperative plan of instruction in terms of its goals, its institutional setting, and the community it serves. For this situation, write the criteria to be used in assisting students in deciding to enroll in the plan.

2. For the plan of instruction described in question 1, what would be your request for adequate facilities and materials? Describe the space and list the materials for the first year, or write a recommendation for a change in an on-going plan you have inherited.

3. Draft a letter of invitation to potential steering committee members. Prepare the draft for your superintendent's signature.

4. Prepare an agenda for an advisory committee that is meeting in two months.

5. List ways teacher-coordinators may cooperate with school guidance personnel in initiating a plan.

6. Brainstorm to get ideas on how you would initially promote the cooperative plan of instruction described in number 1.

For references pertinent to the subject matter of this chapter, see Reading Resources.

8

Coordinator Responsibilities for Adult Workforce Development

A group meeting to explain adult workforce development opportunities available through the school systems.

KEY CONCEPT: Cooperative education is an integral part of lifelong education.

GOALS

After successfully completing the study of this chapter, answering the questions, and carrying out the activities, the student should be able to:

- Explain the role of the teacher-coordinator in adult workforce development.

- Conduct a needs analysis to determine the types of education/training required by an individual or an organization.

- Identify institutional roles in continuing education.

- Describe how to organize an adult workforce development program.

KEY TERMS

adult education

adult workforce development

consulting

continuing education

needs analysis

preparatory instruction

seminar(s)

training and development

upgrading

Adult education, conducted primarily by school corporations, private industry councils, and post-secondary institutions, has expanded greatly since 1970. This expansion is due, at least in part, to funding sources, including state and federal government agencies, as well as private enterprise. The emphasis on life-long learning is becoming a reality in more and more places.

There is a national trend for the *workforce development* of adults, which is being led by business and industry. Most large U.S. corporations have developed active human resource development programs. Presently, U.S. corporations spend an estimated $40 billion annually on the formal training of employees. As of 1995, over 30 corporations were offering college-level degree programs. Much of this effort is directed at retraining as the United States restructures its workforce. Recent studies reveal that 42 percent of the workforce needs retraining at a price tag of $10 billion. Training investment is one of the keys to competitive advantages.[1]

Much of the growth in cooperative adult workforce development programs, however, is in the two-year community colleges. Some community college systems are utilizing cooperative education to support, as well as to make available, those technical programs that are much too expensive because of costly machinery and equipment. These programs are often joint ventures between post-secondary institutions and local industry. For example, in Terre Haute, Indiana, a local laser optics company uses faculty of the Indiana Vocational Technical College to teach course work for its employees while they are using plant equipment for hands-on training. To foster such relationships, many post-secondary institutions employ business and industry training and development specialists.

Experiential (cooperative) learning for adults covers a wide range of learning opportunities. Educational learning for adults is increasingly taking place outside the traditional classroom. Cooperative pre-retirement programs also are becoming more popular within business and industry. Spouses and children may be included in these pre-retirement programs. Cooperative adult programs will be emerging at a greater rate, especially as the lifelong learning concept is accepted by more adults. In addition, by the year 2008, 75 percent of all workers currently employed will need retraining. Some individuals believe that cooperative adult workforce development programs may provide a partial solution to the unemployment problem. Corporations train employees to increase corporation profits. Cooperative education has created a strong relationship between industry and public education.

One might easily wonder why a chapter on adult workforce development is included in a book dealing with using the work environment as part of the school's instructional program. The answer lies primarily in the direct and trusting relationship teacher-coordinators establish with the business and industrial community. Management and school personnel need to recognize and be sensitive to training problems at all levels.

The Adult Workforce Development Division of the Association for Career and Technical Education provides career-technical educators high-quality information and services responsive to the needs of these educators as training and employment professionals. The Adult Workforce Development Division strives to keep pace with the latest trends in technology, education, and legislation in order to produce quality workers for a competitive workforce. The Adult Workforce Development Division was called the Employment and Training Division until December 1999, when ACTE's Assembly of Delegates approved the name change. It focuses on the following adult education needs.

(1) Transitional worker development;

(2) Creating and developing community workforce pools;

(3) Providing comprehensive career services;

---

[1]"Put Quality to Work in Your Own Company," *Training and Development Journal,* July 1991, p. 36.

(4) Customized training for business and industry;

(5) Developing entry level skills for employment;

(6) Upgrading employee skills for future technologies;

(7) Providing skill gap remediation;

(8) Developing employability skills;

(9) Providing the bridge for individuals to maintain economic viability and independence; and

(10) Inspiring an appreciation for life long learning.[2]

The degree of involvement of the cooperative education teacher-coordinator with adult workforce development varies greatly from state to state and from community to community. At one time, teacher-coordinators of secondary cooperative plans were expected to organize, supervise, and even teach in the adult workforce development program related to their occupational area. In such situations, as much as a third or a half of the daily schedule was devoted to adult workforce development. In a small community the teacher-coordinator was, in effect, the director of adult workforce development for the occupational area. The rationale was an obvious one—the coordinator was the best qualified person for adult workforce development in a small community. Because of the person's occupational experience background and personality, and because of daily contacts with the community, the teacher-coordinator had a "handle on" that community's needs.

In most districts, the secondary school coordinator's role is still primarily that of a teacher. In some situations, the coordinator has been assigned responsibility for adult workforce development. Post-secondary coordinators are often involved in lifelong educational programs and have quite a unique role, because of the strategy of the institution being truly

community–service oriented. They are asked to plan and carry out short-term continuing education activities, such as conferences, clinics, and workshops. In addition, they usually have a major role in the planning of classes for part-time students.

Because adult workforce development opportunities continue to expand, the coordinator probably will play some role related to adult workforce development, even with the diversity mentioned, and should possess at least a minimal understanding of the organizational and operational procedures of these programs.

Teacher-coordinators of cooperative plans can, in many cases, look forward to involvement with adult workforce development. Adult education activities may enhance the cooperative plan. For example, development courses for adult sponsors and for supervisory personnel may enhance the on-the-job training given to cooperative student-learners. The contacts with employers on advisory committees for adult courses may result in more employers becoming better acquainted with the coordinator, thus bringing about an increased number of training stations. Other benefits may also accrue to the coordinator, such as increasing knowledge about the field and making contacts that will result in expanding local support for the program.

Enrollments in post-secondary programs, including adult workforce development, are increasing rapidly. Day and evening classes, weekend workshops, and lunch-hour seminars have become an integral part of many in-city located campuses of four-year colleges and community colleges. In this present structure of adult education, it seems reasonable to expect coordinators to accept responsibility.

If a total program of education for the occupational areas is to develop, the coordinator will probably need to assume a leadership role in its development. Involvement in adult programs may help students advance after they enter full-time employment. Confident coordinators, whether of secondary or post-secondary programs, also recognize that they are extremely competent as persons who understand

[2]Adult Workforce Development Division, Association for Career and Technical Education, 1410 King Street, Alexandria, VA 22314, www.acteonline.org.

training and training needs and who have the ability to work with businesspersons. They are cognizant of their ability to fulfill the role of "training consultant" for business and industry of the community. In the smaller town or city, the coordinator may be one of the few people in the community capable of planning and carrying on the training programs.

Even in the larger community, the coordinator's talents are useful both to firms and to trade and professional associations that need advice about training. It is important to recognize that while all CTE teachers should be able to help the community with educational problems, coordinators are somewhat unique because they work daily with businesspersons. Therefore, not only do they have their confidence but they also have greater insight into their problems and needs.

Identifying Goals Through Needs

Education for occupational competency is a lifelong process for most people. For out-of-school youth and adults, increased job competency in their present

occupations or preparation for different occupations to some degree comes from on-the-job training, but increasingly it comes from organized, part-time courses offered by public and private schools, employers, and trade and professional associations and unions. Not only have factors such as technological changes, re-entry of women into the labor force, and geographic mobility made adult training that improves job performance necessary, but they have also made retraining imperative.

PEOPLE'S NEEDS AS A BASIS FOR PROGRAM PLANNING

A basic approach to program planning in adult education is the analysis of people and their occupational training needs at one time or another in their lifetime. The evaluation of training needed by an individual or an organization is called *needs analysis.*

1. **Preparation for a new occupation**—This situation may occur when individuals find themselves dissatisfied with their present jobs. While retaining their current jobs, they may enter an adult course to prepare for another occupation. For example, a factory assembler may wish to become a skilled crafts worker or a salesperson; a cashier to become a word

Many adults wish to upgrade their skills. For example, a retail clerk may wish to learn computer application skills.

processor; or a stockperson to become an insurance agent. Individuals who find themselves unable to continue in their present jobs because of disabilities are another group who need to prepare for new occupations. Both state and private vocational rehabilitation programs have expanded their training services. In these cases, adult workforce development is termed *preparatory instruction.*

2. **Increased performance on the present job**—This situation is one in which individuals are satisfied with their present occupations but wish to improve their performance, typically to enhance their remuneration, but sometimes to increase their job morale and satisfaction. For example, a word processor wishes to become familiar with a variety of software packages; an insurance agent wishes to increase sales and commissions; or a plumber or an electrician wishes to learn new techniques that will improve output. Classes to improve performance on the present job are supplemental rather than preparatory instruction. They are also termed *upgrading.*

3. **Preparation for advancement**—This situation occurs when a person is satisfied with a chosen occupational field but desires advancement to a higher-level position—one of more responsibility. Increased remuneration may or may not result immediately; in some cases, for example, salespersons or crafts workers actually accept reductions in pay to obtain first-level supervisory or management positions. Examples of preparation for advancement would be classes that prepare secretaries as administrative assistants, accounting clerks as computer programmers, salespersons as sales managers, retail salespersons as buyers or store managers, assemblers as supervisors, and automotive technicians as service managers.

4. **Retraining**—This situation is one in which the individual must learn new skills and information because certain aspects of the job have changed, because the individual is unemployed, or will be, or because the skills the person has are obsolete or not up to job standards because the person has been absent from the labor market (as is the case of many women re-entering the labor market). Technological advancements have created the need for retraining programs. Retraining is also necessary when one part of a worker's competency falls below job standards, but such retraining is usually accomplished by the employer.

In addition to adult career-technical needs, it is also possible to think of the needs of adults according to job level. The three useful classifications are: employee (rank-and-file), supervisory, and management. The coordinator who is engaged in adult workforce development will find the "needs approach" very useful in determining whether the adult program is truly one of a total program approach.

The coordinator engaged in adult workforce development should be aware that some adults enroll in CTE classes to meet needs not expressed in the classifications already discussed. Some people enroll because they wish to associate with other people in the same occupations; some feel the need to stimulate their thinking; some wish to increase their status with their co- workers, supervisors, friends, and family, which comes with raising their educational level; and some desire to gain certificates attesting to their accomplishments.

Institutional Roles in Continuing Education

There are many types of educational institutions involved in adult workforce development. While coordinators typically will be associated with secondary schools, area career-technical schools, or community–junior colleges or technical institutes, they should be aware of other institutions, such as the extension divisions of colleges and universities that are involved in adult workforce development. In the role of adult program director, coordinators should avoid duplicating the efforts of others, dovetailing their programs with those of other organizations and agencies whenever possible.

The following is a brief listing of the roles played by various agencies in implementing adult workforce development.

1. **Adult workforce development in secondary school districts**—Many school districts operate adult workforce development CTE programs. Some operate under state plans for CTE; others operate entirely under local school district auspices. These are declining as other educational institutions take over; some high schools no longer see their role as providing adult workforce development, but rather as providing liberal arts–type programs. With the expansion of area career-technical schools, comprehensive secondary districts cannot provide CTE adult classes as cost effectively as the area career-technical districts.

2. **Post-secondary institutions**—Into this category fall the adult workforce development programs of public and private community–junior colleges and technical institutes and of private business and trade schools.

3. **Collegiate extension divisions**—Most public and private colleges and universities operate adult education programs through their extension or continuing education divisions and through their cooperative extension services on both a credit and a noncredit basis.

4. **State departments of education**—In some states the division of career and technical education operates adult workforce development courses through a system of itinerant instructors or by contract with the extension division of state colleges and universities.

5. **Trade and professional associations**—Many trade and professional associations, through their educational directors or through educational institutes operated as part of the associations' membership programs, now offer adult workforce development programs.

6. **Unions**—Labor unions have educational advancement as one of their stated purposes and operate educational programs for their memberships or for prospective workers in the occupations with which the unions are concerned. Unions are also involved with management in the operation of apprenticeship programs.

7. **Non-profit groups**—Some non-profit organizations, such as community action groups, offer adult courses. For example, the Red Cross offers a variety of first-aid courses.

8. **In-house**—Some organizations provide training programs of special interest to their employees such as investments and retirement preparation.

Basic Procedures in Organizing Adult Workforce Development Programs

A well-managed department of career and technical education has a long-range program plan whereby the different groups with which it works may contribute to the realization of the objectives of that program. This type of planning chart would assist in carrying out any adult education program (see Table 8–1).

Among the many advantages of the program approach to adult workforce development education are these:

1. Develops continuity in the training area.

2. Sets definite objectives for students.

3. Provides potential enrollees or individuals with prior preparation.

4. Enables students to upgrade themselves over a long term.

5. Introduces new ideas, skills, knowledges, and concepts in an organized learning situation as technology changes.

Table 8–1
Adult Evening Class Plans for Certified Growers

Objective	Present Situation	Goal	Ways and Means
1. Increase the number of growers of certified seed.	97 certified growers. 15% on registered list.	125 certified growers 50% on registered list.	Conduct evening classes in five centers. Pool seed order for new growers. Sponsor field day on growing new crop.
2. Increase the number of tuber-unit plots.	20% have established tuber-unit plots.	75% with tuber-unit plots.	Include lessons on tuber-unit plots. Conduct demonstration plots. Sponsor tours to demonstration plots and to certified fields with well-arranged plots.

6. Sets high standards for content in carefully planned units.

7. Gives direction and purpose to the training program.

8. Produces a complete series of educational experiences to achieve a goal.

TIMING OF THE PROGRAM OFFERINGS

Planning adult workforce development programs demands looking at the timing of offerings in terms of the needs of adults rather than those of traditional students. Adults are not captive students; they and their employers set their priorities according to the demands of their businesses, as well as according to their own personal values. It is wise to begin first by using the calendar rather than the academic year as a base. The following should be considered: vacations and three-day holiday weekends, peak recreation periods such as fall hunting, shift schedules, hours when working parents would most likely be at home, prime business periods, travel time to the class location, and major sporting and cultural events.

Examples of offerings that reflect these demands on the adult student are: (1) a morning short course on personnel and training techniques offered in the spring before summer hiring; (2) a customer credit course in the fall before the Christmas rush; (3) a secretarial refresher course for the Certified Professional Secretaries' Examination; (4) an income tax seminar in the fall; and (5) a machine technology course operating all year on weekends for upgrading.

SEQUENCING THE PROGRAM OFFERINGS

At state and local levels, experience has shown that the program approach to adult workforce development is advisable. Only in exceptional cases should courses be organized and promoted as ends in themselves. No one course for adults can be exhaustive in the subject matter or problem areas; therefore, it is advisable not to cover areas of instruction too lightly. A program approach (the planning of a sequence of courses and experiences) causes interested groups to ask that courses be scheduled regularly in sequence year after year. This approach will enable the supervisors and instructors to provide continuity and depth to accomplish training objectives. This does not preclude the offering of an isolated adult course if the demand exists. The program approach certainly implies that a sequence of courses is available, with certificates given upon completion. Some certificate programs are given in cooperation with local or national trade or professional associations. Sequences are illustrated by the following examples.

1. **Chartered property and casualty underwriters, program**—This program consists of five yearly courses (parts), each made up of 60 clock-hours of instructions. These are: Part I, "Insurance Principles and Practices"; Part 2, "Analysis of Insurance Functions"; Part 3, "Economics, Government, and Business"; Part 4, "Insurance and Business Law"; and Part 5, "Management, Accounting, and Finance." A national examination is given over each part at the end of the year.

2. **Retail training program**—This is a two-year program. The student attends two classes per week for four semesters, or a total of eight courses.

Required: 4 Courses	Electives: Choose 4 Courses
Retail Sales Techniques	Retail Sales Promotion
Store Operations	Visual Merchandising
Supervisory Techniques	Textile Information
Merchandising Techniques	Advanced Sales Techniques
	Principles of Buying
	Fashion Merchandising

DETERMINING THE LENGTH OF THE COURSES

In some cases, the appropriate course for adults will be the short-unit type, usually less than 20 hours of instruction. A useful approach is the *seminar,* defined as one or two meetings totaling about two to five hours. Examples would be seminars regarding shoplifting conducted for retail salespersons and store managers, telephone courtesy or personal appearance for office clerical workers, and new techniques in advertising for restaurant owners. Another short-unit type of course is that which runs for about 10 two-hour periods; some examples might be "How to Lead a Conference," "Human Relations Techniques for Supervisors," "Textile Information for Salespersons," "Energy and Federal Regulations," "Inventory and

Stock Control," and "Solar Heating." In some cases these 20-hour courses are offered in sequences leading to a certificate of completion.

Besides the short-unit courses, there may be conferences or intensive training given. Many college extension or continuing education divisions and some local adult schools offer courses with intensive instruction all day for a week or two. These courses are particularly useful when participants are sent by their firms and when the courses must serve people in a wide geographic area.

The regular semester-length courses have the advantages of providing more instructional time and of adapting more easily to the schedule of the regular adult school. On the other hand, semester-length courses often conflict with other activities or with evening work assignments of potential enrollees—the evening store hours during the Christmas season are an example.

At one time it was believed the best time to offer adult classes was in the evening. However, with flex-

---

CASE-IN-POINT

Sally Johnson is a marketing education teacher-coordinator in Roanoke, Virginia. In Virginia, adult education has long been a traditional role of the teacher-coordinator. While Sally seldom teaches adult workforce development courses, she does help develop and organize them for her school system.

To get started each year, Sally and the other teacher-coordinators in the city conduct needs analysis of local retail and service businesses. Information from these studies tells Sally what types of training and development are needed. Sally's job then is to determine where the course should be taught, when, what material should be used, what the costs to students will be, and who will teach the course. For her services, Sally is paid a fee based on the number of students enrolled and the cost of the course.

The types of courses offered will vary according to whether a generic course on a topic such as "Selling" or a specialized course such as "Housekeeping Management" for the local hotel association is offered. Other popular courses include "Supervision," "Communications," "Time Management," and "Customer Service."

time becoming more popular, work schedules varying, and lifestyles changing, many adult courses are popular during the day and late afternoon and the "wee" hours of the morning, for example, for those who work the late night shift.

INVOLVING THE COMMUNITY IN PLANNING

Decisions on the timing of the course offerings, the timing of the units within a course, the types of instructional materials to use, the sources of instructional materials, and the suggested sources of teaching personnel should probably be made with assistance from persons in the occupational specialty area. An ad hoc committee is appropriate, with membership made up of from three to seven representatives in the field. After suggested objectives and units of instructions have been formulated, preliminary plans for a course offering have been made, and resource materials have been decided upon, a budget can be determined. The ad hoc committee may take an active role in one or more activities.

Some examples of organizations that might be contacted when establishing ad hoc committees as well as developing instructional courses and materials for adult learners are:

Sources (Local Affiliates/Chapters)	Possible Programs/Classes
American Marketing Association (www.marketingpower.com)	Marketing, sales, customer service, business development, training, and human resources–related courses and information
Data Management Association (www.dama.org)	Data processing program or curriculum
National Association of Executive Secretaries and Administrative Assistants (www.naesaa.com)	Clerical, secretarial, and administrative assistants program; information processing
National Association of Realtors (http://nar.realtor.com)	Real estate classes
National Association of Sales Professionals (www.nasp.com)	Sales techniques, merchandising, retailing, marketing classes
National Joint Apprenticeship Training Committee (www.njatc.org)	Apprenticeship classes in trades and industrial adult offerings
National Society of Professional Engineers (www.nspe.org)	Technical course offerings
Local craft and trade associations	Craft and trade offerings
Local health associations	Practical nursing, home care, elder care
Local police department	Personal and home protection

Other sources of information and advice on the needs and demand for adult education are the state employment service, the state department of education (career and technical division), the chamber of commerce, trade unions, and trade and professional associations.

FINANCING ADULT WORKFORCE DEVELOPMENT PROGRAMS

Almost all educational agencies require that adult workforce development be financed in such a way as to insure that costs will be recovered. The effort must be self-supporting. A few secondary districts perceive adult education to be part of their instructional responsibility and use the tax monies to offer free courses, but such districts are becoming rare as schools fight for increased support.

In some states, public schools cannot charge fees—all aspects of adult workforce development including the use of buildings, heat, and lights are free. Some schools charge a fee for materials used, especially when the materials are expensive, as in classes in welding, visual merchandising, or clothing construction. Some schools charge a nominal fee, which they believe to be advantageous in attracting serious students and holding students through a course once they have enrolled. There is no scientific evidence that these advantages really exist, but experience has taught many supervisors of adult workforce development the advisability of having such a fee.

Persistence in attendance is usually attained as the result of the quality of the program—the motivation it creates and the unity of the courses it offers. An unorganized adult course invites irregular attendance. A course that is obviously planned as a unit tends to hold students because they know if they miss a part of it they will get behind. Charging adult students nominal fees is a common practice. Care should be taken to be sure the fees charged are not so high that they exclude adults who need the program and can benefit most from it.

Some adult programs contract with cooperating firms or government agencies that have a policy of paying part or all of the fees for their employees. The employer is billed by the school for the course and materials fees incurred by the employees. The firms, in turn, may recover any necessary fees from their employees, although some pay all or half if the employees successfully complete the course. Not only does this arrangement simplify billing, it is useful in recruiting students.

SECURING AND TRAINING INSTRUCTORS

There are many sources of instructors for adult workforce development classes: secondary school and college teachers; itinerant instructors employed by state departments of education; personnel from training departments of business, government, and industry; educational personnel employed by trade and professional associations; homemakers with prior teaching experience and occupational experience; and the largest pool of all—employees, supervisors, and managers from industry.

Over the years there has been debate over who is better qualified to teach adults—the individual who knows the job but has no training for teaching or the trained teacher who does not know the job. The argument is fallacious. Adult classes, like any other classes, require instructor competence, both in contemporary subject matter and in up-to-date teaching methods and instructional technology. Adult students vary greatly in their competence in the basics—reading comprehension, computational ability, and writing and studying skills. Some have been failures in school and are apprehensive; some have hearing and eyesight problems; some have emotional problems. The instructor of adult workforce development courses needs to know how to adjust instruction for such students. An instructor of adult workforce development should be:

1. Up-to-date in the technical content to be taught, ranging from broad acquaintance to depth of operational detail.

2. Knowledgeable about the terminology of the industry and acquainted with the climate of the job, preferably from experience.

3. Comfortable with adults, having insight into their needs and their feelings and not feeling superior to them.

4. Competent in instructional methods and technology.

5. Flexible in the approach to course content and structure, understanding the immediate motivation of adults.

6. Licensed in the trade when the courses are part of preparation for licensed occupations; in other cases, recognized for competence in the trade.

7. Willing to engage in meetings or other activities designed to help improve teaching.

8. Willing to have a broad point of view rather than teaching only what is done in a certain firm.

9. Available to see the course through. Other employment responsibilities cannot be allowed to interfere with the teaching responsibilities.

Recruiting of instructors is facilitated by developing a call staff. This is a roster of individuals who are "on call" when an instructor is needed. They are selected individuals who have demonstrated their teaching ability and are listed according to the subjects or courses for which they are qualified. Recruiting is also facilitated by the coordinators, continuing contacts with business and industry. For example, job sponsors and supervisors of cooperative student-learners demonstrate their ability and can be invited to teach classes of adults. Advisory committees can suggest capable instructors and can assist in recruiting them. Presidents of local chapters of trade and professional associations and officers of local unions often know who are the most competent people in their trades.

In any case some instructor training is quite necessary, even if it is only a series of informal meetings devoted to problems raised by the adult workforce development staff. Adult workforce development instructors themselves report the need for orientation sessions in which the course outlines are discussed; furthermore, they want at least short meetings during the time they are teaching to discuss problems that have arisen. Inherent in well-operated programs of adult education is a system for training instructors of adults and certifying their training. These adults from business and industry who learn to teach well also become more effective on their jobs where training and supervision are often among their major responsibilities.

LEARNING ABOUT TEACHING

A functional relationship exists between the abilities and the personality of the instructor of adult workforce development and the degree to which the objectives of the adult offering are reached. In a noncredit career-technical adult offering, the partici-

pants will remain active only as long as they feel good about the instructor and believe that they are gaining in knowledge and skill through attendance.

Most instructors of adult workforce development, particularly those from industry, need some instructor training to help them refine their training skills. Many secondary teachers also need to learn to teach adults because the approaches they use with secondary students will not work with adults. Adults expect instruction that meets their needs now; they feel they have a real investment. They also want instruction that frees them from worry and concern over "passing."

LEARNING TO INSTRUCT AT THE LEARNER'S LEVEL

Adult groups usually enjoy group participation. Instructors of adult workforce development should encourage group interaction and should usually serve as discussion leaders rather than as lecturers. The classroom procedure may be informal, but the planned direction of the discussion is such that objectives are reached.

By learning the experience and background of the class participants at the first class meeting, the instructor can adapt the content to the understanding and experience of the group members. Most adults want to learn those knowledges and skills that can be used in their work. Therefore, the material should be practical. It should be covered at a rate the learners can handle so that they will be able to absorb the instruction sufficiently to use it. The instructor should present ideas to use or to check out before the next group meeting, thus giving the meeting take-home value.

MOTIVATING ADULTS

Without the usual incentive of credits, precise grading, compulsory attendance, assignments, diplomas, and such, the motivation of adults becomes a matter of concern. In adult workforce development courses, probably the main purpose for attendance is the short-run gain of getting a job or performing better on the job or the long-run gain of advancement or promotion. When adults engage in continuing education, they are adding to their personal schedule one additional commitment and one more responsibility. The time and energy required to participate in adult instruction is taken from some other responsibility—family, community, and/or profession, for example.

Adults expect to share in determining the course's goal and in directing their own learning. The wise instructor of adults will utilize the resources that can be offered by members of the class. Realizing the importance of active student involvement, the instructor will invite student help in planning the activities of the course in kind, scope, and sequence. The successful adult class achieves a productive interaction of individuals through the medium of shared learnings and the meeting of minds. Its atmosphere is informal without being disorganized. The group is guided without being dominated. Students feel they are a part of the group without being engulfed by it.

EVALUATING ACHIEVEMENT

The heterogeneity of almost any adult group must be considered in evaluation. The age, motivation, experience, educational background, and sophistication in the content field will vary to a great extent.

Because individual members are in attendance to fulfill self-determined needs, any measurement of success of their effort must be in terms of their own growth. Their success needs to be measured against standards appropriate to their own experience and goals. A comparison should be made of their achievement at the end of the course in relation to where they were at the start of the course. Uniform examinations, exercises, percentage grades, and letter grades are held to a minimum. The best measure of the success of an individual in an adult CTE class is the observed improvement in performance and/or the feedback to the instructor of improved job satisfaction or performance.

Two tangible evidences of successful completion of an adult course often take the form of a pocket-size card (see Figure 8–1) and a certificate (see Figure 8–2). A culminating event for a sequenced certificate program may be an employer–employee dinner; a

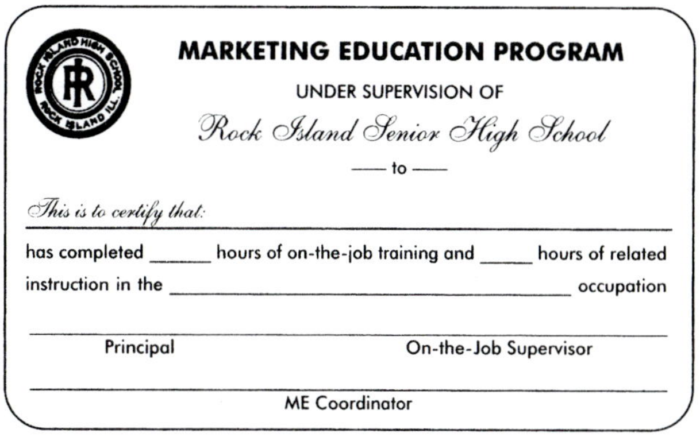
MARKETING EDUCATION PROGRAM

UNDER SUPERVISION OF

Rock Island Senior High School

— to —

This is to certify that: ____________________

has completed _____ hours of on-the-job training and _____ hours of related instruction in the ____________________ occupation

____________________ Principal ____________________ On-the-Job Supervisor

____________________ ME Coordinator

FIGURE 8–1

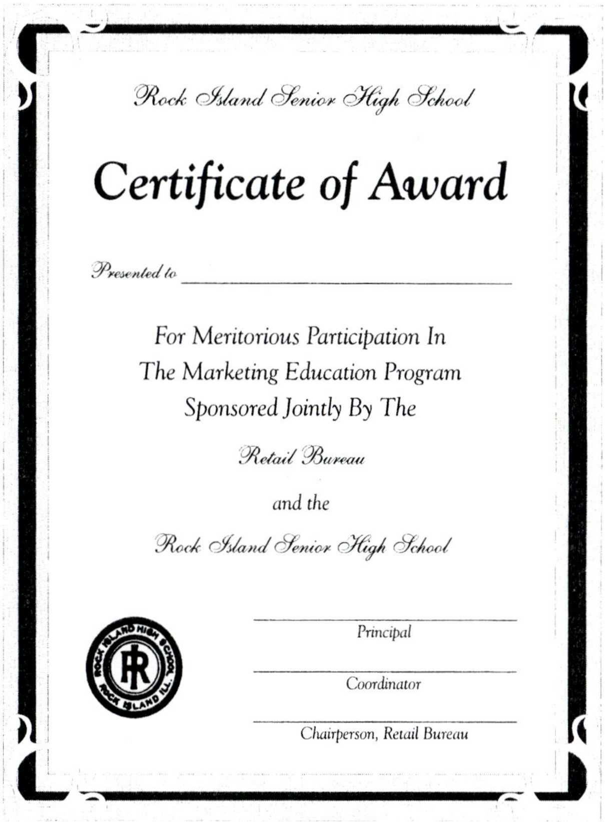
Rock Island Senior High School

Certificate of Award

Presented to ____________________

For Meritorious Participation In
The Marketing Education Program
Sponsored Jointly By The
Retail Bureau
and the
Rock Island Senior High School

Principal

Coordinator

Chairperson, Retail Bureau

FIGURE 8–2

graduation dinner attended by class members, the instructor, the coordinator or supervisor, and the craft (advisory) committee; or even a picnic or a pizza with the instructor.

Questions and Activities

1. Write a paper on suggestions to new instructors of adult classes. Include suggestions from your own experience and from outside references to supplement the textbook reading. Include, among other things, the following:

   a. The types of students enrolled in an adult class.
   b. The desirable characteristics of a teacher of adults.
   c. Suggestions for organizing an adult workforce development course.
   d. Suggestions for physically managing the classroom.
   e. Suggestions for utilizing a variety of methods of teaching adults.

2. Prepare a survey form that might be used to gather information in a community to determine needs for a particular adult workforce development course or program.

3. Prepare a file of forms used in your state by a specific service or division of the state board of career and technical education. Be sure to include forms:

   a. To certify an adult instructor.

   b. To approve a particular course offering.
   c. To gather information on the completion of the course to use as a basis for reimbursement to the local school.

4. Prepare a budget for a specific adult class, including the instructor's salary and an itemized list of supplies, showing the costs, incidental expenses, and estimated available funds for defraying costs and expenses.

5. Find the latest figures available on adult class enrollments and determine what trends are occurring in enrollments in your local community, your state, and the nation.

6. Select an occupational area with which you are familiar. Make a chart showing adult courses that could be offered for that area from rank-and-file through management training.

7. Describe the apparent adult workforce development goals of each educational institution and agency in your community. What programs are offered? To what clientele? What is the apparent timing and sequencing technique? What gaps do you see? What duplication?

8. It is increasingly evident that public school districts are not involving their occupational coordinators in adult workforce development roles. Why is this trend evident? Should it be so? Explain your answer.

For references pertinent to the subject matter of this chapter, see Reading Resources.

9

Coordinator Responsibilities at the Post-secondary and Collegiate Levels

Faculty cooperative advisor discusses potential cooperative placements with college students.

KEY CONCEPT: Cooperative education at the post-secondary and collegiate levels is a key factor in the successful placement of graduates.

GOALS

After successfully completing the study of this chapter, answering the questions, and carrying out the activities, the student should be able to:

- Distinguish between the role of the college cooperative coordinator and that of the faculty cooperative advisor.

- Compare and contrast alternating, parallel, and single-field experiences.

- Discuss the nature of placement problems that are unique to post-secondary and collegiate situations.

- Distinguish between work experience for general education and work experience for professional education.

- Identify and discuss the cooperative professional organizations.

KEY TERMS

alternating cooperative experiences	non-additive credits
debriefing	open-interviewing
faculty cooperative advisor	operational experiences
internship	orientation experiences
learning agreement or contract	parallel cooperative experiences

Work Experience Education in Higher Education Institutions

Higher education institutions are defined, for the purposes of this text, as two-year post-secondary institutions and four-year colleges and universities. Many use the work environment, a part of their curricula, as a way of promoting general education and as a means of furthering professional education goals.

The use of cooperative education in higher education institutions has seen dramatic growth in recent years. Nationwide, according to the National Joint Apprenticeship Training Committee, there are more than 200,000 cooperative education students in more than 1,000 colleges working for approximately 50,000 employers, large and small. The Cooperative Education Association (CEA) conducted a national census of cooperative education programs in 1998. It identified a variety of characteristics associated with cooperative education at the post-secondary level. Seventy-seven percent of the post-secondary institutions responding to the census offered cooperative education as an optional educational experience. The remaining respondents had mandatory cooperative education requirements. Cooperative educational experiences took many forms, including internships, service-learning, and practice. The average number of work experiences completed by students is two. Eighty-six percent of all cooperative work terms are paid. Thirty-three percent of cooperative education students accept offers of full-time employment with the co-op employers.[1]

Post-secondary plans for cooperative occupational experiences are growing in number. Indications are that cooperative education will continue to play an important role in the expansion of higher education in the United States.

---

[1]The Census of Cooperative Education, Cooperative Education Association, 8640 Guilford Road, Suite 215, Columbia, MD 21046, www.ceainc.org.

WORK EXPERIENCE FOR GENERAL EDUCATION

Some liberal arts institutions are well known for their use of work experience education as part of the requirements for graduation. These liberal arts colleges usually include practical work in a productive situation in their curricula for two purposes: (1) to further the aims of general education by acquainting their graduates firsthand with the dignity of work and with the significance of work in a contemporary society and (2) to provide their students with the opportunity to sample areas of professional service and to test their own abilities and interests against the demands of these occupations.

In some institutions, the students' work experience is on campus. Students perform the work necessary for the operation of the college. For example, students may be employed in cafeterias, in the college business office, in maintenance departments, or on the school farm. This "self-help" approach permeates the institution and extends the entire time a student is enrolled. A good example is Berry College in Rome, Georgia.

In other curricula, students are placed in work experiences off campus for one or more periods of time. Students are employed for remuneration by firms, charitable agencies, or government agencies. Students are expected to report on these experiences. With this arrangement the students gain not only the general education advantages of understanding the world of work and of living away from the campus but also a period of occupational exploration.

WORK EXPERIENCE FOR PROFESSIONAL EDUCATION

Work experience is more widely used to further the aims of professional curricula than many educators realize. In general, a professional curriculum may include one or all of three experiences on the job. These are:

1. Experiences early in the curriculum that serve as occupational exploration, acquainting students with opportunities and requirements in the field.

2. Experiences in the middle and latter half of the curriculum that enable students to test theory against practice and that teach those skills the institution believes are best learned on the job rather than in the classroom.

3. Experiences near graduation or shortly thereafter that serve as internships prior to students' entry into professional practice. These provide students with required professional training in the transition from supervision to the full-time role of employed professional workers.

Orientation experiences.—*Orientation experiences* usually occur during the first year of collegiate education because they are designed to give students the opportunity to sample the profession. Through experience as an observer or a paid worker, the student gains insight into the specialties provided within profession. The student also experiences the conditions and demands of employment in the profession and is able to engage in self-evaluation to test personal capacities and motivations against those of the profession.

Such exploratory experiences provide two basic plans. The first plan is to schedule the student part time in a series of observations of the profession while the student engages in the regular courses on campus. For example, the prospective teacher may enroll in a credit course entitled "Field Experience for Teachers" and may participate in a series of group and individual observations in schools. The institutional coordinator arranges experiences germane to the student's interest, schedules interviews with the student to discuss experiences, and evaluates written observation reports submitted by the student.

The second basic plan of providing professional exploration involves one or more assignments of a student to paid employment and enrollment in a credit course for the experience. Some institutions restrict the experience to a regular term, assigning the student full time to the employer; others schedule a series of full-term or partial-term experiences with different units of the same firm or with different firms. In this case the aim is to give the student (for example, the future social worker) the privilege of sampling the different segments of the industry.

Regardless of the variations in scheduling, the experience is part of the curriculum if: (1) the student receives credit, regardless of whether he/she is paid, (2) the institution considers the student's personal interests and abilities before placement, (3) the employer understands the reason for the placement, and (4) the institution provides a coordinator who not only schedules the placement and follow-up but also arranges for student reports, interviews, discussions, and other intellectual activities designed to enable the student to interpret the experience.

Operational experiences.—In some professional curricula, plans of occupational experience, designed to teach certain skills and operational procedures that the institution does not, or does not wish to, teach, are provided. Such *operational experiences* have come about partly because the institutions cannot either physically or financially build laboratories that duplicate the complex and expensive equipment of industry. Examples of such situations are those related to engineering, architecture, and large computer installations. Another reason is that in some curricula, the educational institutions cannot duplicate the human environment. Examples would be the government agencies associated with curricula in police administration, social work, retailing and sales, nursing institutional management, and teacher education. A third reason for including such experiences is the tendency for institutions and their faculties to prune from their offerings those courses that they deem "vocational." Increasingly, facilities expect academic courses to be concerned with theory and principles, with the operational level skills to be learned on the job.

The majority of institutions whose programs have been described in studies or literature appear to use one or the other of two patterns. In the first pattern, a series of full-time experiences in industry are alternated with periods of full-time study. This pattern provides what are often referred to as *alternating cooperative experiences*. Prime examples of this pattern are the cooperative engineering programs, the most famous of which are the one initiated by Schneider at the University of Cincinnati in 1906 and the one initiated by the General Motors Institute in Flint, Michigan. In these long-term programs, students receive a variety of learning experiences in different units of a broad field. A cooperating employer or the unit of the firm is expected to provide careful

supervision and evaluation on the job and to develop a planned training schedule for each student in cooperation with the institutional coordinator. The student earns the full-time salary commensurate with the position. The student gains credit by submitting oral reports and written reports analyzing operations in the firm.

The second plan is to provide one or more periods of experience, often full time for a term/semester, or half time in some schools located in metropolitan areas where school and work can be combined conveniently. The period of experience usually occurs in the latter half of the curriculum when the student has completed the basic or core courses in the major field. In the best programs, careful placement, job supervision and evaluation, and assigned readings and reports are employed to insure learnings suitable to the academic credit awarded. These *parallel cooperative experiences* are found frequently in curricula such as police administration, marketing, accounting, and office administration.

Internships.—The term *"internship"* is used commonly (and erroneously) to describe a variety of work experiences used in curricula in higher education and recently in secondary schools. When the term is properly used, it describes the activity of a person who has completed all the academic requirements for admission to a profession, but who must undergo a period of occupational experience in the profession prior to "certification" (licensing) as a recognized practitioner. Thus, the internship occurs either just prior to the student's graduation and the receipt of the academic degree, as in social work and teacher education, or afterwards, as in medicine, dentistry, dietetics, and other fields such as law and accounting, where licensing examinations are often given after a period of satisfactory experience.

Internships are operated in a manner similar to how cooperative programs are operated. However, one major point needs highlighting—the difference between the intern and the cooperative student described in the previous section. The distinction is

ELEMENTS OF AN INTERNSHIP SYLLABUS

Purpose(s) of internship in the department

Eligibility requirements for undertaking an internship: in the major? minor? class standing? previous course work? GPA? permission? personal qualifications?

Application deadline

Performance requirements—forms (contract?)? time; credit (how much? time ratio?)? written work required?

Some categories for learning goals/objectives:

> job skills/competencies
> career information
> broadened perspective
> interpersonal skills
> environmental knowledge
> research skills
> cognitive skills development
> personal development skills
> experiential learning skills
> other

Site supervision requirement

Confidentiality

Liability waiver

Reporting expectations

Transportation issues

Personal deportment issues

Definition of site supervision needed: frequency, duties, etc.

Definition of faculty supervision required: frequency, duties, etc.

Evaluation/assessment criteria and grading policies

FIGURE 9-1. (Source: Lynchburg College Career Development Center, Lynchburg, Virginia, 1991)

subtle but important, because it affects matters such as placement, supervision and timing of the experience, and credit/pay decisions. The intern is one who has mastered the basic academic content of his/her profession and who has been screened for entry into the profession. The intern is treated during the internship as a member of the profession, albeit a beginner. The internship is under the direct supervision of a practicing member of the profession. From the internship evaluation is derived the recommendation for the intern's acceptance and employment in the profession. On the other hand, the cooperative student is a student-employee who is learning part of the procedures of the profession on the job and is supervised by senior employees or unit supervisors. From the cooperative student's experience comes increased skills and knowledges of the job, a better understanding of actual practice, and, hopefully, a motivation for increased performance back on the campus, a background that can sharpen classroom questioning and discussion and produce a more insightful self-understanding of strengths and weaknesses.

Responsibilities

In higher education institutions, there are typically more individuals involved in the cooperative experience than are found in the secondary setting. Furthermore, the responsibilities of these individuals are sometimes different from those in the secondary cooperative education plan. The responsibilities of each of these groups or individuals are discussed as follows:

RESPONSIBILITIES OF THE FACULTY COOPERATIVE ADVISOR

The *faculty cooperative advisor* is a full-time teaching faculty member who often serves as a liaison between the cooperative education office and the academic department or school. Therefore, the faculty cooperative advisor is the essential element for success in cooperative experiences completed for academic credit. For the most part, the faculty cooperative advisor is involved in internship experiences, although faculty members can sponsor other cooperative students. Although responsibilities may vary, the following are (1) a list of general responsibilities and (2) a list of responsibilities during the cooperative semester.

General Responsibilities

1. Attends orientation/training for faculty cooperative advisors.

2. Attends meetings of faculty cooperative advisors (normally one per semester).

3. Advises cooperative staff personnel of potential placement sites.

4. Promotes the cooperative occupational education plan to students, faculty, and employers.

5. Assures that the student's employment is related to career goals and that the employment qualifies for the cooperative plan.

6. Advises students of job opportunities by referring them to the career center, computer center terminals, cooperative bulletin boards located within various schools, or similar locations.

7. Promotes an understanding of the goals of cooperative occupational education through the development and dissemination of information.

8. Reviews/approves job descriptions for cooperative or departmental credit.

Responsibilities During the Semester

1. Meets with each student the first week of the semester and assumes cooperative responsibility for each student until final grades have been assigned.

2. Discusses student responsibilities and present job.

3. Meets with student and employer at the job site as necessary (two visits recommended). Can use phone calls as a supplement.

4. Encourages students to contact him/her at any time that problems arise on the job or when they need other career guidance.

5. Assigns projects for each student, which include reviewing student progress and assigning grades.

6. Completes all required paperwork, such as travel reports, etc.

RESPONSIBILITIES OF THE COOPERATIVE COORDINATOR

The cooperative coordinator is a full-time staff member who is assigned to the cooperative office. This individual may or may not be given faculty status with academic rank. His/her primary responsibility is to coordinate and supervise the cooperative employment of an assigned group of students. However, the cooperative coordinator serves in a multi-faceted capacity as placement counselor, salesperson, educator, administrator, and referral agent. Some selected general responsibilities include:

1. Acting as a clearinghouse for forms to the faculty advisor from employers and students.

2. Providing lists of students who participate in cooperative occupational education to faculty.

3. Providing travel funds to the faculty advisor for monitoring students.

4. Acting as a liaison between faculty advisor and employers.

5. Establishing and maintaining student records for counseling and placement purposes.

6. Preparing routine correspondence and other paperwork required for daily operations.

7. Developing brochures and other promotional materials.

8. Attracting and selecting students for participation in the plan.

9. Attracting and recruiting employers for participation in the plan.

RESPONSIBILITIES OF THE STUDENTS

College students have much to gain from the cooperative experience. Not only are they earning valuable dollars to aid in paying tuition but they are also gaining work experience that often leads to permanent employment after graduation. The following are some of the primary responsibilities of cooperative education students.

1. Students should remain in the assigned work experience for the duration of the semester or work period.

2. Students are expected to return to the same employer for all work assignments.

3. Students should submit a written evaluation at the mid-point of their work assignments and at the end of each work assignment.

RESPONSIBILITIES OF THE EMPLOYERS

For all employers who hire cooperative students, there is a core of basic responsibilities. These seem to vary little from program to program. These responsibilities include:

1. Providing desirable career training experiences for students.

2. Participating in faculty/coordinator visitations.

3. Providing a written evaluation of student employees at the conclusion of each work period.

4. Providing a safe and appropriate environment.

5. Insuring that job duties remain in accordance with job descriptions.

6. Adequately supervising student employees.

7. Keeping students through the scheduled work period, unless there are compelling reasons not to do so.

Employer-Sponsor Selection

From time to time the faculty advisor will have the opportunity to be involved in selecting potential employer sponsors. The following techniques have proven useful to cooperative coordinators and faculty advisors when they are involved in the selling/selection process.

1. Always identify and contact the "decision-maker" within a company first. This will require some prospecting on your part. Leads can be obtained from sources such as membership directories, alumni, students, and the placement office.

2. Show the employer that you have the company's interests, as well as the trainee's interests, at heart.

3. Discuss specific aspects of a problem with the employer and deal with problems one at a time, for example, hiring restrictions.

4. Explain to the employer that you will assist with any problems that may arise relative to the trainee.

Placement of Students

In most schools, the placement process begins with group orientations to review the philosophy of cooperative education, required forms to complete, procedures of placement, types of jobs, rules and regulations, the art of interviewing, and résumé writing.

THE INITIAL INTERVIEW

Once the student has formally applied, the cooperative coordinator and/or faculty advisor should conduct a formal interview with the student. At this point, the student must understand the importance of mutual trust and communication. It should be noted that the great majority of problems occur because the student did not communicate with his/her coordinator. During the interview, the coordinator should attempt to secure detailed answers to the following:

1. Choice of major—why? Is it definite?

2. Date to begin co-op experience.

3. Personal information regarding citizenship, scholarships, athletic involvement.

4. Placement parameters, such as geographic and transportation restrictions and job preferences.

5. Previous work experience—full-time and part-time.

6. Extracurricular activities.

7. Reactions to courses, grades, and the school in general.

8. Reasons for choosing the school and the cooperative plan.

9. Any questions or concerns that stem from orientation sessions.[2]

In addition, the coordinator should review the student's initial résumé and make suggestions for improvements. The résumé then should be keyboarded and copies made for the coordinator's file. This file should also include notes from each meeting with the student.

STUDENT SELECTION

The process of matching the student's qualifications and interests to the employer's requirements is at best a difficult job. While it is next to impossible to find the perfect match for each student and employer, the coordinator should attempt to find the most appropriate student(s) for each job. The success

---

[2]Barbara B. Porter, *A Review of the Coordinator's Role in Cooperative Education at Northeastern University*, Boston: Northeastern University Publishing Group, 1981.

of this effort will ultimately determine the success or failure of the school's cooperative plan.

When assessing each student, the coordinator should consider the following: (1) grades, (2) personal qualities, (3) geographic and transportation preferences and restrictions, (4) company requirements, and (5) student preferences. Based on these placement parameters, the coordinator will normally guide the student in the selection of several companies to interview with for cooperative positions. Most of these interviews will take place on campus; however, students are usually reimbursed for any travel required for off-campus interviewing. This process of allowing students to interview with more than one organization is called *open-interviewing*. It also implies that organizations can and will interview more than one student for each position.

In the final phase of selection, most students will interview with three to five companies before an offer is made and accepted.

FOLLOW-UP

A primary duty of the cooperative coordinator and/or faculty advisor is the follow-up of students. This requires regularly scheduled personal visits with employers as well as student debriefing meetings.

Visits should be timed so that employers and students have time for appraisal. Normally two to six weeks is allowed for the student on the job before the initial visit. Since some students may be a great distance from the school, the frequency of visits will vary; nevertheless, the telephone is also an excellent source of follow-up information.

Once students have completed their cooperative experience, it is also essential that they be debriefed once back on campus. This *debriefing* gives students the opportunity to express their reactions and impressions of the experience and to review their employer evaluation with the cooperative coordinator.

Evaluation of Students

Each student is evaluated each semester/term. The employer should receive an evaluation form from the cooperative office about three weeks prior to the end of the semester/term. The purpose of the employer evaluation is to provide the student with feedback on his/her job performance and to summarize both strong and weak points. Before the end of the semester, the evaluation form should be completed by the employer and then discussed with the student.

In addition, those students who enroll for credit by registering for a departmental internship should be evaluated by their faculty advisor or a faculty member designated by the department chairperson. Besides feedback from the employer, faculty members will require students to provide some or all of the following academic information.

Organizational Biography

An organizational biography is usually a short (two- to five-page) description of the first two weeks of the work experience. In essence, this description of the organization includes the stated purpose of the organization, the size and major departments, a list of clients, the source of funding, and its relationship to other organizations. One reason for writing such a biography is that it enables the student to get the big picture of the organization right away and to chart its formal structure.[3]

Work Products

A second possible indicator of learning is copies of the work produced on the job. These can vary greatly, from graphics, memoranda, notes, and statistical tables, to reports or articles that the student may have published. In some cases, written explanatory notes are useful to put the work products in context.

---

[3]*Faculty Supervision Manual*, Washington, D.C.: American University, 1984, pp. 15–18.

Journal

One of the most helpful ways to record work experiences is a journal. The journal can be used as "raw data" for analyzing learning over the tenure of the field experience. It can also be used to measure growth toward specific learning objectives. It should not be a log of the student's activities; instead, it should be a collection of the thoughts and insights that the activities have produced. In other words, it should focus on what is learned rather than on what is done. A journal may also be a record of any on-the-job dilemmas or problems and related insights.

Short Papers

While the journal works as a free-flowing device for recording scattered thoughts, it is sometimes useful to make more organized and detailed presentations on topics of particular interest. Short (five- to eight-page) papers, either developing themes or describing other topics related to work, might be assigned.

Readings

Assigned reading that is related in some way to what the student is doing is an excellent way to increase his/her awareness of the purpose of the organization, the problems encountered, and the philosophical questions that should be considered.

Bibliography

If the student is doing substantial job-related reading, one way to demonstrate this is to record the works read in a formal, annotated bibliography. By keeping an on-going list of the literature that is required on the job, he/she will have a reminder at the end of the field experience of the depth and nature of the reading program.

Final Wrap-Up

At the end of the semester, it is often a good idea to require a final medium-size (10-page) paper, which briefly describes the whole work experience and then focuses upon a self-evaluation. In essence, this is similar to an annual report of a company or an employee.

Composing a thorough self-evaluation is difficult but essential. This paper might address questions such as the following:

- What were the most important insights gained during the semester? How were these clarified or modified over the course of the semester? (The journal is a good source.)
- At the close of your program, how accurate does the organizational biography you outlined at the beginning of the semester seem to be? How do you explain the discrepancies, and what insights do they suggest?
- What were the greatest success(es) and/or accomplishment(s) at this job?
- What were the most valuable things learned in terms of educational, personal, and career development?
- What can you identify as weaknesses you would like to correct in yourself or abilities you would like to acquire? What would you do differently if you could start over again?
- How has the cooperative placement affected your academic, personal, and career goals?
- How would you describe the relationship or your cooperative position to what you have learned in class?
- What skills did you tap in your cooperative position, and which ones did you develop during the course of your cooperative experience?
- How might this position help you refine your ideas about the world of work and your work environment preferences?

The Portfolio

The portfolio should include memoranda, reports, statistical analyses, graphics, letters, and other examples of the student's work. Beginning the portfolio with a one-or two-page paper describing its contents and their relation to the work may also be useful. If the material is sensitive or classified, then the general nature of the work in the portfolio can be put on a separate summary sheet. This will give a better picture of progression during the job. It will also be useful for future job placement. Often prospective employers would like to see written work from previous employment, and the portfolio is an excellent record.

Other methods faculty advisors use regularly in supervising cooperative students include:

1. Requiring students to meet in groups to discuss assignments or work-related occurrences or to hear guest speakers who are experts in generally related field.

2. Requiring students to make appointments to meet with the faculty advisor every two to three weeks.

3. Asking students to complete a job information sheet, which includes material useful to both the student and the faculty advisor.

Some students have prepared major research papers (after projects suggested by their job supervisor over and above their cooperative jobs). This is especially true at the graduate level, where some students have developed thesis topics as a result of cooperative research. Whatever methods are used, faculty members have tried to keep in mind the importance of making the student's cooperative experience more than just a job off campus. It is a way to learn what cannot be learned in the classroom. Through this academic reflection and analysis, the student realizes how unique the experience is.

Writing a Cooperative Syllabus

Most cooperative offices follow the procedures of academic departments and ask faculty advisors to submit a written syllabus indicating general student requirements. However, it is important that any structured guidelines not interfere with the ability of the faculty advisor and the student to tailor the academic component to the placement.

Cooperative academic assignments usually fall into four categories: (1) written work (original research or expository writing), (2) reading, (3) portfolios, and (4) meetings. American University offers the following components of a sample syllabus.

For three credits:

1. Read and review one book related to the kind of work your organization does or the world of work in general. (two books for six credits; three books for nine credits)

2. Produce writing that demonstrates an appropriate level of conceptual thinking, research, and/or analysis—approximately 20 pages either in several short papers, two or three medium length, or one long. Topics to be determined. (40 pages for six credits; 60 pages for nine credits)

3. Provide portfolio of work produced on the job.

4. Schedule conferences every two to three weeks either on an individual or a group basis.

5. Have a post-placement session in the cooperative office.

Granting Academic Credit

The awarding of academic credit for the cooperative experience is an increasingly controversial topic. Accrediting agencies have traditionally examined only the curricula of undergraduate academic programs, although some agencies do not recognize cooperative credit as meeting graduation requirements.

A Northeastern University study found that most senior colleges award no credit for the cooperative experience.[4] However, there is a significant trend in community–junior colleges towards the awarding of credit. The study found that 70 percent of all cooperative plans awarded *non-additive credits* in some curricula and additive in others. Additive credit means that credits earned replace academic credits normally awarded when students take campus courses. The average number of credits per experience was 3.75 to 4.2, and the average number of credits awarded was 10.3 to 11.6. The same study indicates that many more junior colleges allow the cooperative coordinator to award credit than do senior colleges. Most senior colleges place the responsibility on the teaching faculty.

AMOUNT OF CREDIT

The most common practice among cooperative plans is to award the equivalent of one course for a one-term work experience. The range, however, is from one credit to a full term of credits. Often community–junior colleges and others using the parallel cooperative plan relate the amount of credit to be earned with the hours worked. For example, Lynchburg College in Virginia relates work and credits in the following manner:

Total Hours of Work per Term	Credit Hour(s)
40–79	1
80–119	2
120	3

Other programs that utilize alternate forms of the cooperative plan have established more flexible credit policies. For example, American University in

Washington, D.C., ties the amount of credit earned to equivalent lab and preparation times found in academic courses. Counting lab time on a two-for-one basis (2 hours of lab/work time counts as 1 hour of class) plus 40 hours of preparation time for scholarly projects and work. American University requires a minimum of 120 hours of work for 3 hours of credit for an undergraduate student. According to American University analysis, the minimum requirements for a standard academic course would be 112.5 total hours (37.5 class hours plus 75 hours of preparation time). Students enrolled in marketing education at the University of Georgia are required to gain 400 hours of occupationally related experience in a marketing position. Students in this program earn 3 semester hours of university credit for their work-based experience.

LEARNING AGREEMENTS OR CONTRACTS

The basis for awarding credit is often the *learning agreement* or *contract.* This contract is an agreement between the student and the coordinator or faculty advisor detailing the goal(s) or objectives that the student proposes to accomplish while on a cooperative assignment. The agreement must also have the approval of the employer.

The purpose of the learning agreement is to encourage students to become proactive in accomplishing their goals so that they will get more out of their cooperative experience.

A typical learning agreement might contain (see Figure 9–2):

1. Objectives to be accomplished.

2. Detailed, time-oriented steps leading to the goal.

3. Anticipated obstacles, resources, and deadlines.

4. Evaluation procedures.

[4]James W. Wilson and Sylvia Y. Brown, *Cooperative Education in the United States and Canada: 1977 Summary,* Boston: Cooperative Research Center, Northeastern University, 1978, pp. 8–14.

THE AMERICAN UNIVERSITY
COOPERATIVE EDUCATION PROGRAM
LEARNING AGREEMENT

Part I

Student _____ SS # _____ Phone _____

Address _____ Faculty Advisor _____

Hours Worked per Week _____ Credit Value _____ Grade Type A–F _____ P–F _____

Employer _____ Address _____

Supervisor _____ Phone _____

Co-op Course Requirements

Readings: _____

_____

Writing: _____

_____

_____

_____

Meetings: _____

_____

Post placement: _____

_____

Other: _____

_____

Important deadlines: _____

_____

How student will be evaluated: _____

_____

FIGURE 9–2. (Source: American University Cooperative Education Program)

LEARNING AGREEMENT (Continued)

■ Part II ■

To the Co-op Student: After reading the co-op handbook on experiential learning and discussion with your faculty advisor and job supervisor, please complete the following:

I have these learning objectives for this semester:

1. _____

_____

2. _____

_____

3. _____

_____

4. _____

_____

5. _____

_____

(This sheet should be sent to the Co-op office. It will become part of the student's file and will become the basis for discussion at the post-placement session.)

Signature of Student _____

Signature of Faculty Advisor _____

Signature of Employer _____

To the Faculty Advisor: Please attach any relevant handouts. (Distribution: white — student, yellow — co-op office, pink — faculty advisor.)

FIGURE 9–2 (Continued)

TYPE OF GRADE

Both letter grades and pass–fail grades are used in awarding credit for cooperative experiences. In either case, the objectives set forth in the learning agreement are an excellent guideline for evaluation. The current trend is for giving pass–fail grades when credit is awarded by the cooperative coordinator for parallel experiences; however, in the case of internships, single-field experiences and other similar forms of cooperative education where faculty members award credit, it is not unusual for better grades to be given. It should be noted that a substantial number of schools that do not award credit or grades do insert a statement of accomplishment in the student record.

An Overview of Professional Cooperative Organizations

Many cooperative coordinators affiliate with one or more professional organizations concerned with cooperative education. Membership in these organizations has many benefits, one of which is the networking of other professionals and employers. The following is a brief synopsis of four major organizations dedicated to furthering cooperative education.

1. **Cooperative Education Association (CEA)**

   Established more than 35 years ago, the Cooperative Education Association is dedicated to offering professional support to individuals in education and business involved with providing cooperative experiences to students. It is the largest cooperative education professional association that supports work-integrated learning. CEA provides professionals with opportunities to enhance skills through training and networking; offers valuable resources and information through its membership network of business and education professionals; and advances cooperative education through research, publications, and programs. Address: 8640 Guilford Road, Suite 215, Columbia, MD 21046. Phone: (410) 290-3666. Fax: (410) 290-7084. Web site: www.ceainc.org

2. **Cooperative Education Division of the American Society for Engineering Education (CED/ASEE)**

   Established as 1 of 51 divisions and committees of ASEE in 1929. Usually referred to as CED. Primarily interested in engineering and technology. Publishes the *Engineering Co-op Directory*. Meets biannually in late June (general meeting) and in January (College Industry Education Conference). Contact: Director of Cooperative Education, Michigan State University, 1410 Engineering Building, East Lansing, MI 48824-1226. Phone: (517) 355-5163. Fax: (517) 432-1356. Web site: www.coop.msstate.edu/ced.

3. **National Society for Experiential Education (NSEE)**

   Established in 1971 for those involved in experiential, field experience, internship-type plans. Published various nationwide internship directories. Meets annually every fall. Address: 9001 Braddock Road, Suite 380, Springfield, VA 22151. Phone: (703) 933-0017 or (800) 528-3492. Fax: (703) 426-8400. Web site: www.nsee.org.

4. **National Commission for Cooperative Education**

   Established in 1962 for the purpose of getting more colleges and universities involved in cooperative education and increasing the number of employment opportunities to match the growing number of schools. Has been instrumental in promoting cooperative education with the federal government and has arranged for testimony in Washington, D.C., on numerous occasions for this purpose.

Offers an annual corporate membership to aid in its efforts to provide various workshops, training centers, employee institutes, and research in the expansion and development of cooperative education nationwide. Address: 360 Huntington Avenue, 384CP, Boston, MA 02115-0596. Phone: (617) 373-3770. Fax: (617) 373-3463. Web site: www.co-op.edu.

Questions and Activities

1. What is the difference, as explained in this chapter, between the terms "interns" and "work experience students," when used in reference to students in collegiate programs of professional education? Is this distinction valid? If so, what are the implications for college educators who organize and carry out such programs?

2. In a collegiate cooperative occupational experience plan designed to teach operational practices and to provide for application of classroom principles, what methods can be used (a) to assure that these outcomes are accomplished and (b) to measure the student's achievement for the purpose of granting course credit and providing a course grade?

3. What devices or arrangements relating to curriculum instructional methods, placement, supervision, and responsibilities of the employing agency are necessary for each of the three types of collegiate professional occupational education experiences as they are defined in this chapter?

4. What type of credit, if any, is given at your college? What rationale is used for justifying the existence or non-existence of credit? Is a letter grade or a pass–fail system used? Are students charged tuition for cooperative education credit?

5. Discuss the types of assignments you could use as a faculty cooperative advisor to evaluate students. How would you weigh those assignments in relationship to job performance?

6. Explain why on-site follow-up visits of students are of key importance.

For references pertinent to the subject matter of this chapter, see Reading Resources.

SECTION THREE

The System of Instruction and Coordination

A SYSTEMS APPROACH TO IMPROVING INSTRUCTION

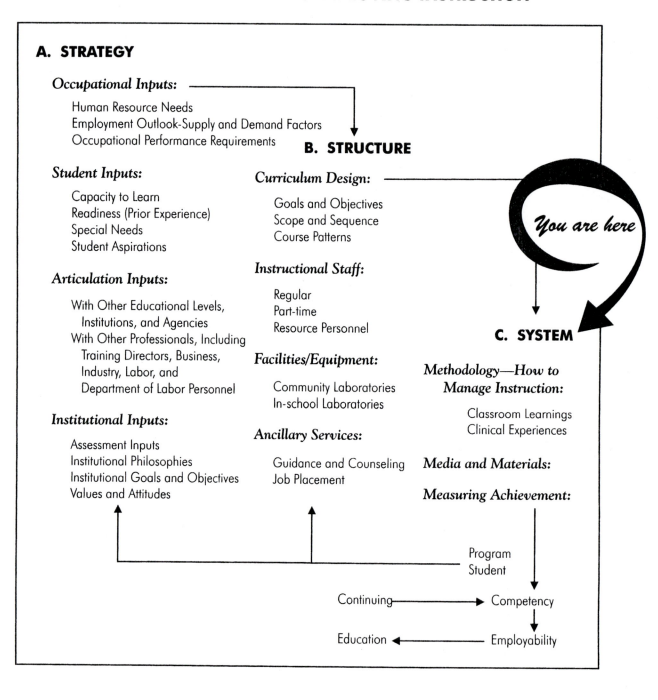

A. STRATEGY

Occupational Inputs:

    Human Resource Needs
    Employment Outlook-Supply and Demand Factors
    Occupational Performance Requirements

B. STRUCTURE

Student Inputs:

    Capacity to Learn
    Readiness (Prior Experience)
    Special Needs
    Student Aspirations

Curriculum Design:

    Goals and Objectives
    Scope and Sequence
    Course Patterns

You are here

Articulation Inputs:

    With Other Educational Levels,
      Institutions, and Agencies
    With Other Professionals, Including
      Training Directors, Business,
      Industry, Labor, and
      Department of Labor Personnel

Instructional Staff:

    Regular
    Part-time
    Resource Personnel

C. SYSTEM

Facilities/Equipment:

    Community Laboratories
    In-school Laboratories

Methodology—How to Manage Instruction:

    Classroom Learnings
    Clinical Experiences

Institutional Inputs:

    Assessment Inputs
    Institutional Philosophies
    Institutional Goals and Objectives
    Values and Attitudes

Ancillary Services:

    Guidance and Counseling
    Job Placement

Media and Materials:

Measuring Achievement:

Program
Student

Continuing ⟶ Competency

Education ⟵ Employability

INTRODUCTION TO SECTION THREE

Section Three covers the components of the instructional system, which consists of (1) the careful initiation of a cooperative occupational education plan, (2) the provision for quality instruction, and (3) the evaluation and improvement that are necessary to enable the cooperative plan to achieve its potential. The system, like any other, is a way of utilizing human resources, methods, and materials in a planned manner to assist students in reaching their goals.

Also emphasized in this section is the important role of the coordinator in unifying and balancing all elements of the plan to provide integrated instructional experiences for the students, specifically in correlating school and job instructional experiences into a unified whole. The lack of this correlation is considered by many career-technical educators as the major weakness of existing cooperative plans.

Those interested in the occupational specialties discussed in Section Four should consult the chapters in this section frequently, for they deal in depth with over-all principles, which are generalizations that must be modified at times to fit the learning needs of student-learners in various occupational fields and the peculiar learning environments in different kinds of businesses and agencies.

A timely topic for those involved in organizing and operating various cooperative plans is the legal and regulatory aspects of cooperative education. These are discussed in Chapter 16.

10

Planning and Carrying Out Effective In-school Instruction

In-school instruction in a well-equipped classroom must be handled by a certified teacher-coordinator. Here a single group is given instruction on handling personal sales techniques.

KEY CONCEPT: Effective related instruction is an essential component of the cooperative plan.

GOALS

After successfully completing the study of this chapter, answering the questions, and carrying out the activities, the student should be able to:

- Explain why correlated instruction is effective in-school instruction.

- Distinguish between the three types of related instruction.

- Explain the career development approach to cooperative occupational training.

- Describe the instructional outcomes expected from correlated instruction.

- Distinguish between cognitive, affective, and psycho-motor skills.

- Explain the career-technical approach to instruction.

- Describe "cycle instruction."

- Identify, in general terms, the key characteristic differences between special needs students and "regular" students.

- Describe the major thrusts of mainstreaming or inclusion.

- Identify the major components that an IEP must include by law.

- Identify resources regarding multicultural information.

- Describe at least six teaching methodologies suitable for providing related instruction.

KEY TERMS

affective domain

career development

career-technical approach

cognitive domain

correlation

cycling instruction

directly related instruction

general occupational competencies

general related instruction

IEP

mainstreaming or inclusion

psychomotor domain

related instruction

specific job competencies

specific occupational competencies

supplementary instruction

A number of times in preceding chapters cooperative occupational education has been emphasized as a plan of career and technical instruction that builds upon learning experiences in school by combining and correlating them with learning experiences at the training station, so as to meet the student's career goal needs. In-school instruction is particularly important in developing occupational competence, because relying solely on the job to provide appropriate learning experiences turns over the educational process to persons who are not professional teachers and who may not have the time to devote to the adequate planning of learning experiences.

The Learning Theory of Cooperative Education

The whole rationale of teaching–learning in cooperative education is quite different from that applied to other areas of the secondary school curriculum. With most in-school instruction, even in career-technical–oriented programs, the classroom work proceeds on certain assumptions. These are: (1) because students will use the learnings *on the job after graduation,* there is time to present topics according to a logical pattern of development; (2) students are committed to preparation for a general family or cluster of occupations and need little or no instruction for a specific job title; (3) there is little or no opportunity to test the application of classroom instruction to business operations because the students are not employed. On the other hand, the theory of cooperative education recognizes the individual student's need for specific instruction. Further, it recognizes that in cooperative occupational education, the instruction of the student-learner occurs in two places, the school and the job, in a correlated manner. The key to instruction is **correlation,** which emphasizes that:

1. Instruction both in the school and on the job is organized and planned around the activities associated with both the student's individual job and the student's career objective.

2. The concepts, skills, and attitudes basic to occupational competence are taught as principles in school but are applied and tested on the job.

3. Instruction is correlated continuously, whereby the sequence of instruction topics in school is related to the sequence of the student's job activities, rather than the topics being presented in the order found in the textbook or in other logical patterns.

4. Each student's learning problems are different; each student's job activities are unique and, therefore, require a different pattern of training. This means that part of the time the teacher-coordinator must individualize the in-school instruction by:

   a. Using group instruction but individualizing the assignment so that each student can apply the learning to the training station job and report on its validity.

   b. Using individual study assignments, such as projects and job study guides, and individual reading assignments.

Competency Areas to Be Developed

As discussed earlier, the cooperative plan may be either the capstone of a curriculum, encompassing two or more courses, or a self-contained instructional program without the prerequisite instruction providing basic facts and skills. In either case, the instruction in school is geared to the career objective, and the training station is seen as a place for learning new skills or for providing practice under real conditions of those concepts learned in the classroom. During the in-school class, three basic instructional needs of student-learners, if they are to be occupationally competent and therefore employable, are *general occupational competencies, specific occupational competencies,* and *specific job competencies.* The proportionate amount of time devoted to each area of need and the question of whether or not each should be covered in a separate course are secondary to understanding the three basic needs.

1. **General occupational competencies**— These are best described as those skills, concepts, and attitudes needed by all workers regardless of their occupations or specific jobs. Some examples of topics in this category are "Employer–Employee Relations," "Job Safety," "Labor Laws," "Labor–Management Relations," "Dress and Grooming," and "How-to-Get-a-Job Information."

2. **Specific occupational competencies**— These include those skills, concepts, and attitudes essential to a broad occupational grouping, those with common usefulness to a family of occupations. For example, certain merchandising and sales principles are applicable to most marketing occupations, while keyboarding is common to a wide range of business occupations. In the industrial field, certain skills such as blueprint reading are common to the mechanical and construction trades but are uncommon in the food preparation trades.

3. **Specific job competencies**—These include those highly specialized skills, concepts, and attitudes that directly relate to (a) the single job classification in which the student-learner is interested, e.g., auto ignition specialist, computer specialist, or shoe salesperson, and (b) the specific requirements of the student-learner's training station position.

Coordinators as Teachers

In the cooperative occupational education plan, the coordinator's primary role is as a teacher. Coordinators design the necessary learning experiences that enable students to progress toward their individual career objectives. It is the coordinators who are suffi-ciently knowledgeable about the students and the training stations to provide in-school instruction that is correlated in time and topical sequence with occupational experiences and that is attuned to career requirements. Some individuals feel that it is not necessary for the coordinator to teach the related class(es) in school. As evidence for this belief, they point to students who are occupationally successful without the benefit of such related instruction. However, in any occupation there are exceptionally self-reliant individuals who learn the occupation without the benefit of formal CTE class instruction. Almost all individuals who have studied and worked in cooperative education believe that it is the coordinator who structures the ideal learning situation as well as teaches the student learning in the related class.

The Nature of Related In-school Instruction

In the cooperative plan of instruction, in-school instruction has been called *related instruction* for many years. Because the cooperative plan refers to either a capstone approach or a self-contained curriculum, the term "related" will be used here to refer to courses or modules of instruction that are taken at the same time as the cooperative job experience and that provide skills, concepts, attitudes, and integrated task practice geared to the student's career goals and immediate training station needs. The related instruction may be in various combinations such as one period or several and may be taught by the coordinator or by other CTE instructors. As used in this text, the term means that the instruction must be geared to the students as individuals and correlated with their training station experiences.

The term "related instruction" has acquired a number of varying definitions, particularly as some states have written their own handbooks and have used the definition they preferred as being indigenous to their own state plan. However, related instruction for the purpose of this text is of three types: *general related instruction, directly related instruction,* and *supplementary instruction.*

1. **General related instruction**—The development of those understandings and competencies that are needed by and useful to all workers, regardless of the specialized occupational skills needed. This instruction, while common learning, is career-technical in the sense that each student can apply it to his/her own particular occupational situation.

Some Units of Instruction for General Related Learnings*

Orientation to the Program

  a. Understanding responsibilities to school, employer and co-workers
  b. Knowing the purposes of the program and becoming oriented to the rules and policies.

Learning to Do My Present Job Well
  (analysis of job)
Personality Development and Adjustment
Employee–Employer Relationships
Co-worker Relationships
Money Management Problems of the Worker
Income and Other Taxes Paid by Employees
Essentials of Social Security
Labor–Management Relations
Insurance for Personal Living
Planning for Progress on the Job
Improving Dress and Grooming

---

*Abstracted from *Manual for Occupational Relations, General Related Study*, Minneapolis: University of Minnesota.

2. **Directly related instruction**—The development of skills, concepts, and attitudes that are directly applicable and meaningful to students, both commonly for a family or cluster of occupations and specifically for the students, career goals and the demands of their specific training station job titles. It also includes remediation, which is the provision of instruction in general education competencies needed for employability or for profiting from advanced instruction in the related class(es).

3. **Supplementary instruction**—The development of skills, concepts, and attitudes that while useful to student-learners, are in the nice-to-know category and are not directly needed for employability. This instruction may also be termed "indirectly related." Such instruction may be either general education courses or applied CTE courses. The distinction is not in its application in the course to a career-technical objective but in its usefulness to the learners as they apply it.

The exact distribution of classroom instructional time among the three areas outlined will, of course, vary from program to program. The wise teacher-coordinator recognizes that to educate individuals only generally for occupations is not to educate them at all, while to educate them specifically for one job title only assures their immediate employment and insures their need for retraining in the future. Enrollment in the directly related class should be restricted to bona fide student-learners—students who are not in training stations cannot profit from the instruction that is geared to occupational experiences. Noncooperative students who are enrolled tend to lessen the effectiveness of the instruction for the cooperative student-learners. Directly related classes should be taught by the coordinators because they know the students and their needs and can most easily correlate school and job.

The distinction between directly related and supplementary instruction can in some way be shown by examples of course titles. In some schools the directly related class is simply titled "Cooperative Related Training" or "Cooperative Business (Marketing, Health, etc.) Education." However, it would seem more to the point to describe the related class by its occupational meaning, such as "Marketing Education," "Marketing Practices," "Business Office Practices," "Health Practices," "Culinary Arts," or "Machine Trades." Supplementary courses would be different for each student, depending on the student's career objectives and needs. Some examples might be:

For trade and industrial occupations: drafting, blueprint reading, shop mathematics, electronics, or a host of other industrial courses.

For health occupations: anatomy, chemistry, advanced biology, keyboarding.

For family and consumer sciences occupations: nutrition, foods, textiles, biology, anatomy.

For business occupations: introduction to business, English, advanced accounting, keyboarding, speech, business mathematics, economics, computer applications.

For marketing occupations: art, speech, accounting, textiles, keyboarding, economics, computer applications.

The Career Development Approach to Content

The forward-looking teacher-coordinator has adopted the career development approach to cooperative educational training. *Career development* is not the same as career education. As shown in the step-by-step training plan, there is a need to train the student-learners for success in their entry jobs while simultaneously preparing the students for advancement in their chosen career fields. This career approach suggests that some of the content in the related class should go beyond the needs of the training station placement and the anticipated entry-level position. Concepts that are useful in advanced positions should be developed. For example, while a student-learner may be placed as a computer operator (based on present skills) and may learn the routines of a word processor (anticipated entry position), instruction related to the job of administrative assistant, which is the ultimate career goal, should also be given.

In furthering students' career goals, related instruction should stress building understanding, which is learning the "why" of the operations. Classroom activities should include many exercises that show a variety of concept applications. In this way, the instruction builds for transfer—the hallmark of depth of understanding. In addition, related instruction should provide many opportunities for developing creativity and imagination in problem-solving. The simple "right-way" approach to problems should give way to many "right-ways," if evidence can be given to support the solutions. Obviously, this approach calls for many cases to be presented.

In designing related instructional patterns, teacher-coordinators should recognize that different occupations require different combinations of skills and conceptual training. For example, in most industrial trades, the manipulative skills are predominant. The training station can be relied on to teach most skills, especially on specialized equipment, but the related technical information must be taught in school. This is also generally true of health occupations and agricultural occupations. On the other hand, manipulative skills in marketing occupations play a minor role; much school instruction is necessary to teach merchandising concepts and judgment skills, while the job serves as a laboratory in which to try out, test, and apply such concepts. Business occupations involve many manipulative skills, but unlike industrial occupations, usually the skills must be developed under controlled practice conditions in school rather than on the job.

Planning In-school Instruction for Cooperative Student-Learners

When planning in-school instruction, teacher-coordinators should:

1. Define the appropriate educational objectives.

2. Define the basic learning tasks facing each student.

3. Use the career-technical approach to instruction.

4. Determine the sequence of instruction.

DEFINING THE APPROPRIATE EDUCATIONAL OBJECTIVES

The various approaches to related instruction are all based on the need to determine decisively the outcomes that are expected from the related instruction. Ideally, student-learners have enrolled in the program because they are interested and because they can benefit from the training, which prepares them for entry employment and for advancement on the job. During the students' participation in the program, the combination of classroom experiences and on-the-job training and student organization activities provides many ways for formal instruction to enhance each student's competency level and to bring about a change in behavior. Thus, four instructional outcomes are sought. They are:

1. Acquisition of facts—basic information or knowledge about the business and industrial world.

2. Acquisition of skills—abilities to apply facts, basic information, or manipulative skills to the job at hand.

3. Development of understandings and concepts.

4. Development of learning attitudes and behavioral patterns.

Skills are abilities to apply basic factual information to various types of problem situations as a necessary follow-up to the learning of such information. In the occupational areas, skills are needed to:

1. Make necessary computations.

2. Read and interpret.

3. Apply basic facts and information in determining the relative merits of alternatives.

4. Exercise judgment in selecting.

5. Draw conclusions.

6. Apply basic information in carrying out job skills or operations.

Understandings might be thought of as the comprehension of *relationships* between *basic knowledges* and *concrete problems*, as distinguished from the knowledge of isolated segments of content material. It is these understandings that characterize the "truly skilled employee." Formed on the basis of understandings and concepts, *attitudes* relate to the position an individual takes on an issue or a problem. Because attitudes involve feelings and emotions and because the attitudes of teacher-coordinators strongly influence the attitudes of their students, teacher-coordinators should strive to reflect positive attitudes gained from their experiences in business and industry. If they exhibit negative attitudes, their students will probably display them also. The attitudes displayed by student-learners on the job or at student organization activities, such as conferences or conventions, may directly reflect the attitudes passed on by their coordinators.

A more sophisticated way of looking at the desired instructional outcomes provided through cooperative education is to classify the outcomes according to Bloom's *Taxonomy of Educational Objectives.*[1]

Cognitive domain includes those objectives that deal with the recall or recognition of knowledge and the development of intellectual abilities and skills— the acquisition of facts and the acquisition of certain skills-abilities to apply facts and basic information.

[1]Benjamin S. Bloom, et al., *Taxonomy of Educational Objectives,* Handbook I: "Cognitive Domain," 1956, and Handbook II: "Affective Domain," 1957, New York: David McKay Co., Inc. See p. 70 of Handbook I.

Affective domain includes objectives that describe changes in interests, attitudes, and values and the development of appreciations and adequate adjustment.

Psychomotor domain involves the manipulative or motor-skill area, including the acquisition of skills-manipulative skills to the job at hand.

Expressing performance goals in behavioral terms.—Educational literature for at least four decades has emphasized preparing instructional objectives for occupational preparation in terms of performance goals, which are expressed in behavioral terms.[2] As a result, teachers think about and plan around objectives expressing what learners are expected to do, under what conditions, and according to what criteria. Such thinking and planning represent the role assumed by teacher-coordinators as they plan purposeful training station step-by-step plans correlated with pertinent related classroom instruction.

DEFINING THE BASIC LEARNING TASKS OF STUDENT-LEARNERS

The philosophy of cooperative training, regardless of the occupational area involved, is based on recognizing that student-learners have made their tentative career choices. Their objectives for employment must be furthered by both in-school instruction and planned experiences on the job in the training firms. Cooperative student-learners have several sets of learning tasks that can be summarized as follows:

The student-learners must:

1. Learn well the skills, technical information, and attitudes necessary for their jobs at the training stations. This is a primary learning task.

2. Learn those concepts and operational techniques applicable to the general occupational fields chosen. Some of these cannot be applied to the training stations because they are more advanced than those demanded by the training station job assignments. On the other hand, students may learn many of these concepts and skills in school and apply them to their present jobs to test their validity and to help reveal how their application to businesses is somewhat different from each individual's own.

3. Meet those needs that are unique to each individual and unlike those of most, or any, of the classmates. This uniqueness of need occurs because work takes place in a *specific firm* on a *specific job* and with a *specific career goal;* thus, each student's needs are likely to be different from those of other student-learners.

USING THE CAREER-TECHNICAL APPROACH TO INSTRUCTION

Effective related instruction is enhanced when teacher-coordinators adhere to the career-technical approach to instruction. They should use several principles as a guide. First, they should make the standards of the job the standards of the classroom—performance in the school is inadequate if it does not meet the minimum standards expected by employers in the community. Of course, the teacher may well wish to pursue standards different from those of the community. On the other hand, the teacher should be aware that student-learners who work daily must appreciate their employer's standards.

The second principle of CTE instruction is developing those competencies that are deemed primary to job success and relegating to later the development of those competencies that might be termed by employers as "desirable to know," or simply "nice to know." In deciding what must be taught first and what will be taught later if time is available, teachers adhere to the employers' viewpoints rather than fulfilling their own desires. This means also that those topics that student-learners *need* to know on the job *early in the*

---

[2]See Robert F. Mager, *Preparing Instructional Objectives,* 1967; *Developing Attitude Toward Learning,* 1968; *Developing Vocational Instruction,* 1967; and *Analyzing Performance Problems,* 1970: all from Fearon Publishers, Lear Siegler, Inc., Education Division, Belmint, California.

school year are taught *early in the school year,* regardless of when they should be taught according to the chapter sequence of the textbook. This principle of the primacy of instruction according to job needs is illustrated in Figure 10–1.

The third principle of CTE instruction is educating workers as a "total package of behavior." In operation, this principle involves recognizing that skill alone is not enough for success on the job. Employees must have human relations competencies, the ability to solve problems, and a host of desirable personal qualities, including initiative, poise, etc. Developing individuals as total employees is to recognize the demands of the contemporary job market in which skills alone are insufficient.

DETERMINING THE SEQUENCE OF INSTRUCTION

In many subjects the sequence of topics is quite logically developed from the content itself, since the students will not apply the learnings immediately. In the cooperative plan, the sequence of instruction is very much dependent upon the students' occupational experiences.

General distribution of time.—In general, the distribution of time in the related instruction period(s) can approximate that shown in Figure 10–2. Note that in the beginning of the year the students' need for specific content is primarily taken care of on the job; hence, the related instruction can be concentrated mainly upon general needs common to all student-learners. On the other hand, as the student-learners progress during the year, they increasingly need, and

NEED TO KNOW
(Technical Related)*

DESIRABLE TO KNOW
(General Related)*

NICE TO KNOW
(Guidance Related)*

FIGURE 10–1. The primacy of instruction. *Note: Industrial education teachers use the terminology shown in parentheses.

Proportionate Allocation of Instruction for a Cooperative Program

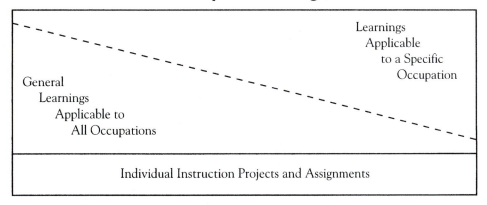

General Learnings Applicable to All Occupations

Learnings Applicable to a Specific Occupation

Individual Instruction Projects and Assignments

September . . . . May

FIGURE 10–2

can operationally apply, the principles of the occupation. Figure 10–2 also points out the need for individual instruction throughout the training period.

In addition to the distribution of instructional time shown in Figure 10–2, teacher coordinators are guided by the need to cycle instruction. *Cycling instruction* refers to teaching a topic briefly once, then returning to it one or more times to add depth of understanding. The need for cycling instruction is pronounced when no preparatory subjects are required. In this case, students on the job need to know many aspects of their occupations, yet there is insufficient time to cover each one in depth prior to the time the students need the competencies. Therefore, topics must be "hit" quickly and returned to later in the school year. For example, most students in marketing education will need some understanding of sales techniques early in the school year in order to perform on the job. However, the teacher cannot afford to spend six to eight weeks on sales techniques because students also should know about human relations, store systems, and the like. Thus the teacher cycles the instruction, covering the minimum essentials of sales techniques, in 7 to 10 days, then going on to other needed topics. The teacher can return to more in-depth instruction in sales techniques later in the year when the need for more complicated techniques is present and when time for that instruction is available.

Number of minutes for related instruction. —There is no magic formula for determining the number of minutes for related instruction nor for estimating the number of months of the course. Judgment of this time element must be based on the complexity of the occupational learnings to be attained; the prior learnings possessed by the students; the type of learning activity, such as laboratory work versus cognitive teaching, which is predominant in the field; and the capacity of the students to profit from instruction.

However, related instruction is likely to be more effective and rewarding if a double or a triple period is used. This is because the block-time is convenient for many laboratory and participatory activities and is useful in helping students become involved with individual projects both inside and outside the classroom.

Planning Related Instruction for Special Needs Populations

As explained previously in this text, the term "special needs populations," is used generally to describe persons who are (1) physically and/or mentally disabled, economically and/or academically disadvantaged, or limited–English-speaking; (2) members of minority groups; or (3) intellectually gifted and talented. In recent years the term "exceptional" has become popular for describing all persons who have difficulty succeeding in "regular" programs, more so than for describing students who are gifted and talented. Therefore, the term "special needs populations" usually refers to individuals who require assistance and/or programs that are modified to accommodate their special needs.

In career and technical education, the term "special needs populations" is being used comprehensively to include the traditional special needs groups (persons with disabilities and disadvantages) as well as to describe adults in need of training and retraining, single parents and homemakers, and displaced homemakers. Thus, the term is being used to describe a broad segment of people who, for one reason or another, are having problems achieving in a regular CTE program and therefore require special teaching–learning techniques and supplemental and supportive services and resources. There is great diversity in the special needs of individuals who make up special groups. Therefore, vocational education for special needs populations must be multifaceted.

The underlying principles and practices used for regular students are basically suitable for those individuals with special needs. All individuals—exceptional and nonexceptional—have many of the same needs and goals. These needs and goals include: (1) feeling good about themselves; (2) exploring and clarifying personal values; (3) emulating successful role models; (4) associating and socializing with others; (5) developing career aspirations; (6) preparing for vocations and avocations; (7) engaging in work that is satisfying; (8) living independent lives; and (9) participating fully in society. Unfortunately, many special needs individuals are not attaining these goals because of the numerous barriers that stand in the way.

With some adaptation, modification, and refinement, the strategy/structure/system model for curriculum development discussed throughout this text can be utilized to meet the needs of special populations.

CHARACTERISTICS OF SPECIAL NEEDS STUDENTS

An essential prerequisite for teaching special needs students is a thorough knowledge of the students—their backgrounds, fears, talents, habits, shortcomings, and lifestyles. However, the characteristics of these students are diverse. The generally accepted characteristics of special needs students can be divided into four categories (general, economic, personal, and educational) and can be used as a guide to help in understanding them and in providing the appropriate instruction for them. These categories are:

General Characteristics

Lack of success in academic competencies.

Tendency toward impulsive decisions and judgments.

Feelings of rejection.

Poor self-concept.

Strong defense mechanisms (aggressiveness).

Unorthodox values system.

Economic Characteristics

Negative environment.

Low income, poverty-level.

High rate of unemployment.

Poor housing.

Inadequate diet.

Improper rest.

Lack of physical comfort and recreational opportunities.

Personal Characteristics

Poor general physical health.

Feelings of hopelessness.

Emotional problems.

Alienation from family.

Lack of imagination and ideas.

Limited experiential background.

Not strong willed—quickly yields to pressure.

Abbreviated communication skills.

Hostile, defensive attitudes; inner rage.

Not willing to work hard (lazy).

High degree of anti-social behavior.

Highly developed coping skills.

Educational Characteristics

Low ability.

Severe underachievement.

Lack of motivation.

Lack of participation.

Poor attendance record; truancy.

Early dropout.

Negative attitude toward intellectual tasks.

Unwillingness to trust adults as sources of information.

Poor logical reasoning.

No reading materials such as books or magazines in the home.

Limited communication skills.

Limited computational skills.

Short attention span.

Major cognitive deficits.

It is important to remember, however, that special needs students are like regular students more than they are different. They laugh, cry, fear, love, and long to become accepted and respected for their accomplishments. The key differences are (1) their own perceptions of self-worth, (2) their own perceptions of failure, (3) their lack of success in the learning arena, and (4) their styles of learning.

Since most teacher-coordinators have had limited professional training in special needs education, the problem of meeting the needs of special populations is immediate especially in response to mandates for mainstreaming or inclusion. The related class must be geared not only toward correlating the students' on-the-job work experiences with the in-school instruction but also toward improving special needs students' reading, studying, and learning skills.

The following tips may be used in the related instruction class:

1. Keep directions simple.

2. Make the beginning assignments short so initial success can be achieved (set up short-term goals for these students).

3. Provide immediate evaluation and feedback on student tasks and assignments.

4. Give short quizzes frequently to evaluate students' progress and to measure their understanding of materials and assignments.

5. Allow flexibility within schedule; vary content sequence.

6. Provide ample praise and encouragement for students with special needs.

7. Have various materials that are interesting and relevant available and accessible to students; design a resource area where students can browse and investigate topics of interest.

8. Modify (but not weaken) curriculum and methods; provide challenges at each student's level.

9. Use professional resources (special educators, guidance counselors, social workers, parents, community members); develop partnerships in education for the benefit of students.

10. Do not underestimate the ability and potential of students; strive to develop their skills and intelligence to the highest degree possible.

11. Learn as much as possible about special needs students.

12. Provide a positive atmosphere of learning in the classroom; create a place where learning takes place and special needs in education are eliminated.

MAINSTREAMING OR INCLUSION

In the past, career and technical education programs for persons who were disabled were generally separate from those for persons who were not disabled. Usually these were special classes within the regular CTE program, such as a keyboarding class and a special class in hotel/motel housekeeping services, or they were CTE classes in special schools or special centers, such as residential and hospital schools, rehabilitation workshops, and sheltered workshops.

The movement away from separate programs for persons with disabilities and toward the mainstreaming or inclusion of these students in regular career-technical education programs in regular settings got its impetus from legislation such as the Education for All Handicapped Children Act. The goal is for persons who are disabled to be integrated within the same educational setting.

The National Advisory Council on Education Professions Development defines "mainstreaming" as:

. . . the conscientious effort to place handicapped children into the least restrictive educational setting which is appropriate to their needs. The primary objective of this process is to provide children with the most appropriate and effective educational experiences which will enable them to become self-reliant adults. Within this objective, it is thought preferable to educate children the least distance

CASE-IN-POINT

Debra Orr is a teacher-coordinator. As a teacher-consultant for students who are mentally impaired, she has several specific duties, which include identifying those students with disabilities who are ready for the cooperative plan and then placing these students on jobs. Debra refers students to the area career-technical center from the five high school districts in the county that are approved for graduate students in special education. She also serves on the educational placement and planning committee (EPPC).

When a student is ready for job placement, Debra locates a suitable placement. She visits the facility to examine the working conditions and to interview the prospective employer. After the student has been placed, she makes periodic visits to the job site to check the student's progress toward meeting the established requirements. She keeps a progress record of the student, in which she includes annual goals, performance objectives and expected achievement, expected time of achievement, criteria for success, and measurement techniques.

away from the mainstream of society. Hence there is a heavy emphasis on movement into the regular classroom whenever possible.[3]

The basic principle underlying mainstreaming or inclusion is that students who have disabilities can benefit educationally and socially from being in programs with students who do not have disabilities. Mainstreaming or inclusion is based on the assumptions that students who are disabled have more similarities than differences in comparison to their peers who are not disabled and that "separate" education can result in "unequal" education. The 1984 vocational education legislation clearly states that students with disabilities are to be in regular CTE programs as much as possible. However, some educators believe that state and federal law should require that

all students who are disabled be placed in regular classes. Legally, these students are to be placed in the least restrictive environment most appropriate to their needs.

Few educators would disagree with the equality and justice of education principles related to mainstreaming or inclusion. As a matter of fact, belief in these principles has led some educators and school systems to expect miracles for students who have disabilities when they are placed in regular classes. In some cases, mainstreaming has been thought to be the panacea for the educational problems of students with disabilities, who have been haphazardly taken from special education classes and placed in regular classes. Many educators are discovering that merely putting students who are disabled in regular classrooms without increased services and support personnel invites failure—even despair. **Mainstreaming or inclusion means more, not fewer, services and personnel.**

Mainstreaming [or inclusion] is not the total elimination of self-contained special education classes. It does not mean that all handicapped students will be placed in regular classes. Further, it does not automatically imply that those handicapped students who are placed in regular classes will be in that setting for the entire school day.

Mainstreaming [or inclusion] is not an arrangement that can be accomplished overnight in a school system. Instant change often leads to unstable classroom environments.

Mainstreaming [or inclusion] is not just the physical presence of a handicapped student in the regular class.

Mainstreaming [or inclusion] does not mean that the placement of handicapped students in regular classes jeopardizes the academic progress of nonhandicapped students.

Mainstreaming [or inclusion] does not mean that the total responsibility for the education of

[3]*Mainstreaming: Helping Teachers Meet the Challenge,* Washington, D.C.: National Advisory Council on Education Professions Development, 1976, p. 7.

handicapped students placed in regular classes will fall to the regular class teacher.

Finally, mainstreaming [or inclusion] does not mean putting educators "out on a limb" and expecting them to accomplish tasks for which they are not prepared.[4]

Instead, mainstreaming or inclusion is the creation of new and different educational alternatives for students who are disabled. It is a student-centered movement. It assumes that the educational system is responsible for meeting the individual needs of all students—that once the students' special needs have been identified, the school environment and instruction must be adapted to meet those needs. Most students who are disabled can be accommodated in regular classes with or without supplementary services. A few may require hospital/homebound instruction. **Mainstreaming or inclusion does not do away with special classes or special schools,** for these settings may be the most educationally appropriate for some students.

All students need some personalized instruction. The best classroom environments for all students are those recognizing every student has special needs and that through individualized instruction these needs can be met.

There is no basis for concern that the presence of these students who are disabled in regular classes will interfere with the progress of students who are not. They can learn a great deal from their classmates with disabilities such as learning to respect individual differences and having an opportunity to gain information about different kinds of disabilities.

THE IEP's

What is an IEP? The Education for All Handicapped Children Act requires schools to develop an **IEP** (individualized education program) for each student who is disabled. This written statement is to be

---

CASE-IN-POINT

Johnny Wong lives in a suburb of upper middle class families in a midwestern city. Johnny has been attending special education classes since he was diagnosed as a slow learner early in his elementary school years. Now at age 17, Johnny has the opportunity to participate in the cooperative plan, which is offered through the area career-technical center as a half-day, shared-time program with his home high school.

Johnny is one of 20 to 25 students who are invited each term to participate in this special program. His parents are sent a letter of explanation, which serves as an introduction to the program and the teacher-consultant. The letter explains that Johnny will spend nine weeks at the area career-technical center in an initial evaluation program. During this period, Johnny will be tested in many different areas and exposed to a wide range of career opportunities through videos, tape recordings, and hands-on experiences. There are 20 occupational programs available to Johnny, each program has from 2 to 8 sub-areas for specific jobs. The evaluation program will help the curriculum planners determine Johnny's maturity level, his hearing and vision capabilities, and his willingness to prepare for a job.

At the completion of this exploratory period, Johnny's performance record will be evaluated by the EPPC so that an educational program that meets his personal needs can be planned. Johnny's parents will be invited to be members of this committee, with equal responsibilities for decision-making. A school administrator, a teacher, and a specialized diagnostician will also be members. Together, they will plan a course of study for Johnny.

In Johnny's county, students who have special needs cannot graduate from high school without having successfully completed a work experience program or a cooperative plan. Thus, the importance of this program is very clear to Johnny and his parents.

---

created by a team of qualified personnel. The law requires that the program be developed in a meeting involving a representative of the local education agency, the teacher, the parents or guardians, and, when appropriate, the student. The law identifies an IEP and its component as:

---

[4]Ann P. Turnbull and Jane B. Schulz, *Mainstreaming Handicapped Students: A Guide for the Classroom Teacher*, Boston: Allyn & Bacon, Inc., 1979, pp. 8–9.

. . . a written statement for each handicapped child developed in any meeting by a representative of the local educational agency or an intermediate educational unit who shall be qualified to provide, or supervise the provision of specially designed instruction to meet the unique needs of handicapped children, the teacher, the parents or guardians of such child, and, whenever appropriate, such child, which statement shall include (A) a statement of the present levels of educational performance of such child, (B) a statement of annual goals, including short-term instructional objectives, (C) a statement of the specific educational services to be provided to such child, and the extent to which such child will be able to participate in regular educational programs, (D) the projected date for initiation and anticipated duration of such services, and appropriate objective criteria and evaluation procedures and schedules for determining, on at least an annual basis, whether instructional objectives are being achieved.[5]

Testing procedures used for placing students who are disabled in instructional programs must be selected and administered to assure racial and cultural fairness. All evaluations must be conducted prior to placing, transferring, or denying placement of students who are disabled into special education programs or regular classes. The law requires the evaluations to include:

. . . testing the child in his native language or mode of communication; using tests that are appropriately validated; having licensed personnel to administer the test; testing specific areas of educational need rather then focusing only on general intelligence; insuring that sensory and/or physical handicaps do not jeopardize their capability to demonstrate their aptitude and achievement level on the test; making the decision for educational placement on the basis of giving a minimum of two tests or types of tests; considering information concerning physical development, socio-cultural background, and adaptive behavior in conjunction with

test scores; setting up a team of persons in the school who possess knowledge about the child, the placement alternatives, and available resource personnel to interpret evaluation data in order to provide an appropriate education; and placing the student in a regular class if, according to the evaluation results, he does not require a special setting for his education.[6]

Public Law 94–142 requires a multidisciplinary cooperating team approach in all areas of program development for students who are disabled—in identifying and locating students with disabilities; in screening, assessing, and program placing; in instructional planning; and in evaluation. In planning IEP's, career-technical educators should be part of the multidisciplinary cooperating teams. Career-technical educators may serve as team members or as resource persons providing input to the teams.

The California State Department of Education, Special Education Unit, has identified seven major parts of individualized education programs in its *Master Plan for Special Education.* They are:

I. **Student Identification**

Student's name and ID number
Parent's name
Language used at home and how determined
Birth date, chronological age, and grade

II. **Assessment Information**

Present levels of performance

III. **Program Information**

Program enrolled in, date and duration
Rationale for placement
Extent of regular class placement
Additional support services, date and
    duration
Type of PE program
Types of preoccupational and career-
    technical programs

[5]*Education for All Handicapped Children Act of 1975,* Sec. 4 (a) (19).

[6]Turnbull and Schulz, pp. 63–64.

IV. *Implementation Information*

Learning style
Learning situation
Behavior strengths
Talents and hobbies
Special instructional media and materials
Personnel responsible for implementation

V. *Meeting Information*

Meeting date
Interpreter required
Signatures of those present

VI. *Long-Range Goals and Periodic Objectives*

VII. *Short-Term Objectives*

BILINGUAL AND MULTICULTURAL RELATED INSTRUCTION

Concerns among educators, including teacher-coordinators, about those students with limited-English or non-English proficiency have become more urgent as the number of immigrants, both legal and illegal, entering the United States each year continues to increase. The impact that the new immigration wave is having on schools is significantly different from years past.

Although many career-technical educators have little or no experience in working with minorities or so-called "culturally different" students (Blacks, Hispanics, Native Americans, and Asian-Americans), it is still possible to have a classroom setting that promotes multicultural education. Career-technical educators must help CTE students to learn to respect cultural differences, to be proud of their origins, to recognize the positive nature of our pluralistic society with its cultural and ethnic diversity, and to appreciate the contributions of all groups to the richness of our national culture. What teachers do in their individual classrooms may serve to break down stereotypes, promote multicultural understandings, and make a crucial difference in the personal develop-

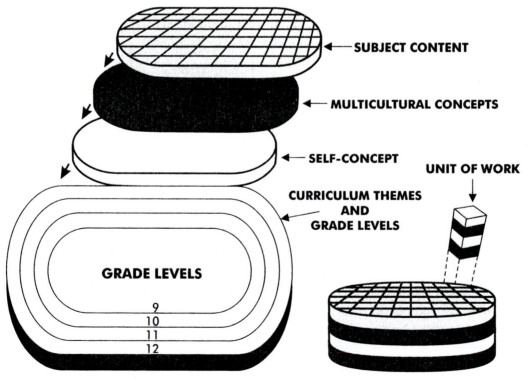

FIGURE 10–3. Framework for curriculum units.

ment of many individual students. Career-technical educators also must help their students to understand that the careless use of language and stereotyped descriptions of people can hurt them and can limit their potential and that all individuals have varied characteristics not limited by sex, race, class, or ethnic background.

Many excellent sources of multicultural information are available in most communities to help career-technical educators create a multicultural learning environment in CTE classes. Career-technical educators should develop a list of possible contacts for materials, resources, and speakers. The local public library may have much of this information, which has already been compiled. If it does not, by using the telephone directory, career-technical educators can begin to develop a list. Most national organizations, such as the National Association for the Advancement of Colored People (NAACP) and the National Conference of Christians and Jews, have branches in large cities. Looking also under the names of various ethnic groups, such as Chicano, Asian-American, Mexican-American, Native American, La Raza, Black, and African-American, can be beneficial in determining what services might be available by these groups in the community. Another source to get in touch with is the U.S. Department of Commerce.

Another important resource is local community colleges or universities. They should be contacted about possible ethnic studies departments or organizations for students of different ethnic backgrounds. Catalogs should be reviewed to determine whether courses are taught in the history or culture of a particular ethnic group. Often, the instructors of such courses can help locate more materials and information as well as provide contacts with other individuals in the community. Of course, these instructors are themselves potential resource persons who can be invited to speak to CTE classes.

In some communities, local service organizations may be an additional resource to address the special needs of minorities. The YWCA, the National Orga-

nization for Women (NOW), the League of Women Voters, the Junior League, community centers, and other similar organizations should be contacted to determine if they offer special programs that might be useful. State teacher organizations, adult training and development programs, and state–county–city human relations commissions are all additional groups that might provide additional resources for bilingual CTE classes.

All students have a need to achieve success. Bilingual students are no exception. Continued failure in school deprives them of satisfying that need. Some students fight back—they become overtly aggressive and rebellious. Others withdraw, stop trying, and become non-learners and dropouts. Vocational education must provide bilingual students with the competencies necessary for success in the world of work.

Most educators would agree that teaching students to value their own ethnic backgrounds, as well as to expand their awareness of other ethnic groups, should be an important objective of education. Teaching respect and appreciation for the cultural diversity that exists in our society should be an important objective in all classrooms.

Employing Suitable Methods, Media, and Materials

Most teacher-coordinators are either experienced teachers or beginning teachers who have had instruction in basic teaching methodology. Those teacher-coordinators who are new to teaching, like the experienced ones, need to keep in mind the participatory nature of CTE instruction. In particular, teacher-coordinators should use those methods that are really applicable to job situations, as well as those that relate to the optimum learning style of the learners. They should understand that CTE students, especially those on the job, want instruction that is meaningful to them in their daily work situations.

Committees.—Student-learners should learn to work with others in small as well as in large groups. Time can be saved and educational value can be derived if the class is broken up into committees for certain phases of the work. Whenever possible, students should be permitted to choose the committees on which they work. However, at times the teacher-coordinator should shift committee members to develop greater leadership and followership.

Discussions.—A class discussion, with emphasis on questions posed by the teacher-coordinator, may yield a list of study topics when a new area of study is introduced. A stimulating class discussion needs to follow brief lectures, reading assignments, and other activities. Generally, a lecture by the coordinator of longer than 20 to 30 minutes will not be productive in the basic instruction period. Informal class discussions should help students to develop tolerance of and respect for the viewpoints of other class members and for the experiences of others at their training stations.

Dramatic presentations.—Attention-getting through unique ways of presenting information appeals to many students. Examples are radio or television skits, mock programs of various types, role playing, and demonstrations. Students should understand that they should not be primarily concerned with the dramatics involved; content is more important than dramatic ability. Many students participate more readily when they know that they will not be judged as actors.

Panels.—Panels should be carefully planned, and students should be instructed in the techniques of panel discussions. A panel should consist of from three to five members. The chairperson introduces the problem, states the issue or problem involved, recognizes the various members, directs the discussion, and summarizes the points made. Each member must be familiar with the topic and be prepared on all issues to be discussed. Each member must be prepared to answer questions from the rest of the group (class).

The teacher-coordinator observes the same rules of audience participation as the other students and, in general, does not "talk."

Reports.—Oral presentations from the floor by individual students help students to learn to speak in a logical and interesting manner and help the group to develop good listening habits. Reports must be well prepared, and an individual giving a report must have something to say; reports should be said and not read; and the group should be held responsible for the information given in the report. Coordinators have found that reports on basic information from current periodicals or trade magazines or observations at work are useful.

Current news items.—A current newspaper item or an announcement on the radio or on TV can become the focal point for a topic of interest to CTE students. A fire in a local business establishment, a negotiated labor dispute, a new technological development, the opening of an industry in the neighborhood might all serve as good launching topics for further study and discussion. Students must be directed toward gaining specific points from the discussion, or much time may be wasted. Therefore, a summary statement and perhaps a quiz on the material are helpful in driving home certain points.

Field trips.—Field trips are not a necessity for cooperative students who work every day, but they do help students see other businesses. They also assist in broadening the experiences of student-learners who may be having rather narrow training station experiences. Field trips must be well planned to places unlike those of the students' training stations. A trip must be a natural outgrowth of classroom discussion, or it must be an activity to stimulate interest in a topic. After the reason for a trip has been identified, the place should be selected. Then a committee of students should either write or visit the owner, asking for permission to visit. The industry or place of business should be alerted as to what will be of interest to the students and what the students should know. A previous visit by the teacher will aid the students in identifying what they should look for. The class members should agree on the factors that will be investigated and should be encouraged to prepare questions. The visit should be followed by a thorough discussion of what was seen, why operations were performed as they were, and what should be the next logical step in the development of the topic. The student committee should write a short letter thanking the business establishment for permitting the students to make the visit.

Resource persons (guest speakers).—

Resource persons broaden the experiences of student-learners. Credit managers, store managers, bankers, social security personnel, farm bureau managers, health center directors, hospital managers, nurses, office managers, and tax accountants are potential resource persons. A student committee should invite the guests to speak to the class, explaining what would be expected of them and what the class would be studying. After an individual has accepted the invitation to speak to the class, he/she should be given a specific subject to talk about and should be alerted to some of the specifics that the class would like described. The class prepares by identifying areas in which its members need information. The class members should be encouraged to write out questions prior to the appearance of a speaker and, of course, to ask questions after the speaker's presentation, which should be followed by a discussion of the points that were brought out and an identification of the implications of the remarks. Shortly after the speaker's presentation, a student committee should write a thank-you note to the speaker.

Classroom reference library.—

Because of the nature of specific instruction and the resultant need for individual study, a classroom library of current periodicals, manufacturers, materials, reference books, and individual resource manuals must be available in the classroom for student-learners. A dictionary, an encyclopedia, the *Dictionary of Occupational Titles,* and the *Occupational Outlook Handbook* should be included among the references. Contemporary collections have trade magazines. Developing skills in judging the reliability of published information and being able to disregard propaganda or unsound ideas and concepts are paramount. Students must learn to take notes on their readings in some systematic and usable form. They must learn to record accurately their data and their sources of information and to make their notes brief and succinct.

Staff resource person.—

The team-teaching approach can accommodate certain needs of student-learners. Another member of the faculty may be invited to present specific concepts or specialized skills. A case in point would be to have the family and consumer sciences teacher present information on color, line, and design to a marketing education class. In turn, the marketing education teacher-coordinator might appear before the family and consumer sciences class to present aids on wise buying.

Chalkboard, whiteboard, flannelboard, bulletin board.—

Teacher coordinators must be ready to use the most practical visual aids to assist in teaching student-learners. The chalkboard is the old standby, too infrequently used by some teacher- coordinators to explain a point. The other two techniques, using a flannelboard and using a portable bulletin board, have been found to fill the needs of teacher-coordinators who must move from room to room or who must share a room with other teachers. By having a flannelboard or portable bulletin board (one that folds together), the teacher-coordinator may arrange displays of materials the day before a class presentation and then carry them in for use.

Overhead projector.—

The overhead projector is a visual aid that motivates, teaches, and grades. There are many methods and devices that, in themselves, motivate, teach, and grade. The projector, however, has an additional advantage. In even the most diversified of situations, an element of routine is likely to creep in; the projector is most effective not only in motivating and teaching students but also in reducing the monotony of routine. The teacher-coordinator gets the job done, and the class enjoys the change of pace. By using a grease pencil for writing or by running the material to be copied and the transparency through a reproducer or a photocopier, the teacher-coordinator can easily prepare transparencies for the overhead projector.

Films, videos, slides, and tape recorders.—For launching a new topic, these help to arouse interest, to provide direction for the development of a unit, to raise important issues, and to give students a common background for analyzing the area to be studied. If carefully selected and previewed for accuracy and recency of material, these can serve as excellent supplementary resources for a unit or a topic of study. Audiocassette recorders and VCRs are especially useful. Because many students have them at home, they can borrow tapes to take home to study.

Video camera.—With the availability of classroom television sets and videotape players in schools, teacher-coordinators can, without much difficulty, prepare and show videotapes within the related classroom. Mock interviews, simulations on human relation subjects, correct operation of various equipment, and field trips to business and industry could be filmed. With teacher guidance, students should be able to plan and tape the video programs.

Interactive video or CD-ROM.—One of the most fascinating technological developments in education today is interactive video or CD-ROM. While not inexpensive, it is a powerful instructional tool. It consists of a computer and a videotape (or video disk) player coupled together or a computer with multimedia capabilities (CD-ROM). Instructional packages, developed by a special authoring system, permit the instructional program to go back and forth between the computer and the video system or CD-ROM. The computer can deliver directions, background informa-

tion, test questions, etc. The sound— voices from a scene, background music—is delivered on the audiotrack of the videotape or CD-ROM disk. The key to the system is interaction—students respond to the instruction, and the program leads them, step by step, through an entire learning process. The most significant feature of the interaction is the branching that is available. A student may respond to a question that follows instruction in a number of ways, giving the right answer or any one of several wrong answers. The program is designed so that each response brings about appropriate but different instruction. In other words, individualized remediation is immediate for each student who gives an incorrect response. The student who gives the correct answer is simply routed on to the next instructional segment.

Computers.—There certainly appears to be no end in sight to the power and versatility of the computer, and prices continue to drop as dozens of new and exciting features are added. Indeed, computers have become an important teaching tool. There are hundreds of programs in which students can drill, practice, and learn basic skills in math, word usage, spelling, reading, punctuation, capitalization, and decision-making. Not only do capable students like to complete work on the computer but also less able students like to drill and practice on the computer because they can set their own learning pace. Many of the skill development programs let students select a level of difficulty from a half dozen available levels. Computer instruction is individualized.

Digital cameras.—Digital photography opens up many new possibilities for teacher-coordinators. With digital images, only the instructor's and the students' imaginations limit projects and educational experiences. Images can be printed, or they can be posted on the Web. Printing is not limited to normal paper. There are heat-transfer papers and papers that are self-adhesive that allow images to be put on everything from t-shirts to coffee mugs. Besides just being fun, many of these projects can be used in fund-raising activities.[7]

Digital video camcorders.—Digital video camcorders are the latest technology for recording and preserving classroom activities and experiences. Quality and ease of use are the two hallmarks of digital video. DV camcorders capture images much like video camcorders. However, they store the images in a high-quality, endlessly reproducible, easily edited, digital format. Digital videos provide better images than analog videos. Digital camcorders costing less than $1,000 record video images equal to or better than professional analog cameras costing 10 times as much. The teacher-coordinator can capture single shots (as with a digital camera), play back tapes from the DV camcorder, or copy them to VHS tapes. Digital videos can be transferred to computers without conversion, digitally edited on-line, and then copied back to digital tapes. Not only is it easy, but there is no loss of image quality as there is in the analog world. Once a video is on the computer, the teacher-coordinator can also easily send short clips as email attachments or post them on Web sites. Digital is the universal format.[8]

INSTRUCTIONAL EQUIPMENT AND FACILITIES

Specialized facilities and instructional equipment are needed in CTE instruction. These will naturally vary, depending on the occupational areas being taught. Recommendations for specialized laboratories for cooperative programs in the various occupational areas are given in Chapters 17 through 22.

INSTRUCTIONAL MATERIALS

For effective instruction in almost every school subject, students should have access to instructional

materials that supplement the text. In an occupational training program, a library of instructional materials is vital to keeping the instruction attuned to the business and industrial world and in aiding students to progress toward their individual career goals. Because each student has individual needs, additional materials that describe operations and products in various occupational areas should be supplied. A cooperative occupational education classroom should have shelving and filing space for instructional materials of the following kind: (1) supplementary textbooks and occupational handbooks, (2) occupational guidance materials, (3) trade journals, (4) product information, (5) job operations information, and (6) videotapes.

A coordinator's filing system will accommodate many of the materials just listed and will fit any occupational area.

Many beginning coordinators wonder how to establish an instructional materials collection. There are several sources of help. First, the coordinator should request from his/her administrator an annual instructional materials budget. A minimum allotment of $100 to $150 per student will finance a basic collection of instructional materials in a new program, particularly if the coordinator will concentrate on obtaining free and inexpensive materials. An annual budget of $75 to $100 per student is needed in an established program. A good source of materials is the employers in the community. Many of them have copies of trade and professional publications, which they would donate to the school if requested to do so by the coordinator. The coordinator could remind students in the cooperative program to ask employers for copies of brochures, product information leaflets, and employee instructional handbooks with their earnings; many times students will donate these to the school when they graduate. On coordination visits, the coordinator should also be alert to acquiring instructional materials from employers, who will probably donate them if asked.

Experienced teacher-coordinators know the job of teaching subject matter that is dynamic and

[7]For more information on digital cameras, visit www.shortcourses.com/how/intro.htm.

[8]For more information on digital video camcorders, visit www.shortcourses.com/video/introduction.htm.

everchanging. These coordinators recognize that a major element of good teaching is to treat the community as their classroom. Effective teachers have learned that the learning of their students in enhanced when they can apply new knowledge to their surroundings. *The development of a resource file is a must for effective related instruction.* A resource file is a systematic file of all the materials, activities, community resources, etc., that the teacher-coordinator might utilize in teaching various topics ordinarily included in the related class. Such a file could contain the audio-visual materials that the teacher-coordinator might use; effective past bulletin board displays that are being saved for the future, as well as ideas and materials for new bulletin boards; the names and telephone numbers of laypersons who might serve as resource speakers; samples of various business forms that are being used locally; newspaper and periodical clippings that have been brought to class by the teacher and students to be shared; and potential sites for field trips.

The resource file should be well organized. The teacher-coordinator should begin with a few broad subject matter topics. After a resource file has been started, it can be expanded and refined over the years. As time and experience permit, the general categories can be subdivided into more specific topics. Once begun, the teacher-coordinator should periodically review the contents of the file to assure that the information is up-to-date.

Questions and Activities

1. Identify three students with career goals, each in a different occupational family. For each student, identify (list) competencies that would fall into each of the three learning categories—general related, directly related, and supplementary. State any assumptions you make in defining the situations of these students.

2. Define the term "correlation" and explain why it is so important to the cooperative plan of instruction.

3. Outline a case of one student—the student's background learnings, career goals, and training station. List under the three headings—"Need to Know," "Desirable to Know," and "Nice to Know"—the learnings that fit this case and each category.

4. Explain the difference between a specific occupational competency and a specific job competency. Give examples under each for the student in the case in no. 3.

5. For the student described in the case in no. 3, prepare an outline of the learning units arranged by a time sequence throughout the cooperative related class for a year. For convenience, try using three-week periods of instruction to show the timing of the student's needs.

6. In planning instruction, how can the teacher-coordinator use performance goals?

7. Prepare a list of reference materials you would put on a requisition list if you were starting a new cooperative plan. Assuming that the whole list will not be approved, which of those do you consider to be imperative?

8. Explain the effect that using the cooperative plan as a capstone method instead of as a self-contained curriculum has on related instruction.

9. Arrange to interview a local career-technical educator at the secondary or post-secondary level. Develop a set of questions to ask during the interview regarding on-going mainstreaming or inclusion practices in the school.

10. Assume you are a teacher-coordinator who has three students with limited-English proficiency enrolled in your cooperative plan. Although they are employed and appear to be doing well on their jobs, they are having some problems adjusting to their new culture. Prepare an instructional unit that you can use in the related class to assist these three students.

For references pertinent to the subject matter of this chapter, see Reading Resources.

11

Developing Training Stations as Instructional Laboratories

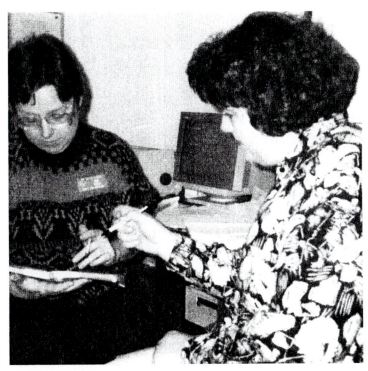

Each job station becomes an educational experience when the teacher-coordinator and the job supervisor exchange ideas on training plan elements.

KEY CONCEPT: The training plan is essential for making the training station experience educational for the student.

GOALS

After successfully completing the study of this chapter, answering the questions, and carrying out the activities, the student should be able to:

- Distinguish between different types of training–learning stations.

- Develop training stations for specific student-learner training objectives.

- Develop training stations that will provide opportunities for student-learners to pursue career objectives.

KEY TERMS

job development

statement of student-trainee
    learning objectives (SSLO)

step-by-step training plan

Targeted Jobs Tax Credit

training agreement

work exploration

The cooperative plan is based on the premise that the world of work has the capacity to provide educational experiences on the job that will reinforce and supplement the in-school instruction. If this premise is correct, the educational institution must take steps to assure that training stations provide more than the normal coaching in company techniques, which they do for all employees. The educational institution must insure that quality controls are built in, which assures parents, the administration, and others that training stations are accorded sufficient school supervision to warrant the conclusion that the students are progressing toward their career goals and are benefiting from the welding together of their in-school instruction and their on-the-job laboratory practice. If these quality controls are not assured, the experience at work becomes only a release of time from school for students to profit from whatever values and income the act of working for pay has for them.

The first task for coordinators is to select and develop training stations appropriate to each student-learner's needs. Coordinators must clearly understand their role in the process of training station development. The word "development" is the key—it is not easily understood by many school personnel.

Developing Different Types of Learning Stations

Anyone who wishes to use an actual job to accommodate a student's learning needs must be a planner-manager. In simple terms, the management exercise is to: (1) *determine* the learning goals, (2) *decide* what type of training station is needed, and (3) *develop* (find or create) such a situation with a minimal investment of resources. Coordinators consider the basic set of learning experiences each student may need and the corollary set of developmental situations available in a given job. These basic types of developmental learning situations are:

1. Work exploration
2. Personal and work adjustment
3. Work-study motivation and income
4. Occupational application and skills development
5. Professional internship

THE CONCEPT OF JOB DEVELOPMENT

The term "job development," used primarily in the MDTA programs of the 1960's, in the CETA programs of the 1970's, and more recently in the JTPA programs, has two basic meanings to the career-technical educator. They are:

1. To find positions in businesses that will provide the learning experiences needed by students and that will convince businesspersons that they will benefit by cooperating.

2. To create jobs for students who are ready after preliminary training to enter the labor market on a full-time basis.

Thus *job development* is the act of creating jobs (training opportunities, part-time employment, or full-time employment) that did not exist or whose existence was not known to the school. Job development is an aggressive, not a passive, function. Coordinators must distinguish it from the mere matching of trainees with existing part-time jobs. Certainly it is far more than the filling of an employer's request for "someone to work part time." The filling of an employer request for a *trainee* rather than a part-time worker can represent much past effort by a coordinator in job development. In such cases, occupational coordinators may have put in much effort, even over a period of two to three years, in convincing employers of the merit of the plan. The distinction just made points out that effective plans cannot exist with so-called "desk coordinators" who are "telephone order-takers," accepting employers' requests at the convenience of the employers and fitting students to those requirements.

A tax credit eligibility for cooperative education employers or a wage adjustment for handling a trainee means that coordinators can be more aggressive in job development activities, such as the development of training plans. If an employer is approved for a tax credit or a wage adjustment, that employer should be willing to provide an educational experience at the training station.[1]

Aggressive job development can assure meaningful cooperation with employers who understand their part in the educational process. Job development obviously occurs when a student's career choice becomes known or when the choice shifts and a move from one training station to another is dictated. But, even after all the trainees have been placed, job development must continue so that training stations for the future can be uncovered. Unfortunately, and all too frequently, overloaded or lazy coordinators do not continue the job development function. Thus, having a new group of trainees will result in one of the following: frantically scrambling at the last minute to find stations, indiscriminately approving any jobs the students can find, or using the same stations over and over again without any real regard for the students' career goals.

It is very important that the heavy resource input to job development be known and understood by the coordinator's administrators. They must be aware of the need for the many hours involved in such an activity, or they will view the coordinator as merely being a placement officer. A descriptive coordinator's log is a useful communications tool with the administration.

TARGETED JOBS TAX CREDIT

Employers can take *Target Jobs Tax Credit* to reduce their federal income tax. Employer tax credits are designed to help people gain on-the-job experience and acquire better employment. This, in turn, will increase U.S. economic growth and productivity.

The Work Opportunity Tax Credit (WOTC) encourages employers to hire people in eight specific target groups: (1) certain veterans with disabilities, (2) veterans who are members of certain families receiving food stamps, (3) recipients of temporary assistance for needy families, (4) recipients of food stamps, (5) vocational rehabilitation referrals, (6) ex-felons, (7) persons receiving supplemental security income, and (8) 16- to 24-year-old residents of empowerment zones and enterprise communities. A business can earn a tax credit of up to $2,400 per new hire.[2]

The Welfare-to-Work (WtW) tax credit encourages businesses to hire long-term welfare recipients. Businesses hiring these individuals can reduce their federal tax liability by as much as $8,500 per new hire: 35 percent of qualified wages for the first year of employment and 50 percent for the second year. Qualified wages, which include tax-exempt amounts received under accident and health plans and educational and dependent assistance programs, are capped at the first $10,000 earned each year. To qualify for the WtW, persons must be employed for at least 400 hours or 180 days.[3]

EDUCATORS' RELUCTANCE TOWARD JOB DEVELOPMENT

Some educators are loathe to engage in aggressive, hard-selling job development. They are reluctant to try to overcome the employer's resistance to the young person who has a poor record, who has had brushes with the law, or who in other ways does not fit the employer's image of what a "good" student should be. Furthermore, many educators are not inclined to try to overcome employers' hiring restric-

[1]Current state and federal laws need to be checked to determine if tax credits or wage adjustments are available to approved cooperative education employers.

[2]For more information on WOTC, visit www.workforcesecurity.doleta.gov/employ/wotcdata.asp.

[3]For more information on WtW, visit www.workforcesecurity.doleta.gov/employ/wtw.asp.

tions on members of minority groups and persons who are disadvantaged and/or disabled. Such persons justify this reluctance by pointing out that (1) the school's role is not to quarrel with local mores or rules or (2) employer support for the school will be lost and the past goodwill be damaged. In contrast, educators must believe that the school exists for the student's needs, which are the coordinator's priority. Thus, coordinators, while respecting employers' opinions, must work very aggressively to bring about change. If individuals do not have the inner toughness and the persistence to engage in job development, they should not become occupational coordinators or work experience directors. There are many other satisfying roles in the school in which individuals do not have to face such problems.

TECHNIQUES OF JOB DEVELOPMENT AND JOB PLACEMENT

The techniques used in job development are similar for work exploration stations, cooperative training stations, and permanent job placement. Only the immediate plan objectives and the emphasis used in each phase differ. Coordinators find the following techniques of job development useful in working with employers.

1. Always contact the "decision-maker" within a company first. A directive should come "from the top down" stating the company's participation in the cooperative plan, rather than requiring you to work up through the ranks. If someone in the personnel department, for example, tells you the company is not interested in being involved, this person will be antagonized if higher management decides that the company will participate. Since the individual in the personnel department will have to carry out the decision, having that person in favor of the plan would be best.

2. Show the employer that you have the company's interest, as well as the trainee's interests, at heart.

3. Always talk in terms of a particular trainee, not the plan.

4. Discuss specific aspects of a problem with the employer and deal with problems one at a time, for example, hiring restrictions.

5. Use all job resources available—employers, state employment service, wholesale distributors, instructors, personal contacts, successfully placed trainees, and the advisory committee members.

6. Convince the employer of the value of employee upgrading.

7. Assist the employee in redefining jobs where applicable.

8. Explain to the employer that you will assist with any problems that may arise relative to the trainee.

9. Relieve the employer of the burden of "firing" the trainee. Before the trainee is "hired," assure the employer of your help in a situation like this.

Caution: Do not "oversell" a trainee; be honest, yet optimistic.

CRITERIA FOR SELECTING TRAINING STATIONS

The following are some criteria to be used in the selection of a training station.

Type of occupation: The training station should provide experience in occupations that require both skills and knowledges.

Career goal: The position at the training station should be directly relevant to the career goal of the student.

Opportunities for rotation: The training station should provide a wide variety of direct experiences associated with the occupation. It should not merely be routine work experience of a repetitive nature.

On-the-job supervision: The training station should be supervised by someone competent in the skills and technical aspects of the occupation. The supervisor should be interested and eager to assist in the training program.

Working conditions: The working conditions of training stations should be safe; the stations should have a good record of accident prevention.

Reputation: Establishments that furnish training stations should have a reputation of ethical business practices in the community.

Hours of employment: There should be a sufficient number of working hours (average 15 hours per week for the school year) at the training station.

Facilities and equipment: Up-to-date facilities and equipment should be used at the training station.

Supervisor and student-trainee: Good supervisor and student-trainee relationships should exist at the training station.

Location: Consideration should be given to the travel time between the school and the training station.

Wages: Consideration should be given to a minimum wage for the student-trainee, based on that paid

other employees of similar experience and training. Tax credit eligibility may be taken into account.

In job development activities, teacher-coordinators should keep in mind the criteria and select the jobs that meet these criteria. The preceding case-in-

---

Top 10 Best Retail Training Stations

1 United Supermarkets
2 Walgreens
3 Brookshire
4 J.C. Penney Company
5 Captain D's
6 Publix
7 HEB Supermarkets
8 Office Max
9 Sack-n-Save
10 Piggly Wiggly

*Rankings are based on 1995 research study completed by Stewart W. Husted and Frank Whitehouse of 343 usable responses from DECA advisors. Using the criteria found in this text, teacher-coordinators rated 763 different companies.

CASE-IN-POINT
"Selling" a Trainee

David Williams is the teacher-coordinator for a marketing education program in a metropolitan area school. His apparent success in his first three years has been due partially to the contacts he has made in promoting marketing education and the trainees in his community. He has given several presentations before community organizations; he has written articles for local newspapers showing the trainees' successes in the program. He visits training stations frequently. He maintains excellent rapport with his students.

During this school year he is the teacher-coordinator for 22 students. One of the trainees in his program is John Turner, an over-age black 19-year-old.

During the past three years, David has been successful in placing four trainees in the cooperative training laboratory of Seth Wyman, the owner of a high-quality men's apparel store in a large shopping center. Mr. Wyman has appeared to be favorably impressed with these four trainees and their contributions to his workforce. Two of them have stayed with his firm in permanent positions. Since Mr. Wyman has never had a black employee, David is aware that this particular case must be handled with care. Before he asks Mr. Wyman to consider John as a cooperative trainee, he feels it is important to have the two meet, perhaps for one or two "exploration" sessions. This would give Mr. Wyman a chance to talk with John, and it would also give John an opportunity to "explore" apparel retailing as a definite career goal.

After two of these sessions, a noticeable rapport seems to be developing between Mr. Wyman and John. When approached with the idea, Mr. Wyman replies that he has already noticed many of the good qualities that John has—interest, punctuality, good attendance— qualities that are important to his business. David and Mr. Wyman agree to a two-week tryout period.

point illustrates one way in which a coordinator can "sell" an employer on the value of having a trainee, no matter what employment handicap that trainee may have. But in a case like this, the coordinator must be prepared and able to deal with any objections the employer might have and be able to "sell" the value of having this *particular* trainee.

DEVELOPING WORK EXPLORATION STATIONS

Work exploration experience provides students with temporary exposure to work, as they are placed in exploratory situations to observe and to assist in occupational areas in which they have indicated an interest or have shown an aptitude. These situations can take place in any combination of time sequences convenient for the employers and for the school. For instance, the trainees may be released from school for two afternoons a week for three weeks, for a week during vacation, for half-days for a nine-week quarter, during a summer session, or other combinations.

There can be no set formula for determining the amount of released time for work exploration, and it need not be during normal school hours or days. Coordinators must provide a sufficient time span for the work situation to allow the trainees to observe and evaluate their jobs adequately. A counseling period or related class to assist students in relating the experience to themselves is advised.

"SELLING" EMPLOYERS ON WORK EXPLORATION

In order to have effective work exploration plans, coordinators must "sell" the idea of developing work exploration stations to employers. Although there will be concern about the cost, the responsibilities to be borne, and the experiences the students should have on the job, most employers are interested in young people and want to give them a chance. Coordinators should stress that:

1. The development of a work exploration station involves no monetary cost to the employers; the

trainees will participate in one or more exploration activities at various times throughout the training program.

2. The "buddy" system will relieve employers of much of the supervision.

3. Work exploration activity guidelines will be prepared and ready for employers to use. These show what experiences are desired.

4. The experience gives employers the opportunity to observe potential employees without commitment; it also provides a way to contribute to the betterment of the community.

5. This kind of experience requires no complicated reports or documents for employers to fill out.

6. Work exploration is one way for employers to really help unemployed adults or young students who do not know where they want to go.

7. The school's coordinator will provide supervision of the trainees as needed.

DEVELOPING WORK EXPLORATION ACTIVITY GUIDELINES

Coordinators must be concerned with the time involved in the work exploration experience as well as with the activities of trainees during that time. Probably the first question (and a possible objection) an employer will have will be "What do you want the trainee to do and see?" Employers seldom have time to sit down and write out work exploration activity plans. Therefore, coordinators must provide employers with at least the skeleton of such a plan. One selling point to an employer, then, will be that the coordinator will help develop an activity guideline so that all that is needed during the work exploration phase is to follow it. By following an activity plan, the coordinator and the employer accomplish another objective—that of assuring the trainee of a complete and informative experience.

A set of exploration activity guidelines for a kitchen worker are suggestions for work exploration experiences in one occupational area. Similar guide-

EXPLORATION ACTIVITY GUIDELINE—Kitchen Worker

To the employer: This trainee is an individual who has expressed an initial interest in your field. Work exploration is designed to help the student make a wise choice, according to his/her interests and aptitudes for future career-technical training. Since this may be your first experience with this type of plan and the trainee's only chance to observe, the following is a proposed exploration plan for three hours each afternoon after school, for four days.

We hope these suggestions will help you in organizing and conducting this exploration.

Day 1: Orientation

1. Tour of kitchen and storage areas.
2. Introduction to those individuals the trainee will be working with.
3. Explanation of the skills needed for an **entry** job.
4. Opportunity for the trainee (if possible) to observe, relate, and work for one person. The trainer should be an interested, patient, considerate individual.
5. Explanation of the importance of cleanliness.
6. Demonstration of the daily routine you go through to keep things sanitary.
7. Opportunity for the trainee to become acquainted with various kitchen appliances—toasters, fryers, dishwashers, etc.
8. Opportunity for the trainee to help with kitchen cleanliness.
9. Opportunity for the trainee to work with the vegetable cook.

DAY 2: Operations

1. Opportunity for the trainee to work with the sandwich maker.
2. Opportunity for the trainee to work with the coffee maker.

DAY 3: Operations

1. Explanation of kitchen safety.
2. Opportunity for the trainee to work with the pastry maker.

DAY 4: Operations

1. Opportunity for the trainee to work with the salad maker.
2. Demonstration in kitchen cleanup.

We appreciate your making this opportunity available. Your trainee will be one of several individuals. Personalities may range from shy and untrusting to extremely aggressive. An individual may fit in quickly or may never feel at home. Whichever it may prove to be, this person is interested in the work and could be a possible future employee.

If you have any questions, please call me at _____

/s/_____

Sue Warren

FIGURE 11–1

lines can be developed for other occupational areas. What is presented here is merely a suggestion for making the work exploration experiences of trainees as meaningful as possible.

Organizing Training Stations

If potential training stations meet the criteria of the educational institution for providing the needed instruction and supervision, the steps in finalizing the selections and organizing them into directed training laboratories can be taken. The first step is the assurance that the employees at the training stations exhibit at least a minimal awareness and understanding of the agreements between the employers, the school administrators, the coordinator, and the prospective student-learners. Everyone concerned must understand that training stations are to serve primarily as a training medium rather than merely as an opportunity for remunerative employment for the students or for employers to obtain part-time help. This is paramount! When everyone involved understands the training role, steps in organizing the training stations can be taken. These are:

1. Solidifying with management and employees the purposes of the cooperative plan and the relative importance of learnings at school and on the job.

2. Setting work schedules.

3. Developing a specific training plan keyed to the short- and long-range needs of each student.

  a. Concluding a training agreement.
  b. Preparing a step-by-step training plan.

4. Designating and training sponsors.

5. Maintaining legal and ethical employment.

SOLIDIFYING THE NATURE AND SCOPE OF THE PLAN

Before employers interview students for possible employment, the coordinator should explain to each employer the following responsibilities and conditions regarding training stations.

1. The cooperative plan is a training plan and not a school employment agency and that the firms will be asked to help the student-learners with some individual study assignments.

2. Employers are considered to be partners in the training plan and should assist the school by providing planned occupational experiences and on-the-job instruction.

3. Training stations are to provide on the average at least 15 hours of employment a week throughout the school year, preferably at least half of this time during the released school time of the student-learners, according to standards set by the school.[4]

4. Frequently, minimum wage, according to state and/or federal laws, will be paid to student-learners, and this amount may not always be increased proportionate to their productivity.

5. Candidates for part-time employment have had career and technical education counseling at the school during which they have determined *tentative career objectives.*

6. Student-learners are enrolled in related classes at the school, where they receive instruction directly related to their work activities and occupational objectives.

7. Student-learners should have opportunities to move from one specific job activity to another in order to gain various experiences leading to their occupational goals.

8. Student-learners should be given the same employment status as that of other part-time employees in matters of social security, insurance, and labor laws (union contracts should be checked).

9. The coordinator will visit the student-learners, observe their job performance, suggest to their

---

[4]The minimum requirement for hours of employment in a training station may differ from state to state according to the guidelines of each state's department of education.

employers or their sponsors sound methods of on-the-job training, and determine the job activities to which classroom instruction should be related.

10. Periodic ratings based on the job performance of the student-learners will be made by their employers or their sponsors and reported to the coordinator.

SETTING WORK SCHEDULES

The on-the-job schedule for each student should be arranged with consideration given to:

1. The student's training needs and an opportunity to participate in various business operations when competent supervisory personnel are on duty.

2. The employer's staffing situation and need for maintaining productivity.

3. The student's need for rest and time for other responsibilities as a student, as an individual, and as a member of a family.

4. Employment laws that regulate student work hours.

5. Opportunity for the teacher-coordinator to observe the student on the job and to confer with the training sponsor, bearing in mind the contractual agreements regarding employment of school personnel.

A specified number of working hours each week should be established for all cooperative student-learners. The suggested minimum is 15 and the maximum 20, because total hours of employment and school attendance should preferably not exceed 40 hours per week and certainly not more than 48 hours per week. Weekend and nighttime hours are counted in this total just as are school-day hours. At one time school personnel believed that the student-learner should not be expected to work earlier than the start of the school day nor later than 5:30 or 6:00 p.m., the usual closing hours in most firms. But work-hour and work-week schedules have changed greatly in some industries, especially those in the marketing sector. And, as the cooperative plan has expanded into new industrial and health occupations areas, the weekend and night shifts are being used.

Furthermore, the issue of how many hours a student should work per week has become a major concern of parents, educators, and policy makers. There is now significant research that indicates that students who work more than 20 hours per week do worse in school than their non-working peers.[5] After studying 1,800 high school sophomores and juniors from California and Wisconsin, Steinberg, Fegley, and Dornbusch report that:

> Before working adolescents who later work more than 20 hours per week are less engaged in school and are granted more autonomy by their parents. However, taking on a job for more than 20 hours further disengages youngsters from school, increases delinquency and drug (and alcohol) use, furthers autonomy from parents and diminishes self-reliance. Leaving the work force after working long hours leads to improved school performance but does not reverse the other negative effects.[6]

The Steinberg studies and others received much national publicity and influenced articles that were published in *Life*, *U.S. News & World Report*, *McCalls*, *Newsweek*, *USA Today*, and other popular press media. As a result of increasing pressure, Congress in 1993 considered legislation that would limit the number of hours teens could work per week to 20. One state, Florida, has enacted a law (1986) limiting high school students from working more than 30 hours a week.[7] Other states such as Tennessee have passed bills to bar 16-and 17-year-olds from working past 10 p.m. on school nights.[8]

---

[5]Maria Mihalik, "The Lure of the Paycheck," *Teacher Magazine*, Vol. 1, No. 4, January 1989.

[6]Laurence Steinberg, Suzanne Fegley, and Stanford M. Dornbusch, "Negative Impact of Part-Time Work on Adolescent Adjustment: Evidence from a Longitudinal Study," *Developmental Psychology*, Vol. 29, No. 2, 1993, pp. 171–180.

[7]Barbara Kantrowitz, "The Kids and Jobs: Good or Bad?," *Newsweek*, June 9, 1986, p. 54

[8]Brett Pulley, "States Step Up Push for Curbs on Teen Labor," *The Wall Street Journal*, June 6, 1990, p. B1

The important point in arranging schedules is not the time or the day; rather it is the aforementioned criteria, as well as the school's primary responsibility to provide a quality educational experience at the training station. Consultation with the school administration and the advisory committee is recommended to determine whether each schedule is in the best interest of the student and is conducive to the student's welfare.

DEVELOPING SPECIFIC TRAINING KEYED TO CAREER OBJECTIVES

One of the most important steps in the proper establishment of a training station is that of preparing a written plan of training. The development of a **step-by-step training plan** for a particular student brings the student face to face with the problem of determining an ultimate career-technical objective and deciding what competencies must be developed. Through the training plan, the employer also becomes more aware of the student-learner's occupational goal and is encouraged to lead the student toward an objective by providing adequate work activities and on-the-job instruction.

Concluding a training agreement.—A *training agreement* would normally include at least the following information. (See Figure 11–3.)

1. Names of student-learner, employer, business establishment, school, training sponsor or supervisor, coordinator, and parents.

2. Dates of beginning and end of training period.

3. Beginning rate of pay, proposed scale of increases during the school year, and working hours.

4. Statement of student's career objective, including a brief description of the skills, attitudes, and information necessary for a worker in the occupation.

5. List of job activities that will contribute to the student's progress toward a career objective (at least the areas of experience and training).

6. Brief outline of the related classroom instruction that will be provided at school (at least the major areas of subject matter).

7. Responsibilities of the student-learner, the employer, and the school to the training plan.

Preparing a step-by-step training plan.— Sound instruction will be enhanced if the coordinator prepares a step-by-step training plan.

1. The coordinator carefully explains to the employer and the student-learner the purposes of a training plan and the procedures for developing one.

2. The coordinator, the employer, and the student-learner cooperatively list the skills, attitudes, and information needed for a successful career in the student-learner's chosen occupation.[9]

3. The coordinator, the employer, and the student-learner develop a list of work activities that should contribute during the school year to the student-learner's progress toward the occupation objective.

4. The coordinator, the employer, and the student-learner list the knowledges and skills needed by the student-learner in performing the work activities referred to in number 3.

5. The coordinator, the employer, and the student-learner cooperatively determine a plan for putting numbers 3 and 4 into action.

A complete training plan is a combination of the training agreement and the step-by-step training plan. It has already been stated that the teacher-coordinator should identify the areas of experience and training available to a prospective student-

[9]In some states, training plans are printed for all occupations, and the activities in the plan appropriate to a given student-learner are checked off.

learner in a prospective training station. Student-learners, because of a wide range of capabilities and career objectives, will vary in the degree of diversity they need on the job.

Expressing training plans in behavioral terms.—A *statement of student-trainee learning objectives (SSLO)* is a possible alternative to the traditional training plan (see Figure 11–2).[10] The SSLO combines the behavioral learning objectives concept employed by educators and the management by objectives tool used by business and industry. This plan changes the role of the coordinator from one of prescribing beneficial, on-the-job learning experiences to one of facilitating the development of learning objectives derived by both students and employers. The student's role changes from a passive one to an active one under the SSLO, as the student is actively involved in determining what learning will take place during a certain period of time.

Under this concept, each cooperative student is treated on an individualized basis. Consequently, each student-trainee has a unique set of learning objectives for his/her job. The coordinator's task is to spend a sufficient amount of time in dialogue with each student-trainee so as to be able to help determine the objectives appropriate to the needs of each student.

Implementation Schedule for the SSLO

First Meeting Between Student-Trainee and Coordinator (This meeting should take place after the student has started to work.)

Coordinator gives the student-trainee two copies of the SSLO, a set of sample objectives from the student's career-technical field, and instructions on how to write meaningful job-oriented learning objectives. The first copy of the SSLO is a working copy for the student's practice use.

Coordinator instructs the student-trainee to review these materials and then to meet with the employer to gather ideas concerning meaningful objectives for the evaluation period.

Coordinator requests that the student-trainee formulate 10 tentative objectives to accomplish during the evaluation period.

Second Meeting Between Student-Trainee and Coordinator

Coordinator reviews tentative objectives with the student-trainee to determine their appropriateness and whether or not they are measurable.

Coordinator must help the student-trainee realize that the objectives identified at this meeting constitute the major criteria for determining the student's grade for this evaluation period.

Coordinator encourages the student-trainee to think of objectives in terms of resultant behavior or what he/she will be able to do after learning has been accomplished (for example, achieve cleaner and safer work areas, be on time to work at all times during the evaluation period, be able to identify at least 10 selling benefits of latex paint, etc.) If objectives are stated in terms of desired results, then the criteria for measurement are incorporated into the statement.

For an objective to be valid, the student-trainee must have control over its accomplishment. The student-trainee should choose objectives that can be achieved within the confines of his/her particular job.

After the coordinator has had input into objectives, the student-trainee is requested to review these objectives again with the employer. This step provides for reality testing.

Coordinator requests the student-trainee to bring back in several days a rather finalized list of objectives that have been discussed with the employer. (Student-trainee uses second copy of the SSLO for these.)

First Meeting Between Coordinator and Employers

Coordinator makes appointment with employers specifically to discuss and articulate the SSLO. The

---

[10]The SSLO material here and the training form shown in Figure 11–2 originated with Dr. William A. Stull, Utah State University,

STATEMENT OF STUDENT-TRAINEE LEARNING OBJECTIVES
FOR
TAZEWELL HIGH SCHOOL
COOPERATIVE EDUCATION

Name of Employer	Name of Student-Trainee

Date	Grading or Evaluation Period

During each grading or evaluation period in which a student participates in the cooperative phase of the program, every effort should be made to insure that actual student learning occurs. The teacher-coordinator, acting as a facilitator, should assist the student-trainee and the employer in determining and establishing what new or expanded learning opportunities may be available through the on-the-job phase of the program. These learning opportunities, stated in terms of measurable learning objectives, enable the student-trainee, employer, and teacher-coordinator to determine whether progress has been made by the student.

The student-trainee, with the assistance of the employer and teacher-coordinator, should establish from five to eight meaningful learning objectives for the specific grading or evaluation period. These objectives should be specific and measurable. They will be used at the evaluation time to determine if the student's on-the-job progress is satisfactory and to determine the areas where improvement needs to be made.

The following rating scale has been devised to evaluate the student's progress on the job. Both the student and the employment supervisor should rate the student's progress toward each learning objective. Prior to the end of the evaluation period, the teacher-coordinator will meet with the student and the employment supervisor to determine the final evaluation.

RATING SCALE

1. Limited Accomplishments
2. Average Accomplishments
3. Better Than Average Accomplishments
4. Outstanding Accomplishments

RATING

Learning Objective	Student	Employer	% Value	Final Evaluation
1. _____				
_____				
_____				

FIGURE 11–2. (Courtesy, Dr. William A. Stull, Utah State University, Logan)

STATEMENT OF STUDENT-TRAINEE LEARNING OBJECTIVES (Continued)

RATING

Learning Objective	Student	Employer	% Value	Final Evaluation
2.				
3.				
4.				
5.				
6.				
7.				
8.				

Signature of Student _____

Signature of Faculty Advisor _____

Signature of Employer _____

FIGURE 11–2 (Continued)

coordinator should probably ask if employers are familiar with the SSLO concept and whether or not they have spent any time helping student-trainees write objectives.

Coordinator reviews objectives with employers to determine their appropriateness, whether or not some possible learning objectives have been omitted, and whether or not employers will give student-trainees an opportunity to accomplish these objectives.

Employers must validate objectives. They must insure that the objectives are measurable and within the job descriptions of the student-trainees. At this point in time the employer may suggest changes in objectives or additional objectives that could be completed. The coordinator should make sure that the employers are aware that these objectives constitute the major criteria for determining the student trainees' grades.

Coordinator and each employer adjust, add to, and generally agree on the objectives for this evaluation period. The coordinator and each employer sign the SSLO. The coordinator requests that the employer review with the student, have the student sign, and request that the student return the SSLO form to the coordinator. A brief explanation of the evaluation process would be appropriate during this visit.

Mid-Term Visit—Evaluation

Coordinator reviews progress on the SSLO with student-trainee and employer. Student and employer are asked to evaluate independently progress to date on objectives.

At a meeting, coordinator, student-trainee, and employer review the accomplishment and progress on the objectives to date. Together they identify weak areas and missing objectives. If necessary, they modify learning objectives, adding new objectives or replacing inappropriate ones. The coordinator determines student-trainee's mid-term grade.

Final Evaluation

Coordinator, student-trainee, and employer go through same process as at mid-term.

Coordinator, student-trainee, and employer discuss possible learning objectives for the next evaluation period.

SSLO advantages.—The SSLO approach to the development of cooperative occupational education training plans provides advantages to all parties involved in the cooperative education phase of the program—the student, the employer, and the school. Potential advantages of the SSLO system include:

1. Students have a greater appreciation for the learnings provided through their jobs, thus greater interest in and motivation on their jobs.

2. Students appreciate the clearer direction that results from the objective-setting process.

3. The procedure enhances communication and understanding between employers and student-trainees. Reality testing occurs.

4. Employers appreciate the business-like and objective approach followed by the educational institution.

5. Employers gain an understanding of a new evaluation technique, which can be incorporated into their entire operation.

6. The procedure allows the school and the coordinator to build a bank of learning objectives (learning opportunities), which may be used by future student-trainees in similar occupational situations.

7. The procedure allows for the assignment of a more objective grade, a grade based on performance objectives rather than on personality characteristics.

DESIGNATING AND TRAINING TRAINING SPONSORS

If the job is to be what it should be—an educational laboratory—then it is reasonable that one person in each firm be designated as the training sponsor. This person is responsible for supervising the learning experiences that, while being worthwhile enough to be the basis for the granting of credit by the school, should relate to the student's needs.

In a sense, training sponsors are the counterpart of teacher-coordinators. They must be job supervisors for the employers, overseeing the trainees' performance and insuring that they are productive employees. At the same time, their role is quite different—they must be counselors to students who may be much younger than they are; instructors who show and tell how an operation is to be carried out; planners of experience that should be varied and related to the students' progress; and evaluators not only of job productivity but also of progress toward defined job goals.

Training sponsors must have the time to devote to the role so that they do not neglect their training duties in favor of other demands. They must be in proximity to the students' work areas to supervise their activities. They must be viewed by their employers as being trustworthy and equal to a unique supervisory task. Above all, they must be individuals who are responsive to young people and who are not in any way threatened by their youth. The sponsors may be rank-and-file employees, or they may be supervisors. As preceding criteria demonstrate, it is the person, not the title or the rank, who counts. If there are several trainees in one firm, there should be several training sponsors, with a senior officer, such as the personnel director, designated as the contact person for the firm.[11] This person can also give direction to the several training sponsors.

The coordinator plays a very important part in orienting the training supervisors (job sponsors) at the participating training stations. The training sponsors will need to know what their responsibilities are as the training supervisors of the student-trainees, and they will need to understand the importance of the related class.

_____

[11]Some firms that employ several student-learners in one location or in several branches use a centralized training technique. A training officer makes up a weekly, bi-weekly, or monthly plan of on-the-job experiences (a departmental rotation schedule), as well as a schedule of group lectures, observations, and other presentations. This technique appears to have considerable merit in a very large organization or in one, such as a hospital, where many specialists are available to teach major concepts or where the student because of inexperience or age cannot actually participate in certain activities. This technique is also recommended for postsecondary programs to give future managers and specialists an over-all view of the firm's operations and to help them and the firm decide their career route within the firm.

Responsibilities of Training Sponsor and Coordinator	
Training Sponsor	**Coordinator**
1. Always treat the trainee with respect—not as a "dummy", or as a "kid."	1. Help the sponsor develop an awareness of what it is like to be economically disadvantaged, a member of a minority group, young, and inexperienced.
2. Recognize the shortcomings of the trainee, but be positive in terms of possible solutions.	2. Don't _talk down_ to the sponsor; _talk with_ the sponsor so that he/she will be able to help the trainee. Avoid using "school" or "teacher" language.
3. Introduce the trainee as an employee, and then help the trainee develop good working relationships with other employees.	3. Encourage the sponsor to praise the trainee, but only for good effort or work performed well.
4. Expect the trainee to abide by the same rules as everyone else. Require the trainee to perform the same kinds of tasks as other employees.	4. Expect the trainee to put forth effort and to perform well. Back up the sponsor.
5. Contact the coordinator immediately if a problem is forthcoming, or if a problem (previously unforeseeable) is present.	5. _Always_ be available when the training sponsor needs your help. Go today … the trainee's problem exists now … it won't wait.

An initial conference with both the employer and the training sponsor (if they are not one and the same) should give the coordinator an opportunity to talk over the problems and possibilities involved in having a trainee in a cooperative work station with that particular business. It may seem like an easy task to list the responsibilities of the training sponsor and the employer, but the real difficulty lies in how to "reach" the sponsor to show what these responsibilities really are. The coordinator must be extremely tactful but helpful in dealing with employers and training supervisors in business and industry.

MAINTAINING LEGAL AND ETHICAL EMPLOYMENT

Educators who direct students in work situations must insure legal and ethical employment situations (see Chapter 16). Teacher-coordinators must know the intent as well as the fine points of the law, for not only are they the representatives of the students, who are usually minors, but they are also the link between the school and the community. They must be conversant and articulate in explaining to all parties the legal language in terms the parties understand. Of primary concern are those laws dealing with work permits, hazardous occupations, working hours and other employment standards, fair employment practices, and insurance compensation such as workers' compensation.

In addition to assuring the students and their parents, the school, and the employers that students are legally employed, coordinators have an equally important responsibility. Because the school controls the student's work assignments, coordinators must insure ethical employment. They cannot allow employment that, although technically legal, would violate what is considered "ethical." This is an important factor, particularly since the mores of students and their parents may conflict with those of the school administration, or even those of their coordinators. Coordinators must advise, caution, and counsel each student to make a decision that everyone can live with and when that decision has been made, to have the final arrangements put in writing.

Guidelines for legal employment.—Coordinators are held accountable for assuring legal employment. In one sense, they enforce the law by complying with regulations, even in cases where they do not agree with them. In another sense, however, they only guide employers by advising them of the law and circumstances in which they perceive the employers might be judged guilty of a violation. Coordinators do employers a great service by calling their attention to such conditions. Coordinators recognize that the failure of anyone—school, student, or employer—to comply with the law results in damage to the image of the plan. If a student-learner's health or safety is endangered, the coordinator is subject to criticism and in some cases to legal action. If a situation is serious, the coordinator should definitely consider terminating a placement or avoiding a placement, along with giving an explanation to the firm about these actions. Legal employment is discussed in detail in Chapter 16.

Orienting Trainees to Learning Opportunities at Job Laboratories

Students in a cooperative plan of instruction are in what is for most of them a very new learning environment and one very different from what they have usually experienced. They are asked to be at one time students in the usual sense of attending classes and at another time individuals who must adjust to a role in the world of work and use it as a learning experience.

These two roles cannot be played without professional assistance from their teacher-coordinators, whom they see every day. The coordinators are counselors, friends, and contacts between school and job. It is the coordinators who help students adjust to their role as individuals outside the school. For mature adults, this adjustment to the scheduling of time and the mixing of roles would be most difficult; for young persons, it is even more complex and can usually be accomplished only if the coordinators have the time to make the process meaningful to all involved.

ACQUAINTING STUDENT-LEARNERS WITH THE VALUES OF THE TRAINING PLAN

Those who enroll in the cooperative occupational education plan should understand that:

1. They must have enough time available for part-time employment and at the same time complete other school requirements.

2. Their employers will be expecting interested, motivated individuals who are willing to learn to produce well on the job.

3. Their learning activities must be carefully planned so as to contribute to their preparation for their chosen careers.

4. Their coordinators will evaluate them during visits to their training stations. They will also help them to overcome any problems in the workplace.

5. Their coordinators will thoroughly explain local administration rules on grades, credits, and student responsibilities.

The orientation process should also teach student-learners how to learn, since the integrated school–job laboratory situation is novel to them. They can be given sample project assignments that show how training station experiences relate to the class in school. They should learn through role-playing how to ask questions of training sponsors and how to explain the training role to co-workers. Other experiences can be those of looking over a step-by-step training plan for career objectives, filling out a sample weekly report, and making an observation of an actual work situation and writing up the report.

FORMALIZING THE STUDENT ORIENTATION PROCESS

In some occupational curricula the cooperative plan is a capstone course or experience. In such cases, the orientation of student-learners to the cooperative experience comes in the prerequisite course, preferably through an intensive unit prior to final placement. When this prerequisite instruction is not feasible or when the orientation is not carried out in such course, some schools use what amounts to a short course or workshop. This is usually carried out during after-school hours, during the summer, or in a daytime series just prior to school opening in the fall.

The objectives of this orientation short course or workshop are:

1. To acquaint students with the purposes, rules, and policies of the cooperative plan.

2. To introduce students to their responsibilities in such a plan.

3. To provide pre-employment training for students *at the time it is needed.*

4. To assist students in developing an awareness of the initial skills, knowledges, and attitudes necessary to meet their immediate employment needs.

5. To help students to understand the importance of further general and occupational education.

6. To encourage students to become aware of occupational opportunities in their chosen areas and to utilize a "career objective" approach to their vocations.

The content of such a short course (five to six meetings, to total 10 to 12 hours) is best determined locally, based upon the most effective ways to arrive at the objectives just shown. The course may be classified as an adult education offering and planned for any of the cooperative occupational areas, or it may be available to all prospective cooperative education students. Preferably the instruction should be handled by the teacher-coordinator or by a team of the coordinators of the cooperative education plans, if more than one type exists in the community. Also, former students, selected training sponsors, and related-subject instructors may be used as resource persons.

Coordination Practices in a New Plan

Some training stations can provide, and some student-learners need, only two or three types of experience. Other training stations can provide, and some student-learners, because they are high achievers and because they are seeking top-level career objectives, will need as many as six or more areas of experience on the job during the school year. Before accepting a particular training station, the coordinator should be satisfied with the possibility of obtaining the variety of experiences consistent with the needs of the student. In the case of a store or an office, a variety of experiences might very well be obtained in a single department or section. In the case of an industrial shop or a laboratory, the student-learner should receive a series of challenging learning experiences and not be kept on minor repetitive operations for excessive periods of time. Accepting poor training stations or "just jobs" for expediency will usually lead to difficulties early in the school year as the result of dissatisfied employers or enrollees and will jeopardize the reputation of the plan in the school and in the community.

When a teacher-coordinator for business occupations has evidence that a business has potential as a training station, he/she should draw up a training agreement form similar to Figure 11–3. Four copies are needed for each student-learner placement (one copy each for the student, parent, training station sponsor, and coordinator). Attached to each copy of the training agreement may be a copy of a step-by-step training plan. The step-by-step plan for a new training station may have to be developed in periodic steps during the school year and on more than one coordination visit. Once a plan is established for a particular training station, yearly review and adjustment of the plan for other student-learners are usually satisfactory.

In preparing complete training plans, coordinators should identify the learning outcomes for the career objective of each of the student-learners. For example, the learning outcomes during the school year for Student-learner X, with a career objective of "Administrative Assistant," might be identified as the broad areas of public relations, telephone technique, mail handling, filing, record keeping, keyboarding,

and word processing applications. The learning outcomes for the school year for Student-learner Y, with a career objective of "Men's Wear Buyer," might be identified as the broad areas of store housekeeping, stock keeping, buying, selling and servicing, visual merchandising, clerical activities, maintenance of inventory control, and fitting. The learning outcomes for the school year for Student-learner Z, with a career objective of "Sheet-Metal Worker," might be identified as the broad areas of fabricating, assembling, altering, repairing, and installing sheet-metal articles and equipment. The first case is a business occupations placement, the second case, a marketing education placement, and the third case, an industrial occupations placement. Similar lists of learning outcomes can be determined for health occupations, agricultural, and family and consumer sciences occupations.

Making Coordination Visits Productive

Anyone who has worked with the cooperative plan knows that administrators have many questions about the time spent on the coordination function. They realize that it is often a one-to-one situation and involves traveling and waiting, and they are inclined to contrast this with the time the teacher could spend with a group of 30 students. Administrators are also faced with the imperative need to justify the uses of funds in their budgets; thus, in attempting to conserve resources, they are inclined to use student credit-hours or some other ratio of teacher–student as a guideline. This appreciation of the administrators' position makes it not only necessary for coordinators to demonstrate why coordination visits are essential to quality education but also to make them as productive as possible. Effective teacher-coordinators should use as a rule of thumb the commitments of at least one-half hour per week per trainee for coordination time. This would involve at

TRAINING AGREEMENT

Name of Student-Learner _____ Age _____

School _____

Business Establishment _____ Address _____

Job Sponsor (supervises student-learner) _____

Dates of Training Period _____

Student's Career Objective _____

Basic skills, attitudes, and knowledges needed in this occupation:

_____

_____

_____

Main areas of related instruction in the classroom during the school year:

_____

_____

_____

The student-learner will consider the job experiences as contributing to career objectives and will perform training station responsibilities and classroom responsibilities thoroughly.

The employer will recognize that a training plan is being followed and that close supervision of the student-learner will be needed. The employer (1) will provide work experiences listed in this training plan, (2) will provide part-time employment of _____ hours a week on the average, (3) will provide wages to the student-learner at the start of $ _____ per _____ , and (4) will consult with the coordinator on any major problems that arise concerning the performance of the student-learner.

The coordinator (1) will provide instruction directly related to the student's job activities and career objective, (2) will suggest ways of supervising the student-learner, and (3) will assist the employer with training problems pertaining to the job.

Additional comments: _____

_____

_____

Student _____ Parent _____

Employer _____ Coordinator _____

Phone _____

FIGURE 11–3

least one visitation per month on the average, but there cannot be a prescribed formula handed out, as each trainee, employer, and occupation will differ in demands. Insecure trainees may require weekly visits until they are more secure.

Visits to the training station will include:

1. **The employer and/or supervisor visitation (evaluation)**

   The employer–supervisor–coordinator conference is the most effective means of evaluating the trainee's work experience. Such evaluations will help coordinators:

   a. Determine adherence of the training station to the training plan for each trainee.

   b. Make adjustments to the training plan as they are needed.

   c. Evaluate each trainee's progress on the job.

   d. Insure that proper training methods are being used.

2. **Observation of trainees**

   The training station visits permit coordinators to observe trainees on the job and can help coordinators:

   a. Determine the appropriateness of the trainee's appearance on the job.

   b. Ascertain how other employees seem to react to the trainees.

   c. See how the trainees meet and serve the public.

Problems That Affect Trainees' On-the-Job Experiences

Attitude problems

1. Feeling that employers owe them a job
2. No feeling of responsibility or obligation to employers for a good job

Inability to follow directions

Misuse of coffee breaks

Alcoholism or drugs

Misunderstanding over employment agreement

1. Wages paid
2. Hours worked
3. Shift changes
4. Fringe benefits

Lack of interest in the occupation for which they have been trained

Inability to accept criticism from employer/supervisor

Attendance and punctuality

1. Frequently absent
2. Late to work

Lack of self-confidence often results in trainees quitting because they:

1. Are not sure they are doing a good job
2. Feel they are not liked by other workers
3. Feel their employers are prejudiced.

Low production

1. Producing fewer units than regular employees
2. Producing fewer error-free units than other employees

Slow learner

Problems with the family

Inability to take orders from superiors

MONTHLY REPORT OF COORDINATION ACTIVITIES
COOPERATIVE PART-TIME PROGRAMS

Send to:
Board of Vocational Education
116 State Office Building
Springfield, Illinois 62706

Cooperative Plan _____

Name of School _____   Month _____, _____

Date	Activities	Date	Activities

(Complete the month on reverse side)

FIGURE 11–4

Date	Activities	Date	Activities

Comments:

Signed by _____ Signed by _____

Coordinator   Superintendent, Principal, or
                                                  Director of Vocational
                                                  Education

FIGURE 11–4 (Continued)

d. Show the trainees that their coordinators are interested in their occupational development.

e. Observe any apparent difficulties that have not been previously brought to their coordinators' attention.

3. **Promotional activities for the plan**

In order to develop new training stations and to assist employers in upgrading existing positions, coordinators will:

a. Make training station personnel continuously aware of training activities and objectives.

b. Keep abreast of current trends in the world of work.

c. Secure publications or other current materials that can be used as source materials in the instructional phase of the training.

d. Solicit participation of employers in training and promotional activities.

e. Distribute and explain brochures that are locally designed or developed by the state staff.

PRE-ARRANGING COORDINATION VISITS

To make coordination time productive, teacher-coordinators should carefully plan their calendar of events, make specific appointments with their various publics for specific reasons, and keep a current log of their activities in a pocket notebook. Local state systems and state departments of career and technical education will ask for coordination reports (see Figure 11–4). It is most helpful if pocket notebook notations of coordination activities can be referred to in these reports. Coordinators might also carry along another notebook on visits as a source for information on items such as school policies, labor regulations, employer responsibilities, student selection procedures, evaluation criteria, school forms, student success stories, and testimonials.

To be most productive, coordination activities should directly or indirectly benefit student-learners.

WHAT TO DO DURING A VISITATION

The specific reasons for visiting a training station will vary, depending on the student-learner, the type of training station, the time of the school year, and the types of instruction being carried out in the classroom and on the job at the time. Coordinators should visit each training station at least once a month, although deviations will be necessary. At the beginning of their experience on the job, some student-learners may have to be visited briefly each week. Some student-learners will need close attention because making certain adjustments is difficult for them. Others will adjust quite well; a visit every three or four weeks will be sufficient for them.

A few do's and don'ts of coordination are listed as follows:

Do's and Don'ts of Coordination

• Do observe the trainees on the job.

• Do consult with the job sponsors frequently about the progress of the trainees and about suggestions regarding how to implement the training plan.

• Don't, as a rule, correct the trainees at the time of visitation without consulting with their sponsors.

• Don't use an extended block of the job sponsor's time without a prior appointment.

Do's and Don'ts of Coordination

- Do consult with top management of the training directors occasionally regarding the cooperative education plan.
- Do be alert for specific instructional materials to have the trainees bring into the classroom.
- Do involve job sponsors in evaluating student projects.
- Do thank employers for their time.
- Do meet where employers want to meet, regardless of your attire.
- Do talk with both the supervisors and the trainees during every visit so that neither one feels left out or insecure about what might have been said to the other.
- Do act with assurance that this is a cooperative arrangement mutually beneficial to all, not a "favor" to the school.

- Don't just drop by for a chat with job sponsors. Have a reasons for asking to see them.
- Don't expect many of the sponsors to be able to teach on the job or to evaluate the trainees' work without some suggestions and guidance from you.
- Don't drop in on employers without first calling.
- Don't ask employers to do any more paperwork than is necessary.
- Don't talk to the trainees' supervisors about the trainees in front of them.
- Don't forget to reassure the trainees before leaving them, if private consultation is necessary with their employers or their supervisors.
- Don't comment on the working conditions unless a change that affects the trainees has been evidenced since the training agreement.

Questions and Activities

1. Explain the concept of job development, and describe the benefits when it is effectively carried out.

2. Why have educators generally been reluctant to engage in aggressive job development? What factors or forces may be emerging now that will compel institutions to overcome this reluctance?

3. Prepare a list of job-development techniques similar to that given in this chapter for a specific community and cooperative plan. Demonstrate in your list certain techniques that are unique to the community, the age of the plan's use, the occupation being served, and other factors such as the level of the institution and the type of plan.

4. Select a job you know well. Develop a statement of learning outcomes for a student-learner in that occupation.

5. Prepare a training agreement for use with training stations in a given locality.

6. Discuss the relative merits, or advantages and disadvantages, of three ways of formulating a step-by-step training plan. They are:

   a. Use the list of skills, attitudes, and information needed for a successful career in the student-learner's chosen occupation, prepared by the coordinator, the employer, and the student-learner cooperatively.

   b. Use a printed training plan for each occupation, and check off the activities in the plan that are appropriate for a given student-learner.

   c. Develop a training station profile. Then, each grade period set down in behavioral terms certain student-learner objectives to be accomplished and evaluated.

7. Prepare a paper listing the advantages and disadvantages of each of the following persons as a job spon-

sor: manager, department head or supervisor, experienced employee.

8. Discuss the work schedules that would be appropriate and necessary for a specific occupational plan in a particular community. If these were not generally during the released school hours, how would you justify this to your administration? What would the trainees do during their released time?

9. To help solve the problem of trainee orientation: (a) prepare a copy for a brochure or handout for the students, (b) devise a topical outline for an orientation workshop, (c) prepare an outline of a teaching unit for the related class.

10. Outline a role-playing situation for three coordination visits, each of which has a different purpose.

For references pertinent to the subject matter of this chapter, see Reading Resources.

12

Correlating Instruction Between School and Job Laboratories

A modern, well-equipped laboratory simulating the work environment becomes an important element in related instruction. Here the teacher-coordinator observes a student getting individual instruction from the auto shop teacher to supplement his study in the cooperative occupational education classroom.

KEY CONCEPT: Individual student needs and on-the-job requirements determine related classroom instruction.

GOALS

After successfully completing the study of this chapter, answering the questions, and carrying out the activities, the student should be able to:

- Explain how and why related classroom instruction is correlated with the on-the-job experiences and career objectives of student-learners.

- Involve students in developing training plans.

- Involve job supervisors in developing training plans.

- Develop the forms necessary for adequate training station evaluation.

- Develop and provide individualized instruction.

- Conduct training station visits.

KEY TERMS

correlation

farm-out system

feedback

related class

sponsor development

weekly training station report

The essence of the cooperative plan of instruction is the sharing of the teaching–learning responsibility between the professionally trained individual in the school laboratory and the occupationally proficient individuals in the work laboratory. But to be shared does not mean that each does a portion of the educational job in isolation from the other. Rather, this sharing of responsibility means a mutual understanding of goals and a correlation of efforts. In a sense this correlation is analogous to the team effort in any sport or activity in which each party relates closely to the other and adapts to the common agreed upon goals.

If school and employer each agreed to do an assigned task in isolation from the other and if each could do this, there would be no need for a coordinator other than for placement and evaluation. But the task is not so simple, for in the cooperative plan there is an interaction between two parties and between the students and the two training sponsors. It is this interaction of the parties that makes the correlation activity so vital to the functioning of the cooperative plan of instruction.

Correlation must be of such priority that coordinators demonstrate it at every opportunity. They begin when they encourage their students to become involved in the planning and management of the learning situation.

A student learner practices skills that she needs in her training station by using laboratory time outside the cooperative classroom.

Involving Students in the Training Plan

The backbone of correlation is a plan of instruction supported by consistent and meaningful follow-up as the coordinator stays on top of the situation. The training plan is prepared, as shown in Chapter 11, primarily by the teacher-coordinator and the training sponsor. However, the students must become very involved with their individual plans. They must review and suggest changes in them. Each student should use his/her particular plan as a guide to gaining experiences by taking advantage of opportunities. Trainees can profit from instruction in school and on the job to the extent that they are willing to be involved, to be responsible for their own learning, and to be active participants in the planning and management of their learning process. Trainees, after many years of directive teaching, may find this freedom disconcerting and difficult to accept and manage. But, they should be persuaded that managing their own learning will be very important during their many years of future employment.

As part of their **related class** instruction, trainees should be asked to do at least four activities that will assist in correlating learnings. These are:

1. List at the beginning of their work experience their career objectives, as well as some of the areas in which they wish to obtain experiences and training during the school year.

2. Prepare a *weekly training station report* that will indicate to them and to the coordinator the areas in which they are gaining instruction on the job (see Figure 12–1).

3. Summarize periodically, preferably once each grade period (whether that be six weeks or nine weeks), the learning outcomes on their job; review their training plan progress.

4. Evaluate at the end of the school year their training by setting up a profile of what has occurred at their training station during the school year. They should indicate the title of their present job, their career objectives at the end of the school year, the experience and training they have received, and their future plans for meeting their career objectives.

Student-learners must get involved in managing their learning; however, they vary in maturity and in their ability to plan and work individually. The less able students, especially those who have difficulty with cognitive learning, will need much assistance in developing their self-confidence and their ability to make decisions. On the other hand, post-secondary students should have developed considerable self-discipline in structuring their own learning.

Making Classroom Instruction Realistic

Often trainees find it difficult to see the relationship between those skills learned within the classroom and those learned on the job (see Figure 12–2).

The progress of trainees depends on how easily they see this relationship. This is not only true of related skills instruction but also of basic education. It may be easier for the trainee who is planning to be a machinist to see the need for learning how to operate a drill press than the need for learning to read.

This trainee needs to be shown examples of how instruction in school and on the job interrelate. The student-learner trainee will gain additional ideas about the applicability of the instruction by engaging in case problem discussions and by undertaking individual projects.

USING JOB PROBLEMS TO MAKE INSTRUCTION REALISTIC

Most students expect their classroom instructions to be relevant. This is especially true of cooperative student-learners. They will expect certain skills and ideas learned in the classroom to be operational immediately on the jobs to which they go. A situation that arises at the job training station or a problem that a student-learner brings to class stimulates interest and suggests to the teacher-coordinator certain areas to be covered in related instruction. Some problems or certain instructional units may require the assistance of another staff member.

A training station problem.—The wise teacher-coordinator carefully studies the weekly training station reports prepared by the students. Such study often brings to light topics for related instruction or the need for individual instruction or counsel. For example, if the teacher-coordinator discovers from reading the weekly reports that several students are not profiting from the job supervision available to them, the teacher-coordinator may use class time to illustrate how an employee can benefit from constructive suggestions from the supervisor. If a weekly report indicates a problem unique to one student, this matter can be handled through a personal student–teacher conference.

An occupational problem voiced by a student-learner.—On occasion, a student-learner may say in class, "But that's not the way we do it down at my training station," or "I certainly had a time of it yesterday when three persons were telling me things to do—all at the same time!" This is a clue to an immediate problem, and without specific names of business personnel being mentioned, classroom time can be spent wisely in the discussion of possible solutions to the problem for the benefit of all the students.

STUDENT-LEARNER WEEKLY REPORT

Week Ending: _____

Trainee: _____   Training Station: _____

Job Title: _____

1. What were your activities on the job this week?

_____

_____

2. How did you apply classroom materials to your job?

_____

_____

3. What new job or procedures did you learn from your work?

_____

_____

4. What mistakes did you make? How did you handle the situations?

_____

_____

5. What problems came up with which you would like help or class discussion?

_____

_____

6. Describe the most interesting experience you had on your job.

_____

_____

7. What related materials did you read outside class?

_____

_____

FIGURE 12–1

STUDENT-LEARNER WEEKLY REPORT (Continued)

Weekly Time and Wages Summary					Comments:
			Total Hours		
Date	**Time Worked**		**Regular**	**Overtime**	
	to	& to			
	to	& to			
	to	& to			
	to	& to			
	to	& to			
	to	& to			
Total Hours Worked					

Hourly Rate: _____   Wages Earned: _____

_____
Signature

FIGURE 12-1 (Continued)

EXAMPLES OF HOW INSTRUCTION IN SCHOOL AND ON THE JOB INTERRELATE

Institutional Training	Employer
1. Trainees who are employed in co-op in different business firms can "share" experiences on how similar tasks are performed in different situations.	1. The employer/supervisor can demonstrate to a trainee how tasks are to be performed in the business in an actual situation.
2. Instructors can devote instructional time to the general study of the interrelationships of jobs.	2. A trainee can see the division of work in one business if the job orientation includes details on the operation of the company. Oftentimes, however, little time is available for such explanations.
3. Instructors may conduct "how-to-do" sessions for the trainees, which feature "extra" task practice in order to perfect performance in certain tasks.	3. In the actual job experience a trainee will perform the task only when it is needed and only the number of times needed by the employer for production purposes.
4. Instructors are specialists in teaching the basics of a broad occupational area.	4. The employer/supervisor is the teacher; the trainee will learn the "one" way to do a task, according to the supervisor.
5. Following certain task exercises with trainees, instructors may make cooperative decisions based on the outcome of their critiques of the trainees.	5. The supervisor will be evaluating a "finished product." Therefore, any criticism may relate to work that may have to be redone if there are any errors. (Extra task practice would help trainee.)
6. Trainees may learn basic operational techniques for a variety of equipment necessary in their occupations.	6. A trainee will learn to operate the equipment of a specific employer and to apply principles to specific job tasks.
7. Instructors may devote instructional time to task analysis and in-depth study of step-by-step procedures employed in a job situation.	7. After an initial explanation of how to do a task, a trainee may have to perform the task under the coaching of an experienced worker or supervisor.
8. The related instruction period permits the trainees to pull together and discuss ideas from many sources—other trainees, coordinator, counselor, instructor, and many businesses.	8. The employer usually limits instruction to the tasks or jobs the trainees will be performing during their experience.
9. The related instruction experience poses the trainees to qualities needed for various job experiences that other trainees are having (punctuality, special dress, attendance, etc.).	9. The on-the-job cooperative experience acquaints trainees with those qualities and traits (punctuality, attendance, grooming) mainly applicable to specific job situations.
10. Instructors can develop materials (related and/or basic) to meet the needs of specific businesses and then instruct to meet those specific needs (at an employer's request).	10. The supervisor uses only those materials that are needed for a specific task or job to be done for the company.

FIGURE 12–2

The teacher-coordinator needs to encourage students to share both their negative and their positive experiences on the job. Because some students believe they may be "tattling" and others may have been conditioned in earlier school years not to discuss their problems, the teacher-coordinator must create an environment in which students feel free to discuss their experiences. One technique used to seek out problems and to aid students in solving problems is to teach them a simple four-step analysis.

1. *Get the facts.* What are the real ones? What are the hidden ones? Which ones are apparent but not relevant?

2. *Consider possible alternatives.* What courses of action could be taken? Which facts support each? What results would occur?

3. *Take action.* Which is the best choice? What others are acceptable? Why? How will this action be initiated?

4. *Check results.* What happened and to whom? Would another solution have been better? Was the solution right but the method of acting on it wrong?

First, students may work on teacher-created problems and then write out their own cases. Students should describe the situation and then apply the four-step analysis, either as a group through teacher-led discussion, through small groups with student leaders, or individually in writing. Occasionally, asking students to write solutions help them have a quiet time to ponder and to consider, without having to feel embarrassed if other students jeer at their spoken thoughts.

PROVIDING EXAMPLES OF CORRELATED GENERAL LEARNING PROJECTS IN THE RELATED CLASS

The immediacy of the job gives rise to many projects for the related class instruction that will demonstrate to trainees the desired relevance to the on-the-job situation. The following examples are illustrative of the areas instructors may consider presenting to their students. It should be noted that most are in the area of general occupational learning, since skill-learning assignments will be specific to the particular job title.

1. How many times each day do you find you must read on your job? (Be sure to count signs, directions, manuals . . . everything you must read!)

2. What kind of arithmetic must you do on your job? Give examples. (*Note:* Be sure to look at your paycheck!)

3. Is there a union where you work? If so, write a short report on what the union does, how many members it has, and what occupational area(s) it covers.

4. Every job has a special language. Keep a list of words (with their definitions) that are used in your field of work.

5. You are the boss. What kind of people would you hire? What kinds of activities would these employees have to perform? Ask your boss to look over your list to determine if your requirements are practical.

6. Will you be able to support yourself or a family on the money you will make in your first permanent job? Will you be able to support your family on the money you will make at top pay? Figure two budgets based on: (1) entry-level pay and (2) top pay scale.

7. What safety and health hazards are connected with your job? What can you do to avoid them?

8. How much does it cost your employer for your coffee breaks? (Figure out the cost for one day, one week, one month, and one year.)

9. What is the arrangement of work space in your working area? How are tools arranged? How are records kept, and where are they kept? What kinds of supplies are kept in the work areas? Draw a diagram of the work area, showing where all work benches, equipment, walking areas, etc. are located.

10. What is the correct way of answering the telephone at your work station (or greeting visitors who enter the area)?

11. If an employee wanted to "check out" a certain record from your working area and go into another, what procedure would he/she have to follow? Is there a "check-out" system? If so, explain what it entails.

12. Are certain kinds of handbooks or reference magazines important in your work assignment? If so, what are they? Why are they considered important? What information do they provide?

13. Is there a special uniform that is required in your working area? If so, prepare a short report describing the uniform and why it is designed as it is.

14. Do you have your own desk or work area? How could you arrange the area so that you could get your work done better, faster, and more efficiently?

DEVELOPING APPRECIATION FOR THE DIGNITY AND VALUE OF WORK

A most important function of related class instruction is to help students who are out in the world of work appreciate the values of work and the satisfaction they can obtain from doing any task well. Students can use this opportunity to recognize that through work they can make a contribution in their own unique way; furthermore, they can relate input to the culture and the welfare of society, to the community, and to the employing agency.

The teacher-coordinator and the job sponsor have frequent opportunities to assist trainees in appreciating the dignity of work. The optimum development of our society depends upon individuals in all walks of life acquiring an appreciation for the concept of work—a feeling that all legitimate work is dignified and respectable. Student-learners have many opportunities to find out that different kinds of work require varying levels of ability and different aptitudes. Every kind of work at every level requires

someone who has mastered the work, if it is to be done most effectively.

USING A TRAINING GUIDE NOTEBOOK, PROJECT MANUAL, OR OCCUPATIONAL PORTFOLIO

The theory of the classroom and the practice of the firm may be welded together if the teacher-coordinator uses a special project manual or some similar form of notebook, workbook, or portfolio to aid students in translating their experiences. This supplementary book or portfolio asks the students to apply theory to the job, it encourages them to analyze job operations, and it provides individual projects that relate to them, their training stations, and their career goals. Student-learners may gain technical information, which the training station through its limitations may not be able to provide. This manual also has the advantage of demonstrating to the job supervisor what each student needs, and it involves the sponsor in actual teaching to help the trainees with their projects. Some coordinators design, write, and produce their own project manuals or portfolio instructions. Some project manuals are produced by state departments, while others are available through commercial sources.

The Related Class Teacher (The Farm-Out System) and Correlation Problems

If the coordinator also has the responsibility for teaching the trainees, problems involved in the correlation of instruction are greatly minimized. The teacher-coordinator is teaching the same trainee visited on the job at the training station. Thus, a go-between has been eliminated. The employer can convey directly to the teacher-coordinator the need for a particular trainee to have extra practice, advanced work, or training in new processes that the trainee will need to know by a given time. Teacher-coordinators do not have to convince the related class instructors of this learning need because in most plans, they are the related class instructors. However, in some plans, the related instruction is assigned to a teacher in an occupational course. Thus, in this *farm-out system,* student-learners are "farmed out" by the coordinator to someone else who will guide their in-school learnings.

Immediately a communications problem arises. The coordinator must communicate the employers' request for trainee instruction to other instructors. This may increase the chance of error and communication breakdown. Nonetheless, correlation of the related instruction with the cooperative experience must be done.

Second, a human relations problem arises. The coordinator must communicate the nature of each trainee's needs and convince the related class instructor to change the sequence of instruction to correspond to these needs. Furthermore, the related class instructor must realize that some learnings not normally in the course outline may have to be introduced for the trainees. The instructor must perceive the importance of the trainees' needing the opportunity to relate job problems, even though most of their classmates are not having an on-the-job experience. The coordinator is asking that the other instructor change a teaching pattern—a most difficult assignment for both parties in many cases. Nevertheless, there are ways of accomplishing this task between two individuals who are adaptable and willing to relate to each other, who see the need to compromise, and who put the students first in classroom planning and management.

OBTAINING INDIVIDUALIZED INSTRUCTION FOR EACH TRAINEE

The ways of providing for the individualized needs of the cooperative student learners in the related class are numerous; the following are a few actual cases that show the possibilities for a creative person—either the teacher-coordinator or the related class instructor.

CASES-IN-POINT

Case No. 1: Trainee X is being trained as a general machine operator. The training agreement states that the employer will have the trainee working on a lathe in a week. The coordinator has checked with the employer and seems satisfied with X's progress. X will indeed be put on the lathe if the 75 hours of classroom instruction needed have been completed. In talks with the instructor, the coordinator finds that even though the instructor is aware of the terms of the agreement, X will not have completed the required number of hours before the next week. The coordinator informs the employer of this fact and makes arrangements for X to be trained in a limited but well-supervised way on the machine at the training station. Meanwhile, the machine shop instructor replaces another student with X in the rotation pattern to give the trainee needed practice.

Case No. 2: Trainee C is stationed as a cashier in a local store. At C's last evaluation, the employer was very unhappy because there were cash shortages. The employer feels that the trainee is honest and that these shortages are caused by mistakes in making change. The coordinator relates the problem to the related class instructor, and C is given intensive work in arithmetic (addition and subtraction practice), counting out change to customers and checking out the register. This practice took precedence over all other classes for the week.

Case No. 3: Trainee B is assigned as a computer operator but has the goal of becoming an administrative assistant. The employer indicates the trainee's keyboarding skills are not yet good enough for B to be assigned these tasks, but the employer will give keyboarding opportunities in a month or so when the trainee is ready. The related instructor continues regular instruction in information processing but arranges to have B given the equivalent of two days a week of practice.

(Continued)

Case No. 4: Trainee M has shown a great weakness in the ability to read. Training as a machine operator and the cooperative job have been making more and more demands on this ability. The employer has complained to the coordinator that the trainee often misunderstands written instructions simply because he/she cannot read. The coordinator confers with the school's reading specialist, and together they decide to use company forms, manuals, and other written items to "spark" the trainee's interest and to improve the trainee's reading skills.

Case No. 5: Trainee W shows an interest in becoming a carpenter and is now employed in a directed cooperative experience as a carpenter's helper. W is having difficulty in mathematics instruction and has not been willing to cope with "textbook study" in order to sharpen math skills. The employer indicates to the coordinator (and to W) that W needs more work in math and helps the coordinator develop a list of examples of computations needed on the job. In the general math class, the instructor agrees to give W additional practice using a programmed text. As W progresses, time is given for W to go to the instructional resources center and work through a multimedia tutorial in the carrel. The instructor also decides to give W extra attention after school one day a week, and the coordinator arranges with the employer for this deviation in W's work schedule.

Feedback from Job to Classroom

The crucial period of adjustment for the typical student-learner is the first five or six weeks on the job. Student-learners must be familiar with the cooperative plan of instruction, must understand the responsibilities they are to assume, and must recognize the place of their teacher and job supervisors in this scheme of activities. Even though the coordinator interviewed each student-learner prior to entry into the program and gave each one at that time information on how the program operates, it is wise to include a two- or three-week unit in related instruction on "School and Business Relationships" as a first unit of the year. For certain students a job visitation early in the school year is a must. The coordinator may find it appropriate at the beginning of the semester to make coordination visitations more frequently than usually scheduled during the school year. Experienced coordinators seem to "get the feel" of when they should visit certain training stations. Prior records of evaluative visitations made in a pocket notebook, on a calendar desk pad, on a state coordination report, or in a job visitation portable index file will assist them in determining the frequency.

Some visitations are routine, while others result from problems reported by students or employers, but the most important after the first six weeks are those made to determine whether satisfactory progress is being made by student-learners. The step-by-step training plan should form the basis for the evaluation and *feedback* process.

A check on the status of this progress benefits the students, the employers, and the plan. If this check reveals that some student-learners have been kept unnecessarily long on certain activities, the teacher-coordinator, the job supervisors, and the employers should review the circumstances to see if beneficial changes can be made. If some student-learners are not performing up to expectations, the probable cause should be determined (see Figure 12–4 for some causal factors). Perhaps the students did not possess the career interest they professed to have at the beginning of the year, or they have discovered through their work at their training stations that the duties and responsibilities are not what they expected or that even after being transferred to new responsibilities, they feel let down. If dissatisfaction occurs on the part of a student-learner or the job supervisor to the extent that the educational objectives of that stu-

STUDENT-LEARNER EMPLOYABILITY RECORD

Name of Student _____

EMPLOYABILITY RATING	COMMENTS

1. Ability to learn the job _____

2. Initiative _____

3. Accuracy _____

4. Dependability _____

5. Attitude _____

6. Courtesy _____

7. Cooperativeness _____

8. Industriousness _____

9. Leadership _____

10. Personal appearance _____

OTHER PERSONAL EMPLOYABILITY NEEDS

1. _____

2. _____

3. _____

4. _____

(over)

FIGURE 12–3a

EVALUATION OF STUDENT

Marking Period	1	2	3	Sem.	4	5	6	Final
Job								
Class								
Days Absent								

EMPLOYER'S COMMENTS

1. _____
2. _____
3. _____
4. _____
5. _____
6. _____
7. _____
8. _____

SCHEDULE OF CLASSES --- TEACHER

1. _____
2. _____
3. _____
4. _____
5. _____
6. _____

AREAS OF TRAINING

1. _____
2. _____
3. _____
4. _____
5. _____
6. _____

STUDENT DATA

School Year _____ Home Room _____

Home Phone _____

Age _____ Average No. of Hrs./Week _____

Employer _____

Bus. Phone _____

Manager or Supervisor _____

STUDENT PHOTO

FIGURE 12–3b

dent, the reputation of the plan, or the successful operation of the business is in jeopardy, a three-way conference should be held between the teacher-coordinator, the job supervisor, and the student. A positive solution is desirable. If that is not possible, the student should be removed from the training station at the most opportune time—perhaps at the end of the semester, to preserve the school credit if the situation will not deteriorate markedly in the estimation of the coordinator.

Employers will appreciate an appraisal of the training stations. On occasion they can make excellent suggestions to the teacher-coordinator and to the guidance counselor on how to make selection and placement more effective. And sometimes the teacher-coordinator needs to remind employers that they may be overlooking a chance to tell the story of opportunities open to student-learners in their businesses. It is unfortunate to find at the end of the year that a graduate of the cooperative plan has been lost to the firm because the employer waited too long to give some individual attention to full-time employment possibilities for the graduate within the business establishment. It is recommended that the coordinator ask the employers to set up specific interviews early in the spring to inform students of their potential future (or lack of it) with the firm. The employer who has an intimate knowledge of an industry can perform a valuable guidance function in helping students determine where their future in the industry lies. After this interview, much of the remaining school year in the classroom can be devoted to individual instruction—remediation or the building of advanced skills.

Measuring the Training Station Experience

Teacher-coordinators of cooperative plans have the task of measuring student performance that is quite unlike that faced by their teaching colleagues.

Their unique evaluation problem is that of measuring performance (achievement) in the occupational laboratory. Not only must they measure progress, because their purpose is to help student-learners advance toward their career objectives, but they must also measure performance in order to determine grades for the student-learners when they are receiving credit for their laboratory experience. Coordinators use several methods for measuring student achievement at the training stations: rating sheets for each grade period, the step-by-step training plan, the training profile, discussions with the job sponsor and observations of the trainees during visitations, and impressions gained in the classroom as the students use (or do not use) job knowledges in classroom discussions and activities.

RATING SHEETS

Almost every coordinator uses rating sheets as one measure of job performance (see Figure 12–5). Unfortunately, all too many coordinators rely solely on this method for grading purposes. In using rating sheets, some coordinators may be troubled by several questions. First, they ask, "Should the report card show a letter grade for each phase of the plan—related class and job—or should it show one grade that reflects both?" Many coordinators report two grades, on the theory that the student is enrolled for two courses. This may be desirable in those situations where the coordinator does not teach the related class. On the other hand, the theory of cooperative education holds that the plan is not one of two separate parts—school and job—but a *total learning experience* consisting of two integrated phases—the school classroom and the occupational laboratory. This calls for one grade reflecting a total achievement. One grade also has the advantage of preventing the image of A's for previously C students.

A second question posed by coordinators is whether or not the employer's rating should be used as the report card grade. It is good practice not to use a rating sheet that requires report card–type letter

REPORTING TRAINEE'S NEED FOR RELATED INSTRUCTION

Date _____

To: _____
(Instructor, Counselor)

From: _____
(Coordinator)

Trainee _____

Job Title _____

Training Station _____

Training Sponsor _____

My recent visitation with the training station sponsor and/or trainee indicates that the trainee needs help with the following.

	Within	
	2 weeks :	4 weeks
1. New job skills or knowledges		
2. Refreshing of old job skills or knowledges		
3. Personal problems, attitudes, or traits		

FIGURE 12–4

RATING FORM FOR THE TRAINEE'S
GENERAL EMPLOYABILITY COMPETENCE*

Trainee _____

Employer _____

Instructions to the Employer: Read each description carefully. For each, place a checkmark on the line over the phrase that describes this trainee most accurately. If you think this individual is halfway between two descriptions, make your mark about halfway between them on that line. Any additional comments you wish to make will be appreciated.

RELIABILITY

Cannot be depended on; requires constant supervision.	Must be reminded of duties; must be carefully supervised.	Satisfactorily performs all assigned duties; requires average supervision.	Is a good, dependable worker; requires little supervision.	Is completely reliable and able to carry on without supervision.

PERSONAL APPEARANCE

Appears slovenly and unkempt.	Often neglects to take care of personal appearance.	Has an acceptable appearance; could make some improvement.	Usually is very careful of appearance.	Always presents an appropriate, well-groomed appearance.

PERSONALITY

Makes a poor impression on others; is inconsiderate.	Is inclined to be indifferent.	Is polite and friendly when approached by others.	Is courteous in dealing with others; has a positive attitude.	Makes a favorable impression on all contacts.

COOPERATION

Is hostile towards others; does not behave as member of a group.	Is a "lone wolf"; works alone and shuns others.	Will cooperate when asked but does not volunteer.	Is willing to cooperate.	Is always very cooperative; has the knack of helping others.

FIGURE 12–5

RATING FORM FOR THE TRAINEE'S
GENERAL EMPLOYABILITY COMPETENCE* (Continued)

ATTITUDE TOWARD WORK

Seems to resent the work; has no desire to learn.	Is willing to work but shows no interest or enthusiasm for job.	Seems to enjoy the work but is willing to "stand still" and not advance.	Shows interest in the work and has a desire to learn.	Shows a keen interest in the work and often takes the initiative to learn.

JOB SKILLS

Has a definite lack of job skills and knowledge.	Has limited knowledge; is lacking in some essentials.	Has an acceptable knowledge of routine and job skills.	Has an above-average grasp of the essential skills.	Possesses all the essential job skills and knowledges.

WORK HABITS

Has to be told several times before doing work.	Has poor work habits and is at times neglectful.	Follows instruction but sees no more to do.	Works efficiently; does more than is required.	Works rapidly and efficiently; is resourceful and finds extra things to do.

What are this student's strongest points? _____

_____

What weaknesses need to be corrected? _____

_____

Comments: _____

_____

_____

_____

Employer–Trainer _____ Date _____

_____

*This rating sheet is general and should be used in conjunction with the step-by-step training plan, which is a measure of specific job skills and knowledges. Chapters 17 thru 22 give examples for each of these specialized areas.

FIGURE 12–5 (Continued)

grades; instead, coordinators should use a rating sheet that describes performance in the language of the trade. The report card grade then becomes a matter of combining the message of the rating sheet, observations, and discussions with job sponsors.

There are many reasons to support this practice. First, employers have varying standards: an A to one is a C to another. Second, many employers tend to overrate student-learners rather than risk being "too critical" on paper. Third, there is the temptation to overemphasize personal qualities and minimize actual job skills and abilities on the rating sheet. Rating sheets usually are not mailed back to the school but are discussed personally with the coordinator. The last question raised by coordinators regarding rating sheets is whether or not they should be shown to the student-learners and their parents. Keeping parents informed is desirable, but coordinators should keep in mind that the ratings are made by employers who may not wish parents to see critical comments. Coordinators know the training stations and can interpret the ratings, while parents might misinterpret. However, parents should be informed of progress, coordinators might prepare brief written evaluations for the parents at the end of each semester.

In summary, evaluation of job progress should be made frequently, and students should be involved in the process so that they can discuss strengths and weaknesses and commit themselves to improvement.

OTHER METHODS OF EVALUATION ON THE JOB

While the rating sheet is commonly used for evaluation, it needs to be supplemented by some of the following:

1. **Observation**—On-the-spot observation of attitudes, skills, and knowledges, through visits to the training stations by the teacher-coordinator, is essential.

2. **Self-evaluation by the student-learner**—Rating scales and self-rating sheets can assist the trainee in self-evaluation.

3. **Job sponsors**—In addition to checking the rating sheet, many job sponsors are quite willing to assist in the evaluation of job manuals or merchandise manuals, if procedures are suggested by the teacher-coordinator.

4. **Co-workers**—With encouragement, student-learners may enlist the assistance of other employees to evaluate their work and tell them how to perform better.

5. **Production standards**—Of course, there are realistic measures based upon production, such as dollars of sales in the retail store, units of production in the industrial establishment, and pieces of correspondence produced or the number of items processed or filed in the office.

Improving Instruction at the Training Stations

In an on-going plan there is all too frequently an assumption by a busy coordinator that the training station personnel know their responsibilities for on-the-job instruction and will carry them out efficiently. Unfortunately, employers and their employees are busy people also and, even with the best of intentions, may neglect their responsibilities as the "downtown faculty." Improving instruction at training stations is a continuous task for coordinators, particularly as new training stations are added, as personnel change in established training stations, and as individual student-learners grow and develop at different rates.

EVALUATING TRAINING STATIONS

It is advisable to review existing training stations annually and to establish new training stations. The composition of the group of student-learners in a given plan will vary from year to year, necessitating perhaps a change in the variety of training stations available. A certain training station that has been used may no longer be available for any number of reasons. Business conditions may have changed so that keeping a student-learner can no longer be justified, or the firm may no longer exist. In certain instances, the job supervisor moves, and a suitable person to handle student-learners cannot be found.

Every training station should be reviewed at the end of the school year. The training plan that was established at the beginning of the year should be compared to the end-of-the-year report on what actually transpired in the areas of training and experience for the student-learner. The list of duties and responsibilities actually provided for the student during the school year becomes a profile of the training station. The coordinator should return to the manager or job supervisor and say, "Here is a picture of the opportunities for learning that you provided our student-learner last year. This was a fine effort. If similar experiences can be provided during the coming year, we would like to cooperate with you by sending you appropriate potential trainees for interview." Or, in some instances the teacher-coordinator may have to return and say, "Here is a picture of the learning experiences that you provided our student-learner last year. This narrow profile needs to be adjusted to a more representative training situation, and I hope you will help use this situation as an educational experience for a student next year." The advisory committee should be consulted when results obtained in the various training stations are being reviewed. The committee members can be helpful in suggesting ways of enhancing the educational experiences of students in various training stations. They can suggest new contacts for the purpose of establishing new training stations.

CARRYING ON A SPONSOR DEVELOPMENT PROGRAM

The business industry will directly benefit from efforts put forth to develop the training abilities of the job sponsors. There are many times when job sponsors are too busy to pay attention to potential learning situations. At other time, however, learning does not take place because the persons under whom the student-learners are working do not know how to relay their work knowledge to the student-learners. Too little is being done by teacher-coordinators to assist job sponsors or supervisors to know how to aid in the training of student-learners. For the purpose of explaining the learning process in simple terms, a meeting called at the beginning of the school year for job sponsors will prove effective. Tactful suggestions by the teacher-coordinator to the job sponsor during coordination visits will be helpful to all concerned. Excellent materials on "how to train" and "how to supervise," useful in working with adults on coordination visits or in formal classroom adult education, are available to the coordinator.

REPORTING ON INSTRUCTIONAL ACTIVITIES TO COOPERATING EMPLOYERS

Some coordinators provide job sponsors, employers, and advisory committee members with a midterm report of instructional activities. This helps members of the "downtown faculty" recognize some of the elements of instruction being carried out in the classroom and helps to enhance their feeling of being an integral part of the training plan (see Figure 12–6). If student study guides, manuals, notebooks, or portfolios are prepared, job supervisors may be involved as resource persons.

SECOND SEMESTER REPORT TO TRAINING STATIONS

I. *What we have done in marketing education in class from January to date:*

   A. Studied textbook: *Retailing Principles and Practices,* 7th ed., by Meyer, Harris, Kohns. (Many store organizations contributed to the materials.)

      Topics covered:

      1. Careers in Retail Marketing

      2. Considering Economic Factors

      3. Enhancing Your Customer Relations

      4. Helping Customers Buy Wisely

      5. Promotion Methods

      6. Looking Ahead

      7. Organizing Your Business Operations

      8. Looking Ahead

   B. Emphasized building a retailing vocabulary.

   C. Reviewed store arithmetic.

   D. Reviewed misspelled words.

   E. Had lectures and discussions on human relations and personal attitudes based on pertinent books and booklets, experiences, and supervisors' comments.

   F. Had each student-learner complete a weekly production sheet on hours, wages, and duties on the job. Student-learner maintains a current and a permanent file containing job sheets, film summaries, manuals, examinations, and supervisor ratings.

   G. Completed project — store manual (including job analysis and merchandise study) by each student, first in rough draft and then in finished form. This project is for open house and *for examination by the job supervisor or sponsor.*

   H. Viewed films. (Student-learner summarizes certain films. Summary is checked for observation, spelling, and grammar.)

II. *What we plan to do in marketing education in class to the end of the school year:*

   A. Study textbook: *Effective Selling,* 8th ed., by Hair, Jr., Netturno, and Russ.

      Topics to be covered:

      1. The Role of Personal Selling

      2. The Job of Personal Selling Traits and Tasks

      3. Buying Behavior

      4. The Communication system

      5. Knowledge Necessary for Successful Selling

FIGURE 12–6. (Adapted from a report developed by Dr. Hebert L. Ross)

SECOND SEMESTER REPORT TO TRAINING STATIONS (Continued)

       6. Promotional Sales Support

       7. Securing and Opening the Sales Call—The Approach

       8. Making the Sales Presentation

       9. The Demonstration

      10. Answering Buyers' Objections

      11. Closing the Sale

      12. Providing Service and Goodwill After the Sale

  B. Continue vocabulary, arithmetic, and spelling review.

  C. Continue discussion on human relations and personal attitudes.

  D. View films.

  E. Complete activities.

       1. Class club meetings based on parliamentary procedure.

       2. Display to be changed weekly. Student secures material from training station.

       3. Trip to St. Louis (guided tour of Famous-Barr Co.).

       4. Trip to Leadership Training Conference.

       5. Written project on advertising.

       6. Bulletin board display.

       7. Tape recording of public relations activities presented on radio.

       8. Speakers: Former ME students now working downtown, a present job supervisor.

       9. Tentative tours: dress factory (re: textiles), stores (re: display), newspaper plan (re: display advertising).

      10. Employer–employee appreciation banquet.

III. *Suggestions for job supervisors:*

  A. Rechallenge student-learners with new and varied responsibilities where potential is evident.

  B. Urge student-learners to make the most of remaining school time, keeping their academic subjects in good order.

FIGURE 12–6 (Continued)

Questions and Activities

1. Explain in terms of the theory of the learning process why the correlation process is fundamental to the cooperative plan of instruction and why the work experience plan suffers from this lack of correlation.

2. Describe students in at least three levels of cooperative plans, and then analyze their competencies in planning and managing their own learning situations. Indicate how the teacher-coordinator might assist each type of student in becoming self-sufficient in learning to learn.

3. Write up on the form of a classroom handout three problems that occur on the job. Also provide a set of suggestions that will help the teacher make effective use of these problems.

4. Describe a student-learner's career objective, present on-the-job situation, and background of readiness. Then, present a series of learning projects for the student-learner that would be relevant and correlated with the occupational experience.

5. Student-learners usually resist the filing of the weekly training report form. They say they "forgot." List some ways of motivating them to turn these in; also describe what you would do as a teacher-coordinator if they "forget."

6. Describe a student-learner's current training assignment, career objective, and readiness and style of learning. For this student-learner, list the specific individual learning activities that you believe would be profitable and explain why.

7. For a given occupational education area using the cooperative plan of instruction, would you recommend or deny the use of the "farm-out" method of instruction. Why?

8. Regarding the use of the grade on the report card as being one for the entire experience or one for each part (school and job), which can you justify? Give reasons in a memo to other departmental faculty as well as to involved administrators.

9. Which is more important—a rating of the skills performance on the job or the personal qualities of the employee? Explain your answer.

10. Describe a sponsor development program by showing the goals, the structure of the program, the timing, and the activities. Who would be engaged in this program as participants and as instructors? Who would provide the financial resources?

11. Describe how the coordinator can help the trainee and the job sponsor be aware of the need for correlation; also describe how they can be involved in the process to make it effective.

For references pertinent to the subject matter of this chapter, see Reading Resources.

13

The Maturing of the Cooperative Plan

A graduate of a cooperative plan gets some advice from his supervisor before being sent on to management training.

KEY CONCEPT: Teacher-coordinators must assess, evaluate, and cultivate cooperative education in order to maintain and improve its quality.

GOALS

After successfully completing the study of this chapter, answering the questions, and carrying out the activities, the student should be able to:

- Identify those maintenance activities that are necessary to preserve a quality plan.

- Identify those activities that are necessary to improve a cooperative plan.

- Plan a schedule of activities within a realistic time frame.

- Justify an extended contract to accomplish the demands of a local cooperative plan.

- Explain why complete and accurate records and reports are important.

KEY TERMS

alumni group

employer–employee appreciation events

extended contract

public relations programs

recurring activities

The cooperative plan, like every other plan of instruction, should grow and develop to its potential. A carefully planned initiation is required, as shown in Chapter 6, and the basic elements of the system must be attended. Evaluation shows how well the basic plan functions and the point at which the improvement activities commence. Students in graduate courses and local administrators at conferences and workshops have often asked, "How long does a program (plan) take to mature to its potential quality?" The cooperative plan, because of its intricate set of relationships both within the school and between the school and the community, requires a period of maturation. Teacher-coordinators should be pleased if the plan is of high quality three years after its initiation. This does not mean they have to subscribe to an initial low quality of performance, but it is a recognition of the time demanded to educate all parties in the plan about their responsibilities. The number of people to be dealt with are many, their values and educational perceptions are varied, and the communications problem is immense. The goal of the plan at the beginning should be *quality* in the basic elements with emphasis given later on to *improving* the desirable conditions.

Establishing Maintenance and Improvement Goals

After a cooperative occupational education plan has been instituted and has been operational for a year or two, teacher-coordinators may discover that a new set of activities requires attention. Just as "green thumb" gardeners pay attention to the maintenance of their efforts, so must teacher-coordinators if their original efforts are to blossom into quality plans that grow and yield extra quality. After the hectic and sometimes emergency nature of efforts in new plans, teacher-coordinators design a set of actions to improve the plans and to clear up deficiencies. Some of these activities require attention but once a year,

while some demand monthly or even weekly attention. What is important is that the plan continue to grow towards excellence with even the minor details attended to. If this is done, the weak plan of the first year will flower, and the strong plan will not decline.

The importance of a continuous plan of *maintenance* as well as *improvement* activities cannot be overemphasized. An educational effort that involves the time and energy of so many groups within and without the school institution can never be static. A plan moves forward, or it begins to slide downhill. It is never resting at an equilibrium; rather it is dynamic and re-active to various groups and their moods and situational changes. Educators know well that unless the effort is given proper maintenance attention, its effect diminishes. Almost every supervisor and teacher-educator can describe case histories of successful cooperative plans that were allowed to degenerate into ineffective learning situations when teacher-coordinators began to rest on their laurels.

All the activities discussed in this chapter might well have been included in previous chapters as essentials for getting started. But no coordinator can do everything the first year; therefore, certain activities were omitted earlier because they were seen as those that could help the plan reach its potential.

Scheduling Improvement Activities Throughout the Year

There are some actions by teacher-coordinators that are connected with the entire year; others have their maximum effectiveness at certain times during the year. Thus, these maintenance and improvement activities are listed in the following paragraphs according to categorization of time. However, teacher- coordinators should recognize that each action is not in every local case the product of time— each is more the product of the on-going situation and should be judged for its cost-benefits. And each should be seen as a generalization that provides a useful classification system for reading and study as well as for the making of a local plan of action.

ACTIVITIES DURING SUMMER MONTHS

The year-long contract is particularly essential as educational institutions change their approaches to service to the community. Some school districts feel that summer school programs are in a sense a "fifth quarter" and that they contribute to the over-all enrichment of the districts' offerings. In some cases, coordinators have the responsibility for adult training and development and need the summer to plan and to promote. But, even more decisive in the argument for the 12-month employment of coordinators is the tendency of districts to offer, in the summer, programs of occupational exploration in which the coordinators with their community contacts serve as admirable resources. In short, districts profit from perceiving coordinators as their community resources for the full year and recognize that their absence in the summer tends to allow the community to see the district as being less than a full-time partner in the local situation.

Some of the major *recurring activities* in which coordinators should engage during the summer months of the *extended contract* are as follows:

Month	Suggested Activities
June–July:	Prepare annual reports for local administration, advisory committee, and state department of education.
	Check on ensuing year's trainees placed in summer jobs.
	Develop the follow-up plan and instrumentation.
	Interview last-minute trainee applicants.
	Improve professional qualifications through additional graduate work, occupational experience, and self-study.
	Teach an exploratory program during the summer term.
	Review and upgrade career training plans for fall students.
	Improve teaching outlines.

Month	Suggested Activities
August:	Attend teacher-coordinators' conference and workshops.
	Recheck training stations with which contacts were made in the spring.
	Establish additional training stations, if needed.
	Start student-learner interviews and placements.
	Check new instructional materials and physical arrangements.

ACTIVITIES RECURRING DURING THE FIRST PART OF THE SCHOOL YEAR

A monthly calendar of activities beginning with the opening of the school term might look like this:

Month	Suggested Activities
September:	Arrange classroom for related instruction.
	Hold orientation meeting for student-learners.
	Start related class instruction.
	Complete student-learner placements.
	Hold advisory committee meeting; present progress report.
	Initiate the student organization program.
October:	Participate in National Careers in _____ Week activities.
	Participate in area student organization activities.
	Visit training stations—plan purposeful visits.
November:	Recognize specific instructional needs of student-learners.
	Speak about cooperative plan before service/civic clubs.
	Participate in National Education Week activities—all groups.
December:	Take first steps toward career objectives.
	Attend convention of the Association for Career and Technical Education.
January:	Give progress report to cooperating employers.

ACTIVITIES RECURRING DURING THE SPRING SEMESTER

In on-going cooperative education plans, teacher-coordinators carry on activities in the spring that are preparatory to handling the new group of student-learners for the coming school year. Some of these are similar to those spring activities described for teacher-coordinators of newly established plans.

Month	Suggested Activities
February:	Training station visitation—check on effectiveness of student-learner selection and placement; encourage counselors to make some visitations with you.
	Check progress of step-by-step training plans.
	Plan employer–employee appreciation banquet.
	Participate in Association for Career and Technical Education activities—all groups.
March:	Disseminate student guidance information.
	Prepare news releases on cooperative plan.
	Visit homerooms and assemblies.
	Arrange for supplementary testing of prospective student-learners, if needed.
	Interview prospective student-learners.
	Assist in class scheduling.
April	Review physical facility needs for coming year.
or May:	Teach a short course on orientation to cooperative education.
	Hold employer–employee appreciation banquet.
	Organize alumni group for the coming year.
	Evaluate adult education needs.
	Follow up last year's graduates.
	Plan National Education Week activities.
	Plan Career and Technical Education Week activities.

Using Advisory Committees to Advantage

All plans utilizing the work environment to furnish CTE experiences need advisory committees to help provide direction and make the best use of resources. In a small community, one advisory committee might serve the business, marketing, and trade and industrial cooperative plans as a joint committee.

There should be a separate committee for each of the occupational plans in a medium-sized city.

Advisory committees may meet as often as once a month if worthwhile projects or activities are planned and are available. The committees will function efficiently if each member has a definite part in planning the over-all operations of the plan. Such activities might include:

1. Speaking for cooperative education plans as a resource person in class, on student organization programs, or at employer–employee appreciation dinners.

2. Helping to obtain proper training stations.

3. Suggesting and helping to secure classroom materials and equipment.

4. Making arrangements for students interested in certain specialties to gain additional insight into those activities.

5. Investigating sources of trade magazines and other trade publications and acquiring these publications.

6. Acquainting coordinators with influential employers and labor leaders in the community and making arrangements for coordinators to talk to various trade groups.

PROMOTIONAL EFFORTS THROUGH ADVISORY COMMITTEES

One of the ways of making advisory committees meaningful for on-going programs is to use the collective wisdom and talent of their membership in a continuing public relations program. Advisory committee members can be involved in activities such as:

1. Developing brochures on employment opportunities that may be used with students, parents, and counselors.

2. Assisting in the school's activities in connection with National Education Week, etc.

3. Making speeches at meetings of the members' trade and professional association.

4. Publicizing the cooperative plan in internal publications of the members' firms.

5. Appearing on radio and television programs with students and coordinators.

Improving Relationships with Other Agencies in the Community

Directors of work experience programs or cooperative occupational education plans within the school institutions must relate to their students as individuals rather than as nonentities who are enrolled for just one period in a given classroom. If they recognize their responsibilities to their students, coordinators know that each student is an individual and each may have need for the services of community agencies beyond the walls of the institution. In short, coordinators realize that students are "whole" persons and that they will or will not learn, depending upon not only their classroom experience and that of their training stations but also their physical and psychological readiness to do so.

THE COORDINATOR'S RELATIONSHIP WITH UNIONS

Union support of the cooperative plan is just as vital as is the support of business and industry, and in some cases, it may be even more important. It is possible to have an industry's support but not be able to implement the plan because its union is not willing to participate. On the other hand, a supportive union may convince a hesitant employer to take part in the plan. Almost every city of size has a joint labor council composed of representatives from all key unions. The support of these group members is essential because of the assistance they can lend in individual companies. If the council is seen as the main focus and source of support, the coordinator should be able

to branch out to union representatives for various trades and hence to those within each company. Operating hand-in-hand with labor and management are the apprenticeship committees. Therefore, communication with the joint apprenticeship committees and with apprenticeship coordinators within the educational institutions in the community is important.

THE COORDINATOR'S RELATIONSHIP WITH THE COMMUNITY

Community ignorance of the school's occupational and career development plan can be the coordinator's biggest hurdle. Most people oppose that which they do not understand, or at best, they are apathetic. Therefore, coordinators must involve many segments of the community in a continuing plan to gain understanding, even if not outright support. The plan must be shown to be a success, because when people are involved in something they view as successful, they become its legitimizers. In effect, they are proud and vocal about their part in something that is a community-recognized success.

Community relationships are not permanent. In most areas, the mobility of people is substantial, and those convinced today move elsewhere to be replaced by others who must be informed and then convinced. Coordinators should constantly remind themselves of the very ephemeral value of impromptu efforts. These often are guised in a moment of goodwill, when everyone is supportive of the school and is dedicated to what is thought to be right. But such moments are transitory and die from the interference of other needs or the later preoccupation with more immediate concerns. The tendency of some coordinators to bask in what is at best a temporary glow of success must be guarded against in their community contacts. A planned schedule of public relations activities is more effective than impromptu attempts to publicize the plan. Publicity should be a continuous process, with the coordinator using a variety of media such as the following:

1. Local newspaper articles on training activities, plan information, and success stories of trainees.

2. Flyers, brochures, and letters directed to trainees and employers.

3. Presentations by trainees (in their sometimes colloquial style) before trainee groups, service clubs, and employer–employee appreciation groups.

4. Career clinics for trainees conducted by cooperative trainees and participating employers.

5. Employer visits to the institutional training site and field trips by instructors to business and industry locations.

6. Displays and exhibits of trainees' work and activities in the institution and in the community.

7. Personal contacts by coordinators with individuals who have interests and concerns related to the plan.

8. Radio and television appearances by cooperative trainees, employers, and coordinators.

9. Seminars and short training sessions conducted by coordinators for people in business and industry. These could be "How to Train" and "How to Supervise."

Participation by community leaders on advisory committees and representatives of social agencies in the development of cooperative training stations is very important to the success of the human resource cooperative plan.

THE COORDINATOR'S RELATIONSHIP WITH THE STATE EMPLOYMENT SERVICE

As trainees enter the final phase of their individual plans—permanent job placement—the coordinators' responsibilities may overlap those of the state employment service (SES) because placement begins with job development.

Because of the importance of placement and follow-up, a cooperative working relationship must exist between the SES and the coordinators. The coordinators can act as field representatives of active job development, and the SES can relay job orders to the coordinators to be used. The SES can also canvass employers for job needs (usually by telephone) and thus provide more job resources for coordinators. Coordinators should report all referrals and results to the SES for its records.

Such a cooperative arrangement will help the coordinators, the SES, the employers, and the trainees.

THE COORDINATOR'S RELATIONSHIP WITH OTHER AGENCIES

All cooperative occupational education plans are designed to meet the personal and occupational needs of students. Therefore, it is necessary for coordinators to secure the cooperation of other already existing agencies that can help the trainees meet their immediate needs (see Figure 13–1).

Coordinators will need to draw upon many resources in order to make their trainees more employable, to provide them with counseling, and to help some with medical and other resources. Students must be treated equally if they are to profit from a cooperative occupational education plan.

Reporting and Accountability

The development of useful records and reports is recognized by effective coordinators as a very important activity. The preparation of useful reports is vital and should not be considered as wasted effort. Effective coordinators recognize early the need for records that reflect an accurate image of the plan, allowing evaluation, back up to requests, and demonstration of effectiveness. A successful coordinator recognizes and understands the administration's need to justify every aspect of the curriculum and to give fiscal evidence to the local board, to the state department of public instruction, and to any other agency that funds the plan. Public relations built upon facts about

AGENCY RESOURCE RECORD

Agency _____ Contact Person _____

Address _____
　　　　　　Street　　　　　　City　　　　　State　　Zip Code

Telephone _____

Request Procedure: _____

_____

_____

_____

_____

_____

TYPES OF SERVICES AVAILABLE

___ 1. _____	___ 8. _____
___ 2. _____	___ 9. _____
___ 3. _____	___ 10. _____
___ 4. _____	___ 11. _____
___ 5. _____	___ 12. _____
___ 6. _____	___ 13. _____
___ 7. _____	___ 14. _____

FIGURE 13–1

the local plan will be more effective than unsupported opinion.

Administrators need to justify every part of their curricula in terms of its contribution to the educational program of their schools in proportion to the budget allocated. Administrators are fiscally responsible to their local boards, to the state department of public instruction, and to any other agency that funds their programs.

A coordinator typically makes reports to two groups: the local administration and, in the case of reimbursed programs, the state department of public instruction.

STATE AND LOCAL REPORTS

Those coordinators who feel that state reports are bothersome probably do not recognize that state government must justify, by evidence, the expenditures made of public funds. They also may not perceive that administrative staff members need the information necessary to fulfill their roles as consultants, providing leadership in program development and engaging in research.

The basic reports required by most states are discussed in Chapter 5.

Continuing Public Relations Programs

Because of the importance of continuing *public relations programs,* coordinators should be aware of the following publicity techniques that can lead to good public relations. Public relations do not begin and end with a specific activity. Public relations are constantly present in every organization and in every individual. Figure 13–2 lists the many activities from which an appropriate public relations program can be chosen.

Competition for the attention of the public is extremely keen. In order for any organization to tell its story to the public through public information media, its efforts must be well planned and executed. Public information media with which coordinators will wish to work include newspapers and radio and television stations. Editors and program directors have the responsibility of bringing news to the people of their area. They must evaluate information made available to them and then utilize that which is of greatest importance to their readers and listeners. Thus, in public information activities, coordinators are in competition with national and international news agencies, as well as with other community organizations and individuals. When information submitted is published or broadcast, it is not a matter of the editor or the program director granting a favor.

Newspapers will be the means for telling the story much more often than any other public information medium. One experienced businessperson suggests six points to keep in mind to make the most of using newspapers for publicity purposes.[1] These are:

1. Become acquainted with the editor or whoever handles the material submitted.

2. Submit only *timely* information. Something that happened a few weeks ago cannot be considered news.

3. Always emphasize the significant and unique aspects of the event or the person involved.

4. Learn and then conscientiously observe newspaper deadlines.

5. Be brief in writing a news story. The reader's time, as well as newspaper space, is valuable.

6. Don't editorialize. The editorial page expresses the editor's personal opinion. News stories consist only of news.

Pictures increase interest and readership considerably. Whether a coordinator takes the pictures or arranges for the pictures with a studio or the newspaper staff, glossy prints should always be used. Student reporters, in many cases, will have full responsibility for working with the school newspaper staff. The principal, faculty advisor, and student editors should be consulted. Public relations represent a two-way street. In order to achieve continuing effective public relations with other groups and individuals, coordinators must express interest in and appreciation of their activities. Representatives of other organizations should be given opportunities to appear in the classroom and at student organization functions. These representatives include administrators and counselors, as well as persons outside the school. Besides contributing to total public relations effectiveness, such activities will provide students with a better understanding of their community and the people and organizations that make it a worthwhile place in which to live and work.

PLANNING AN EMPLOYER–EMPLOYEE APPRECIATION EVENT

Plans for an employer–employee appreciation activity to be given near the end of the school year need to be made early in the new calendar year. A theme must be selected, major elements of the plan decided, the budget determined, and the physical facilities arranged. Student-learners should be involved in the planning from the start to capitalize on desirable educational opportunities—it is their

---

[1] *Public Relations: What's It All About?* by William E. Ramsey, Director of Public Relations, Boys Town, Nebraska, is an excellent reference.

Advance Notices About Coming Special Events

Announcements About Co-op in School and District-wide Bulletins

Arrangement for __ to Issue Co-op News Release

Arranging for __ to Serve on Co-op Advisory Committee

Attendance of Coordinator at Affairs for __

Attendance of Students at Affairs Sponsored by __

Bulletin About Pre-Christmas Training Sent to __

Bulletin Board Displays at __ Meetings

Bulletin Board Displays in Local Stores

Bulletin Board Displays in the School

Contacting __ by Student Public Relations Committee

Contributions About Co-op in School Newspaper

Displays of Photographs of Co-op Activities

Distribution of Bulletins About Co-op Courses

Distribution of Career Information Literature

Display of Co-op Publicity Scrapbooks

Distribution of Co-op Activity News Bulletin

Distribution of Course Enrollment Statistics

Distribution of Descriptive Brochures

Exhibits at PTA, Teacher and Trade Conventions

Exhibit (Posters, Charts, etc.) About Program

Follow-Up News Releases on Special Events

Help of Co-op Students in School Project

Help of Students in Organization's Community Project

Invitation as Guest or Sponsor at Co-op Banquet

Invitation of __ to Accompany Co-op Field Trip

Invitation of __ to Speak in the Co-op Class

Invitation of __ to Sponsor a Co-op Contest

Invitation to Attend or Speak at Career Assembly

Invitation to Attend Showing of Training Films

Invitation to Be Judge at a Co-op Contest

Invitation to Participate in Workshops, etc.

Invitation to Showing of Career Movies

Invitation to Visit Co-op Classes

Invitation to Visit Laboratory Shows

Invitation to Visit or Speak at Co-op Organization

Mention of __ in News Releases

Newspaper News Items, Editorials, Feature Stories

News Release on Big Doings Like "Career Day"

News Release on Follow-Up Study of Graduates

News Releases About Radio Participations

News Releases About Television Participations

News Releases to Publications Sponsored by __

Offering to Help __ Set Up or Publicize a Plan

Participation by __ in "Career Day"

Participation in Contests Sponsored by __

Participation in Curricular Events

Participation of __ in Selecting Co-op Students

Periodic Report on Status of Plan

Personal Conferences with __ by the Coordinator

Presentation of Co-op Assembly Program

Prompt Dispatch of Thank-You's by Class

Prompt Dispatch of Thank-You's by Coordinator

Publication of Articles in Media Read by __

Radio (Spot Announcements, Quizzes, Interviews)

Report of Local Job Openings at Christmastime

Report on Community Adult Training Needs

Representation of Co-op in School Yearbook

Sponsor "Co-op Night" at School; Invite __

Sponsorship of Local Stores

Survey of Christmas Employment Needs

Survey of Community Adult Training Needs

Survey of Public Reaction to Co-op Service

Survey of Public Reaction to Local Industries

Talks by Graduates When __ Are Present

Talks Presented by Student with __ Present

Television Announcements, Interviews, etc.

Window or Counter Display About Program

FIGURE 13–2. Possible activities for a public relations program.

event. The activity is a sound public relations device, but all activities and committee assignments should be student-oriented. Some communities prefer a large banquet that includes participants from all the cooperative plans in the school district. Other schools prefer separate employer–employee event for each school and each occupational plan, to preserve the intimate relationship of a homogeneous business group. In some communities, the students finance the employer–employee appreciation activity with funds raised through student organization activities, from dues, or by the purchase of guest tickets out of their current earnings. An *employer–employee appreciation event* is an excellent opportunity for students to say thanks to their job supervisors and employers. In other communities, the employers, a trade association, or a business association underwrites the cost of the activity. However, the students will probably benefit more if they are responsible for the event themselves.

Traditionally, the employer–employee appreciation activity was a banquet. But, many schools today in addition to, or in lieu of, employer–employee appreciation banquets, plans other social affairs and recognition events. As examples, some schools hold a recognition assembly for all cooperative plans in the school auditorium during May. In addition to appreciation speeches, skits, and other forms of entertainment, cooperative student-learners are presented, by the superintendent, with achievement certificates that record the training completed. Other schools plan afternoon teas or luncheons or breakfast meetings to honor employers or to bring parents and employers together. The point is to plan an activity that will get the largest number of employers together to relate to each other and to the trainees.

ORGANIZING ALUMNI GROUPS

In several cooperative plans across the nation, *alumni groups* reach graduates of 20 years ago or more. Some states have celebrated silver anniversaries of the establishment of cooperative education

and have used these opportunities for publicity. The organization of an alumni group will help the teacher-coordinator see for the first time the importance of the cooperative plan being career-oriented. Those students who entered and participated in a plan with career objectives in mind will be those who now form the nucleus for alumni chapters.

Attendance at an organizational meeting for an alumni chapter may be small, but enthusiastic work by a corps of dedicated graduates, coupled with the publication of an alumni newsletter, will locate many others who are eager to participate. Projects considered by alumni groups might include: activities to promote cooperative education in the local community, professional improvement through sponsoring adult training and development course, service on or assistance to the local advisory committee, assistance to local or state foundations, and research projects. In fact, alumni groups can furnish the enrollment for certain adult training and development classes.

Questions and Activities

1. How do the activities of coordinators of on-going plans differ from those of coordinators preparing for new plans?

2. Explain why an effective public relations program is a continuing program of activities.

3. What contacts in the school and in the community are important for coordinators in continuing public relations programs? Illustrate by preparing a list of publics that must be informed.

4. How might a coordinator proceed to organize a local alumni chapter?

5. As the result of studying appropriate sources of information, discuss some informal research projects that could be carried on by the coordinator and/or students and/or members of the advisory committee.

6. Assume you are the direct administrator for a curriculum using the cooperative plan of instruction. The use of the plan is new in your institution. What goals would you describe as minimal after the first year of operation? After three years? After five years?

7. Describe a case that used the cooperative plan of instruction. For this case make a list of the criteria needed for selection of the advisory committee members. Then, prepare (1) agendas for the meetings for the year and (2) a list of activities in which you would recommend the advisory committee be engaged.

8. Using the problem in number 7, list the local agencies that could provide supportive services to your student-learners. Also discuss the services that the SES can provide to your group and the relationships that might exist with local union organizations.

9. Visit an on-going institutional program that uses the work environment as a plan of instruction. Write in a role-playing situation—a descriptive report of the program that you would send to the administration.

10. For a given situation, which is preferred: a large employer–employee appreciation banquet, which is professionally done and covers all occupational curricula, or an intimate occasion, which is planned and operated by students for their own specialized occupational group? Explain your reasoning.

11. Assume the situation in number 7. Make up a plan for an informational program that communicates to all publics the essence of what is going on and what needs to be done as improvements.

For references pertinent to the subject matter of this chapter, see Reading Resources.

14

Student Organizations as an Integral Part of Instruction

The teacher-coordinator helps her students get ready for a fund raiser to support youth group activities.

KEY CONCEPT: Student organizations are an essential component of the cooperative plan.

GOALS

After successfully completing the study of this chapter, answering the questions, and carrying out the activities, the student should be able to:

- Explain how national career-technical student organizations (CTSO's) operate.

- Identify the general objectives that are common to all CTSO's.

- Integrate CTSO activities into the instructional program.

- Operate and manage a successful CTSO.

- Evaluate the program of student activities.

KEY TERMS

Business Professionals of America (BPA)

career-technical student organization (CTSO)

chapter work/activities

competitive events/activities

DECA—An Association of Marketing Students

Family, Career, and Community Leaders of America (FCCLA)

Future Business Leaders of America–Phi Beta Lambda (FBLA-PBL)

Health Occupations Students of America (HOSA)

National FFA Organization (FFA)

SkillsUSA–VICA

Technology Student Association (TSA)

Philosophically, *career-technical student organizations (CTSO's)* are co-curricular—an integral part of the instructional program. The activities of student organizations have long been recognized as being viable and directly related to the goals of the instructional program. But student membership in CTSO's represents only a small percentage of the total number of students enrolled in CTE instructional programs.

Furthermore, membership in all but the newer CTSO's has stalled or fallen from peak periods in the 1980's. Fewer students ("baby burst" generation) and fewer electives are believed to be the primary reasons for the decline in membership. On the other hand, why are there many students enrolled in CTE programs who are not members of the appropriate career-technical student organization? The reasons are probably numerous, but surely among them are: (1) no one has encouraged students to participate; (2) local chapters are not available for students to join; (3) students believe they cannot afford membership.

The U.S. Department of Education, the Association for Career and Technical Education, and Congress, in the Vocational Education Amendments of 1976 and in the Perkins Acts of 1984, 1990, and 1998, have all recognized career-technical student organizations as being an essential and integral part of the instructional program. Each has provided some leadership in promoting the development of career-technical student organizations. Active membership in CTSO's teaches human relations, leadership, and occupational skills.

A number of student organizations are connected with CTE, occupational exploration, and practical arts. These organizations serve students' learning under the cooperative plan, as well as those learning in other career-technical classes. Teacher-coordinators need to understand the various organizations in the educational system. These organizations include the following:

Business Professionals of America (BPA)—oriented to business and office occupations.

DECA—An Association of Marketing Students—oriented to occupations in marketing and management.

Family, Career, and Community Leaders of America (FCCLA)—oriented to homemaking and to occupations related to family and consumer sciences.

Future Business Leaders of America–Phi Beta Lambda (FBLA-PBL)—oriented to business occupations and to general business principles.

Health Occupations Students of America (HOSA)—oriented to occupations in and related to health.

National FFA Organization (FFA)—oriented to occupations in and related to agriculture.

SkillsUSA–VICA—oriented to occupations in and related to trade and industry.

Technology Student Association (TSA)—oriented to an introduction of students to high-skill technical and industrial occupations.

In some communities there are student organizations that are not affiliated with national organizations but that serve students in the cooperative plans. For example, marketing education students in some schools may belong to a retailing organization. Or, at the local school level, an organization of this type may include all cooperative students regardless of their occupational areas. Usually the local cooperative education organization has the right to affiliate with a national student organization, such as BPA or SkillsUSA–VICA, while retaining its state cooperative organization affiliation. The state cooperative organization typically has a division for each of the major occupational areas. Therefore, a plan of activities is usually organized for these specialized occupational areas. Often an annual state association convention or student leadership conference is held with a general program for all members and some divisional meetings catering to areas of occupational interest. It appears, however, that state-wide cooperative education organizations are on the way out as each CTSO within a given state becomes large enough to support its own meetings.

How National Career-Technical Student Organizations Operate[1]

The student organizations in CTE have several characteristics in common. First, they are found in the nation's secondary and post-secondary schools. Second, each student organization would not exist except for the school's related CTE program. Third, students currently enrolled in the instructional program and graduates (alumni) make up the student organization membership. And fourth, the instructor of the school's program generally serves as the chapter advisor. The organizations are described below in alphabetical order.

BUSINESS PROFESSIONALS OF AMERICA (BPA)

Founded in 1966, the Office Education Association (OEA) became known as **BPA** in 1988. Unlike FBLA-PBL, BPA is the national organization intended for only those secondary and post-secondary students enrolled in career and technical education and in office education. BPA consists of secondary, post-secondary, and collegiate divisions. In addition, alumni members may join any of the three divisions. The aims of BPA are to develop the leadership abilities of its members, to promote interest in the nation's business system, and to encourage competency in business and office occupations.

The *competitive events/activities* program at state, regional, and national conferences is designed to give career-technical business and office education students the opportunity to demonstrate occupational competencies, which include job skills, knowledges, and attitudes, in addition to the leadership and human relations skills needed for success in business and office occupations.

There are three types of competitive events. These are:

1. **Occupational competitive events**—These require students to demonstrate the skills, knowledges, and attitudes needed to gain and maintain employment in a specific business occupational area. These competitive events are occupationally oriented, and students must complete various tasks presented to them as "in"-basket items. Occupational competitive events are grouped into three main areas: (a) administrative support system, (b) information services, and (c) supervision and management.

2. **Specialized competitive events**—These require students to demonstrate the skills, knowledges, and attitudes in leadership and other specialized skill areas related to business and office careers. Examples of these areas are employment skills, legal applications, banking applications, and computerized accounting.

3. **General competitive events**—These require students to demonstrate general knowledge related to business and office careers, such as business law, business math, and economic awareness.

Address: 5454 Cleveland Avenue, Columbus, OH 43231-4021. Phone: (614) 895-7277. Fax: (614) 895-1165. Email: bpa@ix.netcom.com. Web site: www.bpa.org.

---

[1]Logos reproduced with permission of individual career-technical student organizations. Additional information on CTSO's may be found in *Techniques.*

DECA—An Association of Marketing Students

DECA is the national organization of students who are enrolled in marketing education courses. DECA chapters at secondary and post-secondary levels enroll students who are preparing for full-time occupations in marketing. While DECA membership consists primarily of secondary and post-secondary students, the national organization provides for five member categories: high school, community–junior college (Delta Epsilon Chi), college, alumni, and professional (for people in marketing occupations and marketing teacher education). DECA is organized as a private non-profit corporation. It receives its financial support from membership dues, business contributions, and gifts, largely through the DECA Foundation. The national organization is composed of the various state associations, which, in turn, are made up of local chapters. For most students, the local chapter activities have the most meaning because only a small percentage of the members ever participate in state, regional, or national conferences. Opportunities for membership participation exist through career-oriented competitive events on the state and national levels. Opportunities for leadership exist in committee responsibilities for local chapter activities and in the elected chapter, regional, state, and national offices. Address: 1908 Association Drive, Reston, VA 20191. Phone: (703) 860-5000. Fax: (703) 860-4013. Email: decainc@aol.com. Web site: www.deca.org.

FAMILY, CAREER, AND COMMUNITY LEADERS OF AMERICA, INC. (FCCLA)

FCCLA is a national career-technical student organization for young men and women enrolled in family and consumer sciences education in public and private schools through grade 12. The organization's mission is to promote personal growth and leadership development through family and consumer sciences education. Focusing on their multiple roles as family members, wage earners, and community leaders, members develop skills for life through (1) character development, (2) creative and critical thinking, (3) interpersonal communication, (4) practical knowledge, and (5) vocational preparation.

FCCLA functions as an integral part of the family and consumer sciences education curriculum in that it encourages members to extend their classroom learning by planning and conducting chapter projects that have impact on family, school, and community. Projects concentrate on an array of issues facing youth. These concerns include teen pregnancy, parenting, family relationships, substance abuse, peer pressure, environment, nutrition and fitness, teen violence, and career exploration. Through organizational planning, goal setting, problem solving, decision-making, and interpersonal communications, FCCLA members are encouraged to develop their leadership potential and life skills.

FCCLA promotes a variety of student-oriented programs under family-, career-, and community-related headings. Family First is a program that encourages students to become strong family members. It seeks to improve families' abilities to nurture socially, emotionally, mentally, and physically strong, healthy individuals. Career Connections is a program that helps young people explore the important aspects of a career. It endorses a career development process that can be used for a lifetime. The Community Service Day Opportunities program gives students opportunities to make a difference in their communities by participating in local community service projects.

The FCCLA offers several publications to keep its advisors and members knowledgeable about organizational news, events, and activities. Address: 1910 Association Drive, Reston, VA 20191-1584. Phone: (703) 476-4900. Web site: www.fcclainc.org.

FUTURE BUSINESS LEADERS OF AMERICA–PHI BETA LAMBDA (FBLA-PBL)

FBLA-PBL is the international organization of students enrolled in business and business-related courses. However, students need not be in a career and technical education program to belong. The mission of FBLA-PBL is to bring business and education together through innovative leadership and career development programs.

FBLA-PBL contains four divisions—middle, secondary, post-secondary, and professional. The post-secondary division, Phi Beta Lambda (PBL), is made up of business students in two-year colleges and in four-year colleges and universities. The Professional Division comprises alumni, businesspeople, and advisors. Local chapters of FBLA in middle and secondary schools and of PBL in post-secondary schools make up the state chapters. The national organiza-

tion grants charters under which state and local chapters operate. Each chapter adopts programs within the framework of the national organization.

Local FBLA-PBL chapters usually cover all business-related occupations, but they may be composed of special student interest groups studying particular business and office occupations, such as accounting, data or word processing, or information technology. Each chapter has officers, working committees, and one or more advisors. Local projects include those that enhance chapter members' career and technical understanding, such as conducting career conferences and business field trips; running business enterprises at school, including the school store or concession stands; and operating a student employment service. Some chapters provide tutors for children from inner cities, assistance to individuals who are ill, assistance to senior citizens, and financial help to individuals who are economically disadvantaged.

The March of Dimes is the major charity of FBLA-PBL, and more $13 million has been raised by FBLA-PBL since the partnership between the two organizations began.

State chapters of FBLA-PBL, made up of local chapters, function in most states, with all these chapters conducting annual state leadership conferences.

Each year the national organization provides fall leadership conferences in four geographic regions of the United States, a National Leadership Conference, and an Institute for Leaders. FBLA-PBL participates in an ongoing national effort to assist students in learning about the free-enterprise system. Students study entrepreneurship, technology, and productivity. At each level—local, state, and national—there are contest events, projects, achievement degrees, and elected offices that provide opportunities for students to participate. Address: 1912 Association Drive, Reston, VA 20191-1591. Phone: (800) 758-0749. Email: general@fbla.org. Web site: www.fbla-pbl.org.

HEALTH OCCUPATION STUDENTS OF AMERICA (HOSA)

Incorporated in 1976, **HOSA** is the national organization of secondary and post-secondary students who are enrolled in health occupations education. It is the only student organization exclusively serving health occupations students. HOSA has 2,200 chapters throughout the United States and serves more than 60,000 members.

HOSA provides its advisors and members with a variety of programs, activities, competitive events, publications, and services designed to enrich the health occupations instructional program. HOSA offers preparation, practice, and application of skills in parliamentary procedure, leadership, public speaking, interviewing, writing, medical technology, and specific health occupations. Its national competitive events program gives health occupations students opportunities to demonstrate occupational skills in five broad categories: (1) Health Occupations Related Events, (2) Health Occupations Skill Events, (3) Individual Leadership Events, (4) Team Leadership Events, and (5) Recognition for Outstanding Performances.

National HOSA keeps members and advisors informed through an assortment of materials and publications, including a national magazine, a handbook, and a curriculum enhancement guide. Address: 6021 Morriss Road, Suite 111, Flower Mound, TX 75028. Phone: (800) 321-HOSA. Fax: (972) 874-0063. E-mail: info@hosa.org. Web site: www.hosa.org.

NATIONAL FFA ORGANIZATION (FFA)

The National FFA Organization **(FFA)** is dedicated to making a positive difference in the lives of young people by developing their potential for premier leadership, personal growth, and career success through agricultural education. The organization's motto is "Learning to Do, Doing to Learn, Earning to Live, Living to Serve."

The organization changed its name in 1988 from Future Farmers of America to the National FFA Organization to reflect its evolution in response to expanded agricultural opportunities encompassing science, business, and technology in addition to production farming. FFA members, who may enter the workforce directly or pursue higher degrees through technical schools and four-year universities, are preparing for careers in agricultural marketing, processing, communications, education, horticulture, production, natural resources, forestry, agribusiness, and other diverse agricultural fields. Address: P.O. Box 68960, 6060 FFA Drive, Indianapolis, IN 46268-0960. Phone: (317) 802-6060. Fax: (317) 802-6061. Web site: www.ffa.org.

SkillsUSA–VICA

Organized in 1965 as VICA (Vocational Industrial Clubs of America), *SkillsUSA–VICA* is a nonprofit national CTSO for secondary and post-secondary students enrolled in trade and industrial occupation programs. Members in secondary preparatory and cooperative industrial programs have their own division and activities within SkillsUSA–VICA. Post-secondary students form another division. Each division operates on the local, state, and national levels.

SkillsUSA–VICA is an active partnership among people who are involved in the education, training and employment of career and technical education students to influence the quality and relevance of CTE. SkillsUSA–VICA emphasizes respect for the dignity of work and high standards in trade ethics, work quality, scholarship, and safety. It also emphasizes respect for the democratic process through the practice of democracy in the group. Competitive activities in more than 70 occupational and leadership skill areas include air-conditioning and refrigeration, automotive technology, cosmetology, and electronics trouble shooting. Members can participate in individual achievement programs and receive recognition for skill and personal accomplishments. State leadership conferences are held yearly; delegates are chosen to attend the national leadership conference. Address: P.O. Box 3000, Leesburg, VA 20177-0300. Phone: (703) 777-8810. Fax: (703) 777-8999. Email: anyinfo@skillsusa.org. Web site: www.skillsusa.org.

TECHNOLOGY STUDENT ASSOCIATION (TSA)

Founded in 1978 as the American Industrial Student Association (AISA), *TSA* is the national organization of elementary, junior high, and senior high school students who are currently enrolled in or who have completed high-skill technology courses. TSA is the only CTSO to include elementary as well as junior and senior high school students. Address: 1914 Association Drive, Reston, VA 20191-1540. Phone: (703) 860-9000. Fax: (703) 758-4852. Email: general@tsa.org. Web site: www.tsawww.org.

Objectives of Career-Technical Student Organizations

Career-technical student organizations can provide members with general learning experiences as well as those that are directly career-related. While each organization has its specific objectives, there are general objectives common to all CTSO's.

WHO BENEFITS FROM STUDENT ORGANIZATIONS?

Those persons who work successfully with CTSO's believe strongly that everyone who comes in contact with them can benefit.

The students, regardless of age:[2]

1. Gain an opportunity to identify with and be included as part of a peer group.

2. Learn more about the occupational opportunities in the areas for which they are preparing, through activities such as inviting resource persons to speak at chapter meetings, taking field trips, and cooperating in special projects.

3. Learn to plan, organize, conduct, and evaluate chapter activities under the guidance of the teacher-coordinator as chapter advisor.

4. Develop goals and work toward them within a peer-centered environment.

5. Begin to understand the requirements and responsibilities of "leadership" and "followership"—to "lead" in some activities and to "follow" in others.

6. Develop the social skills of cooperation, dependability, sense of humor, and others needed in an occupation.

7. Acquire prestige by participating in an organization that is occupationally oriented and recognized at the state and national levels.

8. Participate in a group that is recognized by employers and other community leaders as part of an educational program of occupational preparation.

The teacher-coordinator:

1. Can become more familiar with the student-learners within the peer-centered setting in which the teacher may be more informal.

2. Can enhance the formal instructional program with additional resources: (a) chapter meetings, field trips, and demonstrations; (b) area or district conferences; and (c) state, regional, and national conferences.

Student organizations allow student-learners the opportunity to develop leadership skills. DECA members in this photo are campaigning for national office at the National Leadership Conference.

3. Can encourage individual student motivation and maturation through competition with others and often more importantly with the student's own past performance.

4. Can direct opportunities for additional responsibilities toward those members who have demonstrated their interest and abilities.

5. Can become better acquainted with the school and the community through organization activities.

6. Has the opportunity to provide information concerning the career-technical student organization and the educational program to the school and to community support groups.

7. Can develop a method of identifying potential enrollees by acquainting students with the activities of the student organization and perhaps inviting them to a meeting.

8. Has the opportunity to follow up graduates by organizing an alumni group.

---

[2]The emphasis in this chapter is primarily on secondary students. At the post-secondary level, the students may be adults in their forties or older. They too like to "belong."

Student organizations permit student-learners to benefit from their contacts with businesspersons in their community.

The school:

1. When describing the activities of the student organization, can inform the public about the school and specific educational programs.

2. Has the opportunity to obtain greater community interest in school activities and to gain support for increased mileage through "visibility" of the programs.

The members of the community:

1. Have the opportunity to become more informed about what the institution is accomplishing.

2. Have the prestige of taking part in school activities, such as sponsoring a field trip or a conference or being speakers.

3. Can see the school in action and can better understand how tax dollars are being spent.

4. Can be proud to say "We helped."

The parents:

1. Can understand what the CTE program is doing for their children, their grandchildren, or their neighbors' children.

2. Have the prestige of knowing that their children are participating in a bona-fide program of school-recognized activities.

3. Can acquire firsthand experience in school activities by serving as sponsors for one or more student organization activities.

4. Can develop a deeper awareness of the general objectives of the school through acquaintance with and participation in the plan of activities.

The Dynamics of Learning Through Career-Technical Student Organizations

Many avenues exist for student learning and personal development through student organization *chapter work/activities.* A well-balanced program of activities supplementing the formal instruction through a variety of experiences, whether at the secondary or post-secondary level, has been found desirable in cooperative education. Teacher-coordinators agree that those activities requiring member participation, member direction, and member evaluation can have learning outcomes relevant to the students' occupational endeavors. Teacher-coordinators can fulfill their role as chapter advisors by *guiding* the progress of the student group instead of *directing* it, as they often might do in a formalized class.

PARTICIPATION

What are some of the activities that call for participation? For example, inviting school and commu-

CASE-IN-POINT

Each career-technical education student organization seeks to train future leaders in business, agriculture, and industry. Each of these programs has produced many success stories. Probably some of the more gratifying successes come from students who, economically disadvantaged as youth, became influential leaders, with the help of a career-technical student organization.

One such case is Harold House, once a shy, economically disadvantaged boy from a small town in Indiana. Harold was one of several children born to parents who lived and worked on a small farm. Recognizing Harold's potential, Jack Stark, a local ME teacher-coordinator, took Harold under his wing. Harold soon became the local chapter president of DECA and district vice-president the following year. With the help of some local business persons, Harold was given the opportunity to attend a nearby college. While there, he ran for student body president and was state president of collegiate DECA. Harold was later elected the mayor of a small town in Ohio and was named as 1 of the 10 Outstanding Young Americans by the United States Jaycees. Today Harold is a career and technical education teacher and SkillsUSA–VICA advisor.

Another example is Kermit Boston, an African American, who was raised in the inner city of Philadelphia. He attended the same elementary school as Bill Cosby and Ed Bradley. Once in high school he became involved in a career-technical student organization at his school. He was elected president of his chapter and later the city-wide president for Philadelphia. Kermit graduated from a state teacher's college and became an elementary school teacher and principal. Wanting to try something new, he switched careers and became a textbook salesman for a major publisher. Kermit became Vice-President of Sales for this company and is now a training and development consultant.

maintaining departmental displays might be the duty of another group. Planning field trips to the business community, presenting a PTA/PTSA program, and offering specialized services to other school and community organizations would call for special committees. For example, a cooperative business and office organization could provide word processing and copying service to the student council, or a marketing education chapter could offer its services as a sales and marketing consultant to those groups planning to raise funds.

For leadership development, elected officers, such as president and vice-president (who is also often the program chairperson), secretary, and treasurer, have many responsibilities that encourage personal growth. Committee chairpersons can provide other students with opportunities to develop leadership abilities.

For increased occupational understandings, participation activities may be presented effectively within the student-centered framework of the student organization. For example, a VICA chapter might cooperate with local industry to survey needs and requirements for employment in the community's trade and industrial occupations. Or, a marketing education chapter, in cooperation with the chamber of commerce or retail merchants association, could participate in marketing products, such as Back-to-School Week, ME Days, or downtown improvement plans.

In general, those participation activities that most effectively support the real purpose of the educational program tend to have the greatest meaning for the individual. For example, the time spent in preparing an office procedures manual as a career-technical business and office education student organization project would help supplement the career objectives of most chapter members to a greater degree than would the same amount of time spent on a speed-keyboarding contest.

nity resource persons to the chapter to speak on topics of current class study might include planning by a number of students. Selecting, ordering, and previewing videos, software, filmstrips, and audio-tapes might be the responsibility of another committee, while

Harold House is congratulated by his former teacher-coordinator, Jack Stark, for being awarded Indiana DECA's Outstanding Alumni Award.

COMPETITION

While some educators have the firm conviction that competition is not healthy, most believe that inherent in competitive activities, in which individuals compete against their own level of ability, is the development of leadership and technical knowledge. While a certain amount of competition with others may be beneficial, competitive activities should be based—as should other educational matters—on the needs of individuals. Some individuals may need competitive activities to motivate them to meet the demands of the work world. If such is the case, occupationally oriented contests are meaningful. The student organization provides the best environment for students to match themselves against others and to seek reward for outstanding achievement. The reward is especially important for those who, having been told they are failures, believe that they are.

Recognizing that interpersonal relationships are vital in the teaching–learning process has shaped new dimensions in education. Typically today, educators possess a better understanding of the emotional growth of individuals and each person's need for increased constructive relationships with others. Objectives have been widened to foster personal security, to expand assertiveness, and to assume self-direction. Among the new dimensions are many theories about what is the basis of motivation for learning and how this motivation can best be acquired by the student. Learning because one anticipates reward is preferable to learning because one fears punishment. The organization contests and awards are considered measures of success.

Almost everyone would like to take "first place" or "be first" in a contest. But, inherent in any contest is the simple fact that only one person takes "first place" or "is first." Teachers have the difficult task of assisting students in developing the ability to compete and the ability to accept defeat—both extremely important competencies in our society. Being able to accept defeats as well as victories and having the ability to convert "losings" to "winnings" are necessary skills for future job successes.

COOPERATION

Much leadership develops in student organizations as well as in the one-to-one student–teacher situation. The teacher whose focus is on helping students integrate cognitive knowledges, skills, and affective attitudes views the group as a laboratory. This laboratory needs control and stimulation—control, lest the group wander off, losing time in unproductive byways, and stimulation to bring in new ideas or different ways of viewing known ideas.

Over and above these, however, is the example set by teacher-coordinators in handling group interrelationships. If, as chapter advisor, a teacher-coordinator listens to a committee report, even though the ideas presented might be too unrealistic or time-consuming for the chapter to undertake, the example of listening might encourage the organization members to listen to one another. It may happen in a group in which those matters that are most immediate, controversial, or exciting becloud the intended purpose of learning. A contest may stir up excitement out of proportion to its worth. Such excitement or preoccupation may easily be mistaken for genuine interest or actual accomplishment by both the students and the teacher. The event may be of value to but a few and beyond the realization of others. It may be seen by still others as an escape from real engagement in learning. Thus, unwittingly, the development of the individual student-learners becomes subordinate to the "fanfare." The competitive activity should be recognized as of little consequence unless it results in learning. If not, it is distinctly a side show to the main tent. It is the teacher, as well as the individual students, who determines whether a student shifts from being other-directed to inner-directed. Students learn through identification with the teacher. If the students and the teacher have mutual respect, some students may adopt many of the teacher's ideas, values, and attitudes. If the teacher is really interested in the student's development, most students will be aware of this, for most are able to distinguish between hollow incentives and temporary aims and the main goal, which is occupational mastery and a continuing desire for learning.

As one student, Billy, said, "Those of us enrolled in cooperative marketing education find that the program has three parts: (1) the classroom, where we receive group and individual instruction geared to our career goals in marketing; (2) the job or training station, where we gain actual supervised experience and practice in an occupation that interests us; and (3) the organization, where we learn more about opportunities in vocations and where we can enjoy meeting with other CTE students."

Billy, the state president of a high school organization, understood the basic components of a successful cooperative education plan and the interrelationship of these components. He had discovered that the formal classroom instruction was organized by the teacher-coordinator. He had learned that at the training station new learning experiences as well as continuing duties and responsibilities were presented and supervised by the training station sponsor. He understood also that successful student organization activities were organized by the members of the local chapter, with guidance from the chapter advisor. As the year progressed, Billy came to realize that the organization's members became leaders in some activities and followers in others, and by the end of the year, he had learned how to build a cohesive team spirit among his peers. When the year was over, Billy also had discovered that some of his peers believed the classroom instruction had been the most useful to them, others felt they had attained their greatest achievements at their training stations, while still others thought they had realized their greatest successes through the student organization program. The leadership role that Billy worked to achieve in the group and the results he was able to accomplish provided personal growth, which is difficult for many of his age to obtain. These would be of value in his personal life as well as in his work situation.

Keys to Successful Operation of a Career-Technical Student Organization

There are several keys to operating a career-technical student organization successfully. These are methods that knowledgeable advisors have found useful in organizing chapters, correlating student organization activities with instruction, guiding the chapters throughout the year, understanding legal responsibilities, and evaluating the activity programs.

ORGANIZING THE CHAPTER

Because the responsibility of introducing the idea of the CTSO rests with the teacher-coordinator each year, the following procedures may be helpful.

1. Near the beginning of the year, introduce the idea to the related class. Describe the organization program on all levels—local, state, and national. Use examples of past student organization activities— newspaper articles, magazines, conference programs, brochures, and emblems. Invite some past officers or some other alumni to this meeting.

2. Discuss with the class the values of the organization. New chapters might visit an established chapter in a nearby school.

3. Suggest that a student committee be formed to review the matter further and to obtain the views of all concerned. Later, the committee can report its findings, and the entire class can vote on the findings. (Many coordinators require membership since CTSO activities are considered a part of the instructional program.)

4. If the vote is affirmative, the group elects a temporary chairperson, who appoints a nominating committee. The responsibility of the nominating committee is to provide candidates for the offices of president, vice-president, secretary, treasurer, and others.

5. After the election of permanent officers, the president appoints those committees determined necessary, in consultation with the executive committee and the teacher-coordinator as chapter advisor. These committees might include occupational development, financial or fund-raising, civic, service, public relations, and social. The new chapter will need a committee to develop a constitution and by-laws that meet the chapter's objectives. The president should also request the chapter advisor to determine the proper procedures for affiliation with the state and national organizations.

6. Encourage each member to serve on at least one committee. It is important that no one chapter member be asked to give a disproportionate amount of time to the organization by serving on too many committees. The committees confer with the executive committee and the teacher-coordinator as chapter advisor to plan suggested activities for the year, since these activities would relate to the content of the instructional program.

7. Each committee then reports to the chapter, presenting its plan for the quarter, semester, or year. The executive committee—with the advisor— can take on the responsibility of synthesizing all committee activities and suggesting a general plan for the year. This plan would need to be flexible and changeable to provide for special occasions and emergencies.

8. As chapter advisor, periodically review with the officers and committee heads the responsibilities they have undertaken, their progress, and their plans.

9. Evaluate the progress of the chapter on a continuing, student-directed basis, emphasizing the constructive suggestions of the group. Some questions for evaluation purposes are given at the end of this chapter.

In guiding the career-technical student organization, the chapter advisor must make sure that the activities are related to the objectives of the total program. Also, it is the advisor's responsibility to be certain that the initial projects and activities attempted are successful in the eyes of the members. In this way a foundation of enthusiasm is established, enabling the group to reach the more complex objectives as the year goes on.

Naturally, the teacher-coordinator ascertains that student organization activities are within school policy and have school authorization. Before undertaking a chapter activity, the advisor clears it with the principal or with a designated school administrator. Some matters that usually need local approval include: (1) procedure for establishing the time, place, and content of chapter meetings; (2) amount of local chapter dues; (3) number and kinds of fund-raising activities; and (4) participation in conferences and meetings away from school. Keeping the school administration informed about chapter activities helps the advisor to be certain the chapter is following school policy and can also build goodwill in the school for the student organization.

To conduct a meaningful program of student organization activities, the teacher-coordinator and the student members must make a time commitment. Chapters are expected to use class time for meetings that provide meaningful learning experiences. Some chapters within the same school district meet together once in a while to accomplish more than each chapter could do alone. Perhaps the most frequent example of such cooperation is the parent or employer–employee appreciation banquet; another example may be a social event such as a dance or a hay ride.

CORRELATING STUDENT ORGANIZATION ACTIVITIES WITH THE INSTRUCTIONAL PROGRAM

If the CTSO is truly to be an integral part of the vocational program, its activities need planning, just as all instruction is planned. Ideas for chapter activities can be noted when the major learning outcomes and areas of study are prepared for the year. Some teacher-coordinators find a course content and chapter activity chart useful in providing student committees with suggestions for chapter activities. When general content areas are listed along one side and types of activities across the top of the chart, both the teacher-coordinator and the students can perceive relationships between instructional content and activities (see Figure 14–1).

A course content and chapter activity chart can facilitate the work of each student committee as it plans chapter activities. When students plan the program, the instructional program takes on more meaning for them. In planning the program activities, the chapter advisor and members will want to keep in mind these questions: (1) What would be a suitable, well-balanced plan of chapter activities for the school year? (What is meaningful to us—what makes sense?) (2) What projects can we get done as a group? (3) What projects might individual members wish to assume? (4) How much can we get done (are we realistic about time and money)?

GUIDING THE CHAPTER THROUGH THE YEAR

Organizing the year's activity program is a major effort, but other important matters also arise as the year progresses. Among the first is the question of when chapter meetings should be held. Another question is how chapter activities should be financed, and a third, what the major thrusts of state regionals, over-all state, and national conferences should be.

Providing for chapter meetings.—The time and place for chapter meetings should be selected to fit into the student's schedules and the school's administrative policies for all organization meetings. Few schools still set aside a portion of the regular school day for co-curricular activities. Some schools plan for organizational activities immediately following the normal schedule of classes, but busing may negate this and employed students may not have the opportunity to participate. Because the activities are part of the instructional program, meetings are often held during a portion of the related class time.

Obviously, successful chapter meetings need advance planning by the student officers, in cooperation with the teacher-coordinator as chapter advisor. Some chapters find that a business meeting every two weeks, possibly in conjunction with a resource speaker, a planned video, or a short field trip, meets their needs. Other chapter meetings may occur outside school and most employment hours to allow for more extensive field trips, social events, and district (regional) or state meetings. Participation in district, state, and national student organization activities when possible enables chapter members to broaden their horizons through: (1) learning more about career opportunities in the fields of their interest, (2) gaining more thorough understanding of the related educational program, (3) exchanging ideas with other students having similar career interests, and (4) learning to live with students from other cultural settings.

COURSE CONTENT AND CHAPTER ACTIVITY CHART

General Content Areas	Types of Activities (Questions)	Business and Professional	Financial

School, business, and community relationships	As chapter members, what are your responsibilities to your training station and your community?	1. Films
		  a.
		  b.
		  c.
		2. Speakers
		  a. Employment or personnel managers
		  b. Chamber of commerce executives
		3. Projects
		  a. Demonstration job interviews
		  b. Survey of job application blanks
		  c.
		4. Other
		  a.
		  b.
			1. What group activities during the school year will require funds?
			  a. Local activities
			  b. Area, state, and national affiliations
			2. Will fund-raising activities be needed to meet organization expenses?

Civic	Service	Public Relations and Social

1. Study various civic groups and their objectives. How do these groups relate to CTSO's?
 	
2. Have group representatives attend local business and civic organization meetings.
 	
3.
 	
4.
	1. Identify some of the existing approved service organizations:
	  a. In school
	  b. In the community
	  c. Nation-wide
	2. Are there any organizations established by businesses?
	3.
		1. Invite adult guests, advisory committee members, school administrators, parents to attend a "get-acquainted party."
		
		2. Recognize outstanding training stations and/or sponsors.
		
		3. Conduct Careers Through Career and Technical Education Week
		
		4.
		
		5.

FIGURE 14–1

COURSE CONTENT AND CHAPTER ACTIVITY CHART (Continued)

General Content Areas	Types of Activities (Questions)	Business and Professional	Financial

Business organization and operation	All businesses are not organized alike. What are the advantages of the organizational structure of the business for your training station?	1. Creative projects
		  a. Organize a clinic in which businesspersons discuss the management, control, operation, personnel, and marketing of their types of businesses
		  b.
		  c.
		2. Displays
		  a. Posters and displays of types of business organizations
		  b.
		3. Speakers
		  a. Representatives of various divisions
		  b.
		4. Films
		  a.
		  b.
		5. Other
		  a.
		  b.
			1. How is the chapter fund-raising activity to be organized?
			
			2. Name who has responsibility for:
			
			  a. Management
			  b. Purchasing
			  c. Receiving and marking
			  d. Sales and promotion
			  e. Accounting and control
			3.

Civic	Service	Public Relations and Social
1. How are local civic groups organized? 2. 3.	1. Organize and conduct activity for a service organization of your choice. 2. 3.	1. Invite local civic and business leaders to area, state, or national conferences. 2. Organize and hold an employer–employee activity, such as a banquet or luncheon. 3. 4. 5.

FIGURE 14–1 (Continued)

Financing chapter activities.—The activities that a chapter plans for the year determine the amount of funds needed. Basically, four ways are employed by local chapters, state groups, and national student organizations to raise necessary capital. They are: (1) dues, (2) savings plans, (3) fund-raising events, and (4) special assistance projects. The first method is used at all levels—local, state, and national; the second and third are more successfully utilized by local chapters; and the last method is used at the national level and by many state organizations.

1. **Dues**—Chapter dues, paid by each member, should provide for most of the operating expenses of the chapter. The amount of dues should be within the local school's policy for chapter dues and within the income the members are earning. A chapter might plan a field trip to another city or attendance at a state or national meeting in its program of activities. In this case, the members agree to establish dues sufficient to provide for the major cost of the trip before it takes place.

2. **Savings plan**—Since chapter members of many CTSO's are employed part time, a savings plan can be successful. Under the plan, each member saves a predetermined portion of the weekly or monthly earnings for a special purpose. The use of savings plan funds may be for a field trip, major social event, a conference or a meeting, or a gift to the school.

3. **Fund-raising events**—Special fund-raising events may be necessary from time to time. Funds for special CTE, civic, or service activities, such as state and national meetings, scholarships, or contributions to a special cause, usually cannot be met through membership dues. A chapter's fund-raising events may be a continuing service, such as holding a car wash regularly, maintaining a word-processing service for students and teachers, or operating a school store. Or, a chapter may sponsor one event or a series of single events, such as a fashion show, a candy sale, or a concession stand for athletic events.

4. **Special assistance projects**—Some chapters, but more often state and national organizations, request special assistance from interested groups, such as community organizations and industry employers. In certain instances, special assistance organizations have been established for this purpose. For example, the FFA and DECA foundations have as their respective goals the support of programs of student activities. On the national level, these organizations obtain financial assistance from business and charitable organizations interested in CTE. Locally, active solicitation of donations of merchandise or money by students should be cautiously done, if at all. The potential ill-will such solicitations may create may be detrimental to both the chapter and the school. Naturally, any fund-raising event that depends on support from the community in general should have prior approval from the school's administrators responsible for such activities.

Financial activities that are thoughtfully planned can enhance the general content of the educational program. For instance, when a family and consumer sciences class is studying family finance, a practical reinforcement would be for that class to assist the FCCLA chapter in planning its anticipated budget; or when a career-technical business education group is practicing production methods, the chapter could produce and sell a student directory or a newsletter; or when a marketing education class is studying marketing practices, the chapter could conduct, as a fund-raising project, a survey of the potential consumer market for a fast-food franchise or a survey for the retail committee of the chamber of commerce.

In taking trips to state or national meetings, delegates are expected to pay for at least a portion of their own expenses. A successful pattern in many cases has been for one-third of the total expense to be supplied by each of the following: the delegates, the state organization, and the national organization.

Participating in the state and national organizations.—The activities of the local chapter have the greatest meaning for the individual members when these activities are student-planned, student-

A capstone of instruction for career and technical education programs is the annual student organization Leadership Conference Awards Ceremony.

directed, and student-centered. However, state and national meetings have other purposes: they are designed to inform the public of the goals of the student organization and the educational program and are organized to encourage adult interest and support. In fact, they are designed partly to be a "showcase" for the schools.

The annual regional, state, and national student meetings have great meaning for those members who can attend. But, effectively reported at home in newspapers and on TV, these meetings can raise the status of the program for those members who cannot participate. The organizations are so large in most states that attendance is limited to a percentage of the total membership— usually those members who are contest winners and/or chapter officers.

Typically, regional and state meetings follow the pattern set by the national organization. A national leadership conference usually consists of several types of events: (1) registration of all delegates; (2) student officer executive committee meetings; (3) a general opening session for all delegates and advisors; (4)

various competitive events, ranging from those requiring a beginning level of occupational competency to those demanding a high degree of competency; (5) general sessions featuring leaders from business, education, and/or government; (6) discussion groups, workshops, and project sessions of educational and occupational interest; and (7) nomination and election of officers. The climax of the state and national meetings is traditionally an awards banquet and program, where the competitive events trophies and other major awards are presented.

Advisors experienced in conference leadership procedures state that students gain a greater understanding and appreciation of the total organization when time is planned for conference orientation for all members, regardless of whether or not all are to attend the regional, state, or national meetings. In this way those who stay at home can take a greater interest in the post-conference reports of the delegates. For all conferees, daily or more frequent meetings or buzz sessions of the local delegation, led by the chief student delegate, can be meaningful for these reasons:

1. Conferees have an opportunity to review fully the goals of the meeting.

2. The delegation leader designates time for questions from the delegation members and for clarification of individual responsibilities.

3. Members can report to the delegation on the sessions attended, since the delegates all probably cannot attend each discussion or work group.

4. Conferees can discuss with other members of the delegation and with the advisor the new ideas that they gained for occupational and personal use, as well as salient conference information for post-conference reports to those at home.

5. The delegation members have an opportunity to evaluate the meeting in the light of the goals originally discussed.

With frequent conference buzz sessions, the advisor also maintains an up-to-date check on the morale of the delegation, the general health and well-being of each member, and the educational significance of the meeting.

UNDERSTANDING THE ADVISOR'S LEGAL RESPONSIBILITIES

Legal liability for students away from campus for a chapter activity is the concern of every chapter advisor. Naturally, the chapter advisor cannot assure the safety of the advisees; nevertheless, the chapter advisor must take steps to limit the potential liability imposed by the law on the chapter advisor, the school district, school administrators, and even school board members. Chapter 16, "Legal and Regulatory Aspects of Cooperative Education," discusses in some depth the potential liability of teacher-coordinators, including advisors, and it gives some guidelines for limiting liability. One guideline, for example, for minimizing liability, and perhaps even avoiding personal liability, is to obtain written parental permission for each student. Under the law, the advisor assumes the responsibility of acting with reasonable professional care

and taking positive steps to protect the students from injury. To minimize the likelihood of students being injured, the advisor should:

1. Plan the activity, giving attention to the safety and protection of the student.

2. In writing, fully advise the school, the students, and their parents of the organization's activities.

3. Supervise the students in every way consistent with the planned activity; anticipate the consequences of misjudgment by enthusiastic, energetic students.

4. Establish rules and regulations for the excursion and be sure that all students and their parents are aware of them and the penalties that will be imposed should any student violate any rule or regulation.

5. Acquaint students with the risks that might arise, especially in a large city, such as walking alone on a city street, being pickpocketed, being solicited for drugs, alcohol, sex.

6. Teach safety education along with program activities, and insist that all group members follow good safety practices not only when they are in the classroom but also when they are away from school for a chapter activity.

7. Instruct students to report any safety problems to you, the chapter advisor, and/or the proper authorities, and to refuse to use any defective material, equipment, or facilities.

8. In the event a student is injured, get competent medical aid at once. Depending on the seriousness of the situation, it is generally advisable to contact the student's parents or guardian once medical assistance has been obtained.

Evaluating the Program of Student Organization Activities

After defining the extent of interest among chapter members, determining the amount of time available for student organization activities, and obtaining suggestions from members for a meaningful activity program, chapter advisors evaluate the recommended student activities in relationship to their contribution to the total instructional program. In this connection the following questions should be considered.

1. Are the proposed activities student-centered; that is, are they based on the career needs of the students as identified in the instructional program?

2. Can the activities be organized and conducted by the members without the assistance of the chapter advisor?

3. Does the activity program present worthwhile learning experiences without exploiting the time or energy of the chapter members?

4. Is the available chapter time best spent on a specific suggested activity, or are there other projects of greater value to the chapter members?

5. Is the teacher-coordinator providing a commitment of time for adequate follow-up of the student program?

6. What recommendations could the teacher-coordinator as chapter advisor offer for improving suggested projects to give them more meaning?

Some of the questions to be considered in evaluating the career-technical student organization might include the following:

1. Is the CTSO an integral part of the instructional program?

2. Are all enrollees in the instructional program included in the student organization?

3. Is there a written statement of objectives for the student organization?

4. Does the program of student activities meet these objectives?

5. Is the student organization peer-centered?

6. Is the program of student activities planned cooperatively on a long-range basis?

7. Does the program of student activities complement the instructional program?

8. Are there balance and variety in the activities?

9. Are all members, regardless of individual differences, provided with opportunities to plan for, take part in, and gain recognition for their participation in the student organization?

10. Are supportive guidance and adequate follow-up provided by the teacher-coordinator in the role as chapter advisor?

11. Is participation in the affiliated area (or district), state, and national CTSO encouraged?

12. Is the program of student activities evaluated on a continuing basis by the members as well as by those adults concerned with the instructional program?

Questions and Activities

1. Describe each CTSO. For each, state the organization's full name, initials, and structure.

2. What are some of the major values students can acquire from membership in a CTSO? Are there any values to be gained by the school or school district from a program of student activities?

3. What are the advantages and disadvantages for the teacher-coordinator in organizing a CTSO?

4. Who, other than the members, might benefit from the organization? In what ways?

5. Describe how you would go about encouraging the establishment of a local CTSO. What methods would you use to attain a program of chapter activities that would enrich learning for the greatest number of students?

6. List five steps chapter advisors can take to reduce the risk of liability when students are away from the school for a chapter activity.

7. In what ways can the CTSO be an integral part of the instructional program? Give examples of chapter activities that match specific learning outcomes.

8. In what ways does the role of the teacher-coordinator differ from that of the student organization advisor?

9. Assume you are the advisor of a local chapter. Given the related course content and the areas of organizations work, construct a course content and chapter activity chart for a specific CTSO.

10. Assume you are the advisor of a local chapter. Write a lesson plan in which you introduce the topic of the student organization to the related class.

11. Make a list of sales projects or other ways a CTSO might use to raise funds to conduct its activities.

For references pertinent to the subject matter of this chapter, see Reading Resources.

15

Accountability Through Evaluation

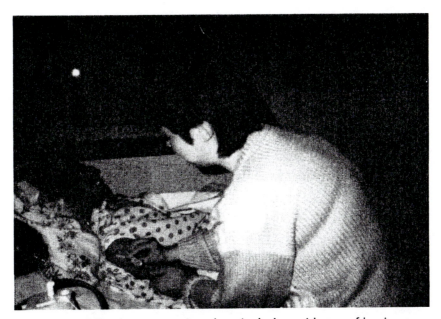

Accountability in cooperative plans includes evidence of having produced graduates with employability, as shown by psychomotor skills, cognitive learnings, and the necessary affective skills. These qualities are apparent as a health occupations student demonstrates how to use a hypodermic needle.

KEY CONCEPT: Accountability necessitates assessing results and then refining the structure and system of the local cooperative plan.

GOALS

After successfully completing the study of this chapter, answering the questions, and carrying out the activities, the student should be able to:

- Explain the differences between the three types of evaluation used for the cooperative plan.

- Describe the evaluation mandated under the Carl D. Perkins Vocational and Technical Education Act of 1998.

- Describe internal evaluation and third-party evaluation.

- Distinguish between process- and product-oriented evaluation.

- Describe the critical factor technique as a form of evaluation.

- Prepare an outline that includes the major components of a cooperative plan that should be included in a program review.

- Outline accountability measures identified by President George W. Bush.

KEY TERMS

accrediting associations

cost-benefit studies

critical factor technique

engaged-time analysis

internal evaluation

learning-time analysis

mandatory evaluation

process-oriented evaluation

product-oriented evaluation

program review

third-party evaluation

time-on-task analysis

time-use analysis

value-added assessment

"Accountability," "evaluation," "assessment," "cost-benefits," "cost-effectiveness," and "time-on-task analysis" are all popular and controversial concepts. To some people, these terms suggest financial and business operations; for others they mean instruction and student learnings; and for still others they mean getting the job done. To most educators, all these terms represent a new era for education—an era when being held accountable for results is demanded.

Fed by federal funds and foundation grants and nurtured by the vested interests of various groups, private learning corporations, concerned taxpayers, alarmed administrators, and school board members, the concept of accountability manifests itself in a variety of forms. Accountability focuses on results. The accountability policy shifts the principal focus in the school from input to output, from teaching to learning.

It is common for educators to ask: "How are things going?" "Am I getting through to these kids?" "What's happening to the students in my classes?" "Are they getting the job done?" "Are they in positions that relate to their training?" "Have they advanced toward their career objectives, and, if so, to what extent?" A commitment to being accountable is to know what should be accomplished. The accountability process requires comparing outcomes to previously stated objectives. Accountability in the schools is everyone's job—teachers, students, parents, school board members, and administrators. The responsibility for quality education—learning—does not rest upon the shoulders of the teacher alone.

President's Accountability Measures

In *No Child Left Behind*, President George W. Bush provided a blueprint for his agenda on educational reform. The President's plan calls for increased accountability structures influencing states' school systems and districts. Schools will be held responsible for the successes and failures of their student populations. Federal dollars will be attached to schools' abilities to meet specific performance goals and ensure improved results. Key components comprised in the President's educational reform include achieving equality through high standards and accountability, improving teacher quality, moving limited-English-proficient students to English fluency, promoting parental options and innovative programs, encouraging safe schools for the 21st century, increasing funding for impact aid, and granting flexibility and expecting accountability.[1]

Achieving equality through high standards and accountability.—Literacy, a foundational building block of successful educational experiences, will be improved through state-wide reading initiatives and programs designed to ensure that all children will be reading by the third grade. High academic standards must be established by all schools and met by all students, including those who are disadvantaged. Schools will be held responsible for improving student achievement. All students in grades 3 through 8 will be assessed annually in reading and math. Results of assessments must be comparable from year to year and be submitted to parents and the public. Disadvantaged populations must show improvement on state content and performance standards. Technical assistance may be provided to states and districts needing improvement so that established standards may be increased. Schools not showing progress will be identified as needing corrective action. Schools will be rewarded for narrowing the achievement gap, and federal sanctions will be mandated for schools failing to show yearly progress.

Improving teacher quality.—Teachers will be vital to achieving improvement in student and school performance. Therefore, federal grants will be

[1]President George W. Bush, *No Child Left Behind*, U.S. Department of Education, 400 Maryland Avenue S.W., Washington DC 20202, www.ed.gov/inits/nclb/index.html.

provided to states and districts for developing high-quality teachers. Professional development for teachers must include high standards, promote innovative teacher reform, and ultimately improve the quality of teaching performance. Excellence in teaching will be measured using gains in student academic performance and will be rewarded.

Moving limited-English-proficient students to English fluency.—Performance objectives will be established to improve the English fluency of all students regardless of their background. School districts will be held accountable for the English proficiency of students. Federal monies will be tied to performance-based grants, and sanctions will be imposed if performance standards are not met.

Promoting parental options and innovative programs.—Under this title, charter schools will be promoted, educational savings accounts will be broadened, school choice will be implemented, more federal dollars will be sent directly to classrooms demonstrating innovative program structures, and public-private partnerships in school construction will be encouraged and expanded. Parents will be given data that will permit them to make informed choices regarding the education of their children.

Encouraging safe schools for the 21st century.—Funding will be provided that establishes educational programs for students and communities about the prevention of violence and drug use. Schools will be held accountable and must offer reports that demonstrate school safety. Teachers will be empowered so that they might have options to remove persistently disruptive students from their classrooms. Funds will be provided to implement a variety of programs that prove effective in ensuring safer school environments.

Increasing funding for impact aid.—The federal government will be held responsible for serving students and rebuilding schools for Native American and U.S. military children. The quality of these public school systems will be improved through availability of federal funding.

Granting flexibility and expecting accountability.—Unprecedented flexibility will be given to school districts regarding the spending of federal dollars. However, results in the form of student achievement will be paramount and expected. An accountability system will be created for the improvement of student achievement. School systems will be expected to establish student performance objectives that will be viewed as charter agreements with the Secretary of Education. Monetary rewards as well as sanctions will be based on state assessment results.

The Evaluation Process

Quality implies a systematic and continuing evaluation and a search for data, both subjective and objective, that will help identify the strong and weak points of the effort. The evaluation plan begins with a review of the written goals that form the strategy of the operation.

Evaluation continues with the use of those measurement tools that furnish evidence of goal attainment. The primary goal of any assessment is to provide information to improve the decision-making process.

Evaluation is a continuous process of collecting and interpreting information in order to assess previously made decisions. This definition has three major implications: (1) Evaluation is an on-going process, not something that is done only at the end—or once. (2) The evaluation process is not haphazard; instead, it is directed toward achieving specific goals and toward finding answers about how to improve—how to make better decisions. (3) Evaluation requires using accurate and appropriate measuring instruments to collect information needed for decision-making. The evaluation process involves collecting information to enable decision-makers to determine how the career and technical program is progressing, how it turned out in the end, and how to do better next time.

Types of Evaluation

There are three types of evaluation for the cooperative plan and the work experience plan in the curriculum. The first type is *mandatory*, in which the state and/or federal government that provides funds requires evaluation. The second type of evaluation is called *third-party*, in which the educational institution contracts with other sources to review the program. The third type is called *internal evaluation*, in which the institution uses its staff, students, and advisory committees to review the plan.

MANDATORY EVALUATION

There are several forms of **mandatory evaluation**. One form is specified in federal legislation, such as the Carl D. Perkins Vocational and Technical Education Act of 1998. Another form is the comprehensive program review required by some state departments of education. Many states have developed state-wide evaluation systems for career and technical education.

Federal-level mandates.—The Vocational Education Amendments of 1976 (Public Law 94-482) mandated that programs be evaluated by the assessment of planning and operational processes, the achievement of students, and the success of program graduates. The Carl D. Perkins Vocational Education Act of 1984 (Public Law 98-524) required that vocational programs address the changing content of jobs through the improvement and assessment of general occupational skills and academic foundations.[2] The

Carl D. Perkins Vocational and Applied Technology Education Act of 1990 (Public Law 101-392) required each recipient of financial assistance under Part C of Title II to evaluate annually the effectiveness of the program based upon the standards and measures identified in Section 115.[3]

The Carl D. Perkins Vocational and Technical Education Act of 1998 (Public Law 105-332) requires that individual states develop and submit to the federal government state performance accountability systems based on identified core indicators (student attainment of academic, vocational, and technical skills; student attainment of school diploma or equivalent; student placement in post-secondary education, advanced training, military service, or employment; and student participation in, and completion of, vocational and technical programs). Under this new legislation, states are required to maintain an evaluation structure for their career and technical education programs that must be put forward for continuous federal funding. States exceeding their agreed-upon performance levels will be awarded incentive funding, and states not meeting their agreed-upon performance levels may have their federal funding reduced. In some cases, states may even loose their federal funding for career and technical education.[4]

Evaluation and the current definition of "vocational and technical education."—The Carl D. Perkins Vocational and Technical Education Act of 1998 provides the definition of "vocational and technical education":

> The term "vocational and technical education" means organized educational activities that (a) offer a sequence of courses that provides individuals with the academic and technical knowledge and

[2]Carl D. Perkins Vocational Education Act of 1984, Sec. 113(a)(D)(11).

[3]Carl D. Perkins Vocational and Applied Technology Education Act of 1990, Title II, Sec 117(a).

[4]Carl D. Perkins Vocational and Technical Education Act of 1998, U.S. Department of Education, Office of Vocational and Adult Education, 400 Maryland Avenue S.W., 4090 MES, Washington DC 20202, www.ed.gov/offices/OVAE/VocEd/InfoBoard/legis.html.

skills the individuals need to prepare for further education and for careers (other than careers requiring a baccalaureate, master's, or doctoral degree) in current or emerging employment sectors; and (b) include competency-based applied learning that contributes to the academic knowledge, higher-order reasoning and problem-solving skills, work attitudes, general employability skills, technical skills, and occupation-specific skills, of an individual.[5]

This definition more restrictively defines what vocational and technical education is and narrows the discipline's focus. Many believe it clearly identifies the purpose of vocational and technical education. Therefore, program evaluation can be more measurable. Few career and technical education leaders would question that the intent of the law is to measure the results of CTE—with more emphasis on what students can do after completing a CTE program than on what is done to the student in the program. The intent of all evaluation as mandated by the law focuses on results—the effects the programs have had on students.

Advisory councils' role in evaluation.—In the Vocational Education Act of 1968, Congress mandated the formation of a National Advisory Council on Vocational Education (NACVE) and State Advisory Councils for Vocational Education (SACVE). Among their responsibilities is the evaluation of career and technical education, including the cooperative plan. Evaluations are generally made through the council members' reading of reports and the holding of hearings, although in some states the councils commission research. In these ways they become third-party evaluators. The impact of the councils can be strong because one of their charges is to make input into the development of the state plan for CTE.

Local advisory committees were mandated in the 1976 Act to assist in the evaluation process. These were omitted in the 1984 Act. Even though current career-technical legislation does not mandate local advisory committees,[6] they can be effectively involved in the evaluation of cooperative education. Generally, the local council (committee) will use secondary data supplied by the local education agency (LEA) in looking at the program and in giving advice. The committee members rarely engage in any evaluation based on gathering their own data through formal research procedures. Their evaluations are usually subjective—based on their perceptions.

State-level evaluation—program reviews.— Many states have developed comprehensive statewide program review systems. The Vocational Education Act of 1976 and the Perkins Acts of 1984, 1990, and 1998 required that local CTE program had to undergo a comprehensive evaluation. This comprehensive evaluation included examination of the planning and operational activities, student achievement, the employment status of graduates, and the effects of support services provided for special populations. To comply with federal mandates, comprehensive statewide evaluation programs have become common. For example, Michigan uses a six-step evaluation system model (see Figure 15–1). This model identifies a plan for collecting, analyzing, and interpreting data, which is used to measure the effectiveness of career and technical education programs in the state.[7]

Most state-wide program reviews involve self-assessment by the LEA. A self-study by the school's career-technical educators prior to the on-site visitation by the evaluation team can be very helpful to the school. Program Review for Improvement, Development, and Expansion in Vocational Education (PRIDE), a state-wide evaluation system for voca-

---

[5]Ibid.

[6]An extensive text on advisory committees, which contains both principles and operational procedure, is *Advisory Committees in Action: An Educational–Occupational–Community Partnership*, by Leslie Cochran and L. Allen Phelps, Boston, Allyn & Bacon, Inc., 1980.

[7]Jane Coviello, "Evaluation of Vocational Programs," *Key Concepts in Vocational Education: A Monograph Series*, East Lansing: Michigan Department of Education, Vocational–Technical Education Service, 1979, p. 6.

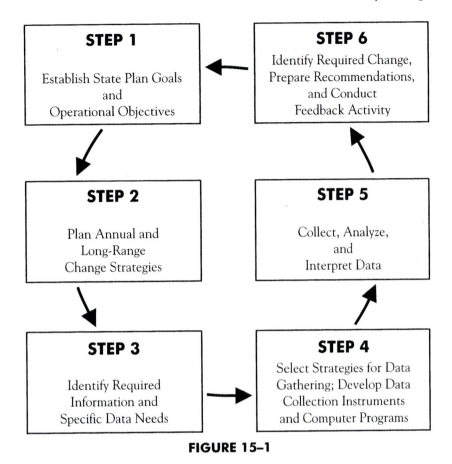

FIGURE 15–1

tional education programs in Ohio, and District Review of Vocational Education (DROVE), California's evaluation system, are both examples of state-level program review plans that have been used in those states. Both evaluation systems require a self-study by the local personnel and an official on-site review by a selected committee of persons knowledgeable about career and technical education. In addition to having the self-assessment report, the visiting team is able to make judgments based on what it observes. Many career-technical educators are concerned about the evaluation mandates stated in the law. Some believe mandated curricula, state-wide testing efforts, state and even national standards for student performance will soon be additional outcomes of the mandated evaluation requirements. But, no matter what one's beliefs are about federally mandated evaluation efforts, in response to the thrusts promulgated by the Carl D. Perkins Voca-

tional and Technical Education Act of 1998, states are continually refining and modifying existing assessment systems.

THIRD-PARTY EVALUATION

Third-party evaluation is done by contract with a private consulting firm, a consultant from a university or another institution, a government agency, an accrediting association, or perhaps a graduate student. Third-party evaluations are required by some state departments of education. Some contracts with the federal government require third-party evaluation. Because the evaluator is not directly involved in the program's scope of work, the third-party evaluation is relatively free from bias and prejudiced judgments. Third-party evaluations are being used because of concerns about sex stereotyping and discriminatory practices.

Accrediting association.—Another evaluation method for the cooperative plan is the accreditation program of the regional *accrediting associations* for secondary and post-secondary education. Visitation teams from accrediting associations come to secondary and post-secondary schools periodically. Coordinators are called upon to make self-evaluation studies prior to each visit. In the early years of cooperative education, much of the assessment was built on quantitative data—number of students, size of facilities, etc. Although quantities in assessment studies are important, they remit the comprehensiveness of the study. Evaluations conducted by accrediting associations have become more qualitative in nature.

INTERNAL EVALUATION

Internal evaluation occurs daily as each teacher-coordinator asks, "How did I do today?" It continues as the teacher-coordinators check with the school's administrators, the employers at the training stations, and, most important, the students. It culminates with end-of-the-year reports to the administration and with the students' final grade.

The critical factor technique.—The *critical factor technique* involves effort by local personnel. It assesses a host of factors thought to be vital by those in the local district. This assessment is beneficial to the professional development of the staff. All too often the formal evaluation reports gather dust on shelves. The teacher-coordinators do not put the results into use, nor do they let their publics know the results. It would be better if the local coordinator and the administrator were to use the critical factor technique, which specifies that if a complete picture of an operation is warranted, there are a few factors that are crucial to quality. If any crucial factors are not present or are badly handled, the plan may be judged as being in trouble no matter how high the quality of a host of minor factors turns out to be.

The critical factor technique is very valuable to busy administrators who need to use their time because they may look carefully at only a few items to judge the quality of the operation. This technique is a form of management by exception and tells administrators whether they should investigate further. Among the crucial questions or factors are:

1. Is a certified teacher-coordinator on the job?

2. Is the related class teacher responsible for the coordination visits and does that person have adequate coordination time?

3. Is related instruction that is appropriate to each student-learner's career goals provided?

4. Is the classroom for related instruction adequate in space and equipped and laid out to facilitate individualized instruction?

5. Are training plans developed and used?

6. Are training stations selected and developed by the coordinator?

7. Are advisory committees for each occupational area operative?

8. Are there plans for helping the special needs students?

Process- and/or Product-oriented Evaluation

Should the process and/or the product of career and technical education programs be evaluated? Which should it be—process or product? These questions and many others like them create much discussion among career-technical educators. *Process-oriented evaluation* requires looking at the *procedures* (instructional strategies; instructional facilities, including laboratories; and instructional materials) used in the career and technical education programs to teach occupational skills and knowledges. Evaluating CTE's product—the student—requires assessing the *outcomes* (skills and knowledges, job placement, and program costs).

The Perkins Acts of 1990 and 1998, as well as President Bush's blueprint for educational reform, prescribes *product-oriented evaluation*. The law does require some data collected from process assessment—condition of facilities and equipment, teacher qualifications, and instructional procedures. Congress has clearly mandated that CTE must prove its effectiveness in terms of measurable outcomes. Most career-technical educators would agree that product-oriented evaluation is important. Career and technical education's success should be measured based on the number of students trained, the number placed on jobs, and the effectiveness of the occupational training, measured by how well students perform on their jobs. However, many career-technical educators would argue that there is danger in focusing only on product-oriented evaluation, especially job placement data. They feel evaluation of CTE programs should be based on both process and product criteria. Product-oriented evaluation emphasizes the easily quantified elements of a program, while it neglects the more subjective, less tangible points that may be more directly related to program improvement. Student achievement may be an increased poise and confidence, success in learning compared to previous failure, and better command of the education basics. Such achievements are difficult to assess except in the most subjective terms.

Evaluation of occupational cooperative education plans should be in terms of both process and product. If the primary purpose of the evaluation is to look at specific aspects of the instruction that should be improved, then *process-oriented evaluation* is what is needed. If, on the other hand, the assessment's primary goal is to determine the general efficiency and effectiveness of all the CTE instruction included in a

Process/Product-oriented Evaluation Model

INPUT	PROCESS	OUTPUT
(What goes into the program)	(What goes on in the program)	(What the clients are like when they leave the program)

Resources
   Budgets
   Facilities
   Equipment
   Materials
 	
Community belief system
 	
Curriculum structure
 	
Characteristics of human
   resources
 	
   Students
   Teachers
   Administrators
	Teaching-learning process
	
	Teaching-learning styles
	
	Teaching-learning strategies
	   World-of-work experiences
	   Individualized study
	   School laboratories
		Student achievement
		   Occupational skills
		   Job placement
		   Personal traits and
		     characteristics
		   Basic skill development
		   Economic accomplishments
		
		Student attitudes/values
		
		   as workers
		   as citizens
		   as family members

FIGURE 15–2. Process/product-oriented evaluation model.

particular program, then *product-oriented evaluation* should be used. In practice, however, it is imperative that both process- and product-oriented assessments be employed.

In the final analysis, how well cooperative education works is judged on the basis of how well the product—the student—does. Many evaluation schemes measure the degree of plan effectiveness based only on job placement. If "output" is defined broadly to include more than acquiring occupational skills and knowledges, to include a wider range of knowledges, attitudes, and behaviors as well, one will have "product-oriented" approach that will be sensitive to all the outcomes of education—attitudes, values, and behaviors.

Before judgment is passed on the degree of effectiveness of a cooperative education plan, the evaluation scheme should consider the systems inputs (what goes into the educational program), the schooling process (what goes on in the program), and the outputs, or outcomes (what the clients are like when they leave the program). The following process/product-oriented evaluation model, while not neglecting the program's outcomes, draws attention to other important program components.

MAJOR EVALUATION QUESTIONS

Those who are policy makers and influencers with regard to career and technical education ask some major questions about its effectiveness. These questions are designed to assess both process and product. While the following questions are broad, they can be asked about the total cooperative plan in an LEA, as well as about the plan in one occupational area in one building.[8]

1. **What evidence do you have that CTE graduates have advantages in the market place over non-CTE graduates?**

Placement rates, salary levels, the amount of time graduates spend unemployed, and promotability statistics on graduates provide the best responses to this question.

2. **Does the course content of CTE actually reflect changes in the demand for workers and in the needs of the work place?**

The proportion of students employed in the field for which they were trained, occupational survey data that show a demand for graduates that is greater than the number completing programs, a survey of workers, attitudes concerning the relevance of the training they received, descriptions of curriculum revision activities designed to keep content up-to-date, and reports on courses that have been discontinued in favor of more appropriate ones provide good responses to this question.

3. **What evidence do you have that women are now enrolling in non-traditional career-technical areas that offer better than average pay?**

Comparisons of enrollment in key fields by women today and 10 years ago offer the best response to this question.

4. **What evidence do you have that CTE is enrolling students who are members of minority groups in career-technical areas that offer better than average pay?**

Comparisons of enrollment of students who are members of minority groups in engineering technology and other skills trades programs today and five years ago offer the best responses to this question.

5. **What evidence do you have that CTE is effective in assisting existing workers in efforts to upgrade their competencies? To increase their productivity and promotability?**

Figures on the number of adults who have been promoted after participating in a CTE

---

[8]This material was adapted from a handout at a national conference, circa 1978. The author(s) is not known.

program, personal testimonies from participants, or a survey of employers provides the best response to this question.

6. **What evidence do you have that CTE is working effectively with students who are physically and/or mentally disabled?**

Enrollment data for students who are physically and/or mentally disabled and placement data today compared to 10 years ago provide good responses to this question. The teacher-coordinator may wish to write a case example to illustrate efforts in this area.

Evaluation Schemes and Strategies

Available to those who are responsible for assessing cooperative occupational education plans are many evaluation schemes and strategies. Selection of the proper evaluation tools should be based on questions such as:

1. What is being evaluated? (*e.g.*, process and/or product components)

2. What resources are available? (*e.g.*, monies, facilities, and personnel)

3. Who will be conducting the evaluation? (*e.g.*, trained evaluators, teachers, members of the local advisory committee)

4. Why is the evaluation being done? (*e.g.* for accreditation purposes, for the state department)

5. How will the data be analyzed? (*e.g.*, computer assistance, statistical models, the evaluation design)

6. What are the intended uses for the data collected? (*e.g.*, curriculum revision, program expansion)

Of all the strategies, schemes, and tools available for program assessment, there are three evaluation schemes being widely used to assess career and technical education programs. They are: (1) follow-up studies, (2) comprehensive *program reviews,* and (3) *cost-benefit studies.* A fourth evaluation scheme, *time-use analysis,* has been gaining much attention. This evaluation strategy, typically used in connection with the three widely used evaluation strategies, is also called *time-on-task analysis, learning-time analysis,* or *engaged-time analysis. Learning-time analysis* has gained significant favor in response to the national thrust for school excellence. Engaged time is the time students work on tasks directly related to the subject they are supposed to be learning. The basic premise of the strategy rests on the research findings that show that student performance can be improved by:

1. Increasing the time students spend actively learning and practicing a skill.

2. Having students engage in learning tasks that provide them with a high rate of success (successful learning experiences are motivating).

3. Involving students in learning activities that require them to interact with others.

4. Tailoring instructional strategies and materials to students' learning styles and achievement levels.

5. Providing students with frequent feedback. Learning-time analysis focuses on the first research finding. It provides methods that assist coordinators in increasing the amount of time students must apply themselves to learning both in the related class and on the job.

LEARNING-TIME ANALYSIS (TIME-ON-TASK ANALYSIS)

The actual amount of time a student spends during the instructional period, both in the related class and on the job, has become important in assessing the teaching–learning process. The National Commission on Excellence in Education's report and a significant number of other educational reports have called for longer school days and years. The National Commission on Excellence in Education's report strongly recommended that the actual time students spend in learning activities needed to be increased.

Most educators would agree that simply increasing the school day and school year, thus creating "more sit-time," would have little if any positive effect on student achievement. But, several research studies do support the premise that the more efficiently time is spent in learning activities, the greater the students' achievement. Thus, the more efficiently time is spent in learning experiences in the cooperative related class and at the training station, the more likely it is that students' achievement will increase. The more efficiently coordinators manage the time students actually spend engaged in learning experiences in school and on the job that further develop technical know-how, basic skills, and general employability skills, the greater the chances are that students will be better prepared for entry into the world of work.

The terms "learning-time analysis," "time-use analysis," and "time-on-task analysis" are frequently used interchangeably. Each relates to a method for determining how much related class time and on-the-job training time students actually spend engaged in learning.[9] Each method requires observing how much time students actually spend learning; identifying which students are off task (and what they are doing instead); and selecting teaching–learning methods to increase the time students actually spend learning. The basic goal is to increase the amount of time students actually work on any assigned activity that builds the desired skill.

[9]For a publication that is designed to provide coordinators with detailed step-by-step procedures for evaluating the training station and for determining the amount of a student's time actually spent in learning activities, see *Assessing Learning Times at the Co-op Training Station*, by Jeanne Desy, Lee Norton, and Stephen J. Franchak, Columbus, Ohio: National Center for Research in Vocational Education, The Ohio State University, 1984.

FOLLOW-UP OF LABOR MARKET BEHAVIOR

A follow-up of graduates and non-graduates in programs reimbursable from federal and state funds is required. A follow-up study should ascertain the employment status of each graduate, any additional education being undertaken, and the relationship of the graduates' employment to their career objectives and training station placement. A follow-up made about nine months after graduation provides information regarding the graduates' employment patterns that have followed temporary first jobs or college entry. The longitudinal study of two to five years helps to determine the graduates' current employment status and to illuminate patterns of advancement, besides discovering relationships between their occupational areas of original training and their current occupations.

1. What proportion of the graduates (cooperative trainees) are employed approximately three, six, nine months after graduation?

2. To what degree do graduates find employment in occupations the same as, or similar to, the ones for which they were trained?

3. To what extent do employers of cooperative trainees retain them as full-time employees?

4. In what ways does the trainees' scholastic ability, as measured by rank in class, affect labor market behavior?

5. What is the duration of the trainees' unemployment following graduation?

6. What additional education have cooperative trainees completed after graduation?

7. To what extent do cooperative trainees enrolled in a post-secondary school concentrate in a field comparable to their cooperative training?

8. To what extent are cooperative trainees experiencing greater job satisfaction?

PROCEDURES IN MAKING
FOLLOW-UP STUDIES

Various techniques can be used in making follow-up studies. Some schools rely on the mail questionnaire exclusively. The problem with this technique is that some graduates simply do not respond and must be contacted personally. Telephone interviews are quite satisfactory in communities where most graduates remain. An interview is more reliable in ascertaining the exact nature of the graduate's present job or educational endeavor and provides the personal touch in maintaining good relations.

Whatever the method of contact, the key to follow-up is an accurate file on each graduate. The file card should be made up before the student's graduation and should include names and addresses of relatives, a code number showing the job title of the student-learner, rank in graduating class, career goal, and other data to be used later as criteria against which to measure success. This file is essential in accomplishing longitudinal studies covering periods of three to five years or more.

In most schools the follow-up studies are made by the coordinators. If a school has a CTE director, that person is usually assigned the responsibility. Large school districts often centralize follow-up studies by conducting them out of the LEA's central office, with the assistance of counselors and chapters of career-technical student organization. Public school districts have never exhibited much interest or investment of resources in the follow-up of CTE students except for "nose-counting" to see who obtained what job. However, there has been interest in the premise that occupational education is not complete without systematic and personal follow-up to aid the students with whatever transitional problems occur. Ideally the school—secondary or post-secondary— should provide follow-up assistance to graduates of the school's cooperative plan. Few would argue that a follow-up visitation program is beneficial for the students and the school. However, because so many schools are facing severe financial crises, they cannot afford the assistance, even though they may be in philosophical agreement. Many states have developed state-wide follow-up systems. Because the Vocational Education Act of 1976 emphasized evaluation, state departments of education and other agencies developed various assessment models. A state-wide follow-up study standardized the information collected, validated follow-up information collected by teacher-coordinators, and established a reliable and valid data base that could be used for state-wide program improvement efforts. Not only did a more comprehensive follow-up system measure skills development and job placements, but it also included attitudes and feelings about work.

Employer follow-up studies.—Those who have hired cooperative education students should be an integral part of the evaluation system. The information they can provide may be helpful in assessing the program's success in preparing students for the world of work. Employer follow-up studies can provide data about the on-the-job performance of cooperative students they have hired and the level of occupational skills possessed by the students. Information from employers regarding their satisfaction with the employees' training and job performance can be a significant input for improvement efforts. Analyzing the employer satisfaction data collected can be most helpful when the program is being refined and modified to improve the cooperative plan's quality of instruction, the instructional content included, and the training stations used.

PROGRAM REVIEWS

The *program review* has been the major evaluation technique in cooperative career and technical education. Through it, components such as staff qualifications and development, availability of instructional materials, appropriateness of space and facilities, and recruitment and placement practices are reviewed and judged. Assessing these program components is important, for without needed staff, materials, and media, instruction is likely to suffer. But, the program review that concentrates on quantity factors, such as square feet of space available and number of students enrolled in the program, is not truly assessing the cooperative plan. Any review should be equally concerned with quality measurement. For example, how well are students achieving? Are they learning the skills needed for entry-level employment? Are they developing positive work attitudes?

Program reviews should be a combined endeavor by the local school and state consultant staff in CTE, by teacher-educators, and by employers. In some states, brief program reviews are conducted annually; in other states, they are sporadic, because some states may have only one consultant for 200 or more secondary programs. To reduce the work load of overburdened staffs, some state consultants are now using the technique of management by exception. They use reports and brief visitations to uncover those situations that seem to have major weaknesses. For these, they schedule full-scale program reviews.

How to conduct the review.—The purposes of a program review are to involve the local school staff in looking at itself and its program, to pinpoint the strengths and weaknesses of various program components, and to provide administrative support for the correction of deficiencies. Only the local school can improve a program, when and if it wishes to do so. A "heavy-hammer" approach from an agency such as a state department can cause resentment and can create little desire to improve the quality of the program—which is what evaluation is all about. The review should consist of the following: (1) visitations to the classroom; (2) interviews with students and graduates; (3) discussions with training station sponsors and advisory committee members; (4) meetings with counselors and teachers who relate to the CTE students; and (5) examination of training memorandums and agreements, step-by-step training plans, instructional materials, and student organization records. Among the criteria for a program review are:

1. Time requirements of the appropriate occupational area, especially those requiring licensing exams, must be met.

2. Each student's career goal, application, and personal interview must be on file as a matter of written record.

3. Textbooks and instructional materials, including written and multimedia programs such as audio and video tapes and tape–slides, must be up-to-date, readily accessible, and used. Self-study and reference material should be available and used in conjunction with specific instruction based on each student's career objectives.

4. Minimum and maximum class enrollment must be met.

5. Business, industry, and labor leaders in the community must be consulted as resource persons in the instructional program.

6. The selection of students for the program must be done by the teacher-coordinator and guidance personnel working together to determine the student's interests, needs, and career objectives.

7. The professional growth of the instructor must be evidenced by his/her attaining updated occupational experience; becoming a member of professional education associations; and attending workshops, special training sessions, and regional and/or national conferences.

8. The teacher-coordinator must be on an extended contract for at least four weeks.

9. Data from follow-up of graduates must be used in program planning.

10. An adequate budget to purchase equipment, consumable supplies, and instructional materials must be maintained. Adequate monies to cover travel costs for on-the-job coordination visits must be allocated.

11. An active local career-technical student organization must be operative.

12. The related class must be taught by the teacher–coordinator.

13. A training agreement for each student must be on file. Training stations must comply with state and federal laws pertaining to student employment.

14. A schedule of on-the-job training experiences (training plan) for each student must be drawn up with each employer and must be on file in the teacher-coordinator's office.

15. Safety must be taught in the related class and on the job.

16. A minimum of one-half hour per week must be spent by the teacher-coordinator with each student on coordination activities.

17. Credit toward graduation must be given students for their related class instruction and on-the-job experience.

18. An adequate number of professional leave days and travel monies should be available for the teacher-coordinator to engage in professional development activities.

19. Training stations must be visited regularly by the teacher-coordinator, according to the students' needs and the employers' time and day preferences.

20. The teacher-coordinator must have adequate office space and access to a telephone.

21. A program of public relations must be consistently carried out.

The primary purpose of a program review is to determine whether the ends (goals) have been accomplished. Each element of a cooperative plan is only the means to an end—one part of the entire set of operational functions that make for quality programs. Judgments about a program's strengths and weaknesses should be formed, based on all the data collected in the whole review. To determine a program's successes or failures based only on the results of a few isolated elements is to violate sound evaluation principles.

EVALUATING A WORK EXPERIENCE PLAN

The criteria used for the evaluation of a general work experience plan should differ from those used in the evaluation of the cooperative occupational education plan, since the goals of each are significantly different, and the operational strategies are not similar. The primary criterion in evaluating the success of a general education work experience plan is: "Have the objectives (goals) of the plan been realized?" However, the evaluation process must go beyond this, using criteria related to operating procedures and seeking evidence from the groups affected—businesspersons, counselors, teachers, students, and parents. The criteria in the list that follows may be used by curriculum planners and program coordinators to plan procedures for organizing and operating work experience plans.[10]

1. Does the coordinator (supervisor) have sufficient time for placement, supervision, counseling, and other program responsibilities? In a general work experience plan, a coordinator can usually supervise from 40 to 60 students if allotted half a school day. On the other hand, work experience plans for persons with special needs may require a full-time coordination load to exceed no more than 25 students.

2. Are sufficient visits made to the job placement situation, to assure evaluation of each student's progress, maintenance of good working relations with employers, and early detection of student problems?

3. Are adequate records that reveal number of hours worked, job progress, and observance of labor laws maintained?

4. Is there close cooperation between the coordinator and the guidance department personnel with regard to selection of students whose needs can be met by the plan? Are records of each student's progress in

[10]The criteria in this list have been adapted from Work Experience Education, a handbook for California secondary schools, Sacramento: California State Department of Education.

the program reviewed and discussed with the individual student?

5. Has a class been organized, at least one day a week, or several hours a week, in which the coordinator–work experience supervisor meets with the students to discuss their job problems and to provide group and individual counseling?

6. Do the school and the community really understand that the plan is general education in purpose, and not vocational? Are differences in plan procedures, student selection and school-related instruction between the work experience plan and the cooperative occupational education plan clearly understood by the groups affected by both plans?

7. Is an advisory committee used for each type of work experience plan? Do these committees meet regularly, and have they been continually involved in counseling with regard to the plan and in assisting with public relations, in locating work stations, and in evaluation?

8. Are there a functioning evaluation process and a follow-up of students?

9. Are employers briefed as to the nature of the work experience plan, the type of students they will be supervising, and their responsibilities for students on the job?

10. Are there standards established to be used in approving work stations and work assignments?

11. Is the plan individualized enough to accomplish its desired outcomes, rather than standardized to fit the regular school program?

12. Is full advantage taken of community resources in planning and carrying out the plan?

13. Is academic credit applicable for graduation given only to students in plans in which the school is assured that learning, rather than earning, is the primary outcome?

14. If the plan deals primarily with persons who have special needs, is the coordinator professionally prepared to deal with these special needs?

COST-BENEFIT STUDIES

Accountability thrusts by local, state, and federal agencies and the Vocational Education Act of 1976 and the Perkins Acts of 1990 and 1998 clearly reflect the need for career and technical education to try to prove its effectiveness (its worth to society). **Cost-benefit** (or cost-effectiveness) **studies** move beyond a description of the occupational status of graduates of CTE programs. They get at the economics of input–output by asking: Was the CTE program worth its cost to the individual? Was it worth its cost to the taxpayer and to society?

If the appropriate data are collected and analyzed, cooperative education will have few if any problems in convincing interested publics of its value and contributions. The financial impact of cooperative education on the economy of the local community, the state, and the nation may be more far-reaching than some people recognize. More studies about the cost-benefits of cooperative education are being conducted. Cooperative education programs must give priority to evaluation schemes that document the need for the cooperative plan.

Cost-benefit evaluation is the least developed evaluation strategy.[11] Cost-effectiveness studies are complicated because inputs and outputs are not absolute. To determine the rate of return on investment requires having absolute figures. In business and industry, cost-benefit studies are relatively simple to conduct because cost figures are definite; in education, costs are sometimes obscure and the benefits can never be expressed purely in financial data. Many of the benefits of cooperative education are intangible—better learner attitudes, positive self-concepts, established work ethic values, and improved achievement in school are just a few of the ways individuals and society in general benefit from cooperative education, but these benefits are difficult if not impossible to quantify.

---

[11]Frederic Welch, *Cooperative Education, A Review,* Information Series No. 116, Columbus, Ohio: national Center for Research in Vocational Education, The Ohio State University, 1977, pp. 35–53.

VALUE-ADDED ASSESSMENT

Another scheme for measuring performance is the "value-added" model. This evaluation model searches for the positive difference made by an existing program. It assesses the ability of the teaching–learning process in influencing students favorably by making a positive and identifiable contribution to their development. *Value-added assessment* focuses on the degree of improvement in terms of knowledges, skills, and attitudes. Basic to this evaluation scheme is the notion that "value-added" causes change. The improvement, the margin of effectiveness, presupposes the existence of change—change that has taken place as a result of the cooperative plan.

A Summary View

The concept of accountability implies precise educational goals. A careful assessment will tell "how we're doing" compared to "what we want to do." A systematic approach (a management approach) to education includes identifying goals and objectives, planning and carrying out processes for their attainment, monitoring and reporting results, and continuing to refine the total endeavor.

All career and technical education, including cooperative education, needs to be more in tune with program planning and evaluation. Steps need to be taken to define more carefully where career-technical educators want to go (*goals*), how they will get there (*educational process*), and how they are doing (*assessment*). Evaluation schemes and strategies that will allow career-technical educators to be more definite about outputs need to be developed. The end product—adults equipped with relevant knowledges, interested in being contributing members of society and with the competence to do so—is what educators must be accountable for.

Questions and Activities

1. In outline form, prepare a follow-up plan for use with cooperative education graduates. Assume you are a teacher-coordinator and will be using the system without computer assistance.

2. Design an evaluative checklist for a cooperative education plan in your chosen discipline. Include criteria to cover goals and objectives, organization, nature of offerings, physical facilities and equipment, staffing, instructional materials, budget and outcomes.

3. Compare the evaluation requirements included in the Carl D. Perkins Vocational and Technical Education Act of 1998 with those required by the 1976 Vocational Education Act. How often must periodic evaluations be conducted and which elements of a program does the law require to be evaluated? Are there any special evaluation requirements for cooperative education? If so, what are they?

4. Contact your state department of education, career-technical division, for information about evaluation. What evaluative criteria or checklists does the staff use when evaluating a local cooperative plan?

5. Make arrangements to visit a local cooperative or work experience plan. After the visit, describe the plan and list the crucial questions you asked or the factors you looked for during your visit.

6. Assume the role of counselor, businessperson, an advisory committee member, another CTE teacher, a principal, a CTE director, or a student. For the role, list the crucial questions you would have about a cooperative plan. Be sure the questions or factors can be measured in some way.

7. Oppose or defend the proposition that the crucial evaluation question for cooperative education (or any CTE offering) is: "Do the graduates obtain jobs in the fields for which they are trained?"

8. Assume you are a state supervisor of an occupational program and your task is to conduct a new series of program reviews of cooperative plans. Prepare a paper that explains how local personnel are to be involved, what materials are to be available for inspection, and what factors are to be assessed.

9. Assume you are responsible for making arrangements for a third-party evaluation. Make a list of the quali-

fications you would want the third-party evaluator(s) to possess.

10. Assume you are a teacher-coordinator. What do you perceive your responsibilities to be for increasing engaged-time, learning-time on the job? How would you present the issue to the training sponsor?

For references pertinent to the subject matter of this chapter, see Reading Resources.

16

Legal and Regulatory Aspects of Cooperative Education[1]

Student-learners are affected by local, state, and federal laws in relationships at school and at training stations. At a food center, health is protected by standards of cleanliness.

---

[1]Parts adapted from a paper prepared by Lorraine T. Furtado as fulfillment of the requirements for Independent Research, Capital University Law School, October 1985.

KEY CONCEPT: Teachers are expected to know the law. The courts will not accept ignorance of the law as a defense.

GOALS

After successfully completing the study of this chapter, answering the questions, and carrying out the activities, the student should be able to:

- Identify those topics that have direct impact on career and technical education teachers and their students.

- Explain the importance of case decisions in establishing law.

- Enumerate basic student rights that must be recognized in the classroom.

- Explain how teachers act in loco parentis when they are working with students.

- Explain how schools can legally regulate student behavior at school functions.

- Discuss key labor laws and regulations that apply to teen workers.

KEY TERMS

academic sanction

administrative laws

age certificate

due process

employment certificate

equal educational opportunity

in loco parentis

From Classroom to Courtroom

Over the past four decades, state and federal courts have had a profound impact on how the nation's public classrooms operate. As teachers and administrators will testify, the application of law to school conflicts has frequently changed the way decisions are made and the role of professional educators in developing educational policies. Three decades ago, the legal concerns of education generally related to rather mundane issues, and school board decisions were commonly accepted as the last word. Today, schooling is no longer regarded as a take-it-or-leave-it proposition. As never before, the legal fulcrum on which the student–teacher–school relationship balances requires the interpretations of constitutional, statutory, and case law.

While it is generally true that the courts are reluctant to interfere with the judgments of educators, except in cases of clear professional abuse, the courts have decided questions embraced with the human rights issues contained in the U.S. Constitution, as well as curriculum mandates; teacher contracts; student personnel issues, including discipline; and the vast area called "equal protection of the law." The courts increasingly are being called upon to deal with school-related issues that represent individual and group social, economic, political, and personal rights.

TEACHERS HARBOR MYTHS ABOUT "THE LAW"

Even though most educators, including career-technical educators, are aware of the burgeoning litigation and legislation and have some degree of familiarity with a few of the landmark Supreme Court cases, they may harbor misunderstandings and uncertainties about the legality of the daily decisions they must make in their classrooms. Only those topics with a direct impact on career and technical educa-

tion teachers and their students are addressed here. But, each topical area is documented so that the reader can explore specific cases and/or points of law in more detail.

Keep in mind several propositions. These are:

1. Laws have a powerful force in shaping educational policies at the national, state, and local levels.[2]

2. Laws are created by people; and human biases and prejudices impact the decisions made. Thus, laws and legal principles are not always objective; instead, they reflect personal considerations.[3]

3. Laws are not static, nor are they created in a vacuum. But rather, they are continually evolving as courts redefine constitutional provisions and legislatures enact new statutes.[4]

4. Laws on some school issues are far from clear, and some questions confronting teachers have not yet been addressed by the courts and the legislatures.[5] But in spite of these problems, certain legal principles have been established and can be relied on for direction in making daily classroom decisions.

IT'S A LITIGATION ERA

Most educators would probably agree that the classroom is not what it used to be, nor is teaching what it used to be. In an era of unprecedented litigation—we are probably the most litigious people in history—the law is more and more the means for attempting to resolve school problems that were once either resolved within the school's walls or left unresolved. Many teachers have discovered, some too late, that there are legal as well as educational consequences in the smallest and seemingly most innocuous daily decisions. Career-technical educators

---

[2]See, *e.g., San Antonio Independent School District v. Rodriguez*, 411 U.S. 1, 93 S. Ct. 40 (1973).

[3]See, *e.g., Snow v. State of New York*, 98 A.D.2d 442, 469 N.Y.S.2d 959 at 961 (1983), affirmed December 18, 1984.

[4]See, *e.g., New Jersey v. T.L.O.*, 105 S.Ct. 733 (1985).

[5]See, *e.g.*, Former Justice White, writing for the majority in *New Jersey v. T.L.O.*, 105 S.Ct. 733 (1984).

should understand the legal dimensions of their decision-making, the current application of the law to classroom decisions, and their rights and responsibilities in making these decisions. Not only does preventive legal planning tend to decrease the likelihood of a lawsuit actually being filed but it also gives better assurance of good defense if one is filed.

The Classroom and the Legal System: An Overview

As educators become increasingly involved in legal actions, at minimum, some basic knowledge about the judicial system and how it operates is essential. Having some awareness of what "the law" is as it applies to career and technical education, and how "the law" functions, can be helpful in avoiding lawsuits. Lawsuits in which teachers are either plaintiffs or defendants are on the increase.

WHAT IS "THE LAW"?

The term "the law" is used to describe broad, general rules that govern the actions of society. Legislative enactments such as congressional acts, state statutes, and city ordinances create laws. State boards of education develop rules and regulations, which are laws. These laws are commonly called **administrative laws.** A great deal of law results from court decisions. These laws, along with opinions of attorneys general and similar rulings, are generally called *common law.* The president of the United States can issue executive orders, which are also "the law." Taken together, actions by the judicial, legislative, and executive branches of the government are "the laws."

Since *Brown* v. *Board of Education*, most school laws are the result of court decisions—common law.[6] From case decisions has emerged "case law," or common law, establishing the base of proper conduct in

our schools. Thus, many of the principles governing what schools do and do not do are not the result of statutory or constitutional provisions, but rather the result of court decisions, the common law.

Laws are not static; most laws change as society changes. Laws need to be definitive enough to allow people to know what is a violation and what is not. In addition, they need to be clear enough so as to predict what the courts might rule when they are violated. On the other hand, the laws must be responsive to changing societal values. The civil rights movement brought about many court decisions impacting on individual rights. Racial segregation was no longer acceptable to most people. Rather than wait for legislative bodies to change the laws, people turned to the courts for new laws.[7]

ARE TEACHERS RESPONSIBLE FOR KNOWING "THE LAW"?

Like every citizen, teachers are expected to know the law. The courts will not accept ignorance of the law as a defense. Actually, the courts recognize that it is impossible for every individual to be aware of all his/her rights and duties in every instance; but, if the courts allowed ignorance of the law as a defense, and there was no way to refute such a claim, ignorance of the law would most likely be claimed not only by those who truly were ignorant but also by others as well.

Thus, teachers are expected not only to know the law but also to abide by it. Because elementary and secondary teachers deal with the lives of the young, the next generation, they are, as professionals, vulnerable to being sued. Quite understandably, juries demand that teachers practice a higher standard of care than ordinary citizens. It is probably true that most teachers have faced the potential of being sued. Consider for a moment these typical career and technical education activities and their potential to generate lawsuits:

---

[6] *Brown v. Board of Education (I)*, 347 U.S. 483, 74 S.Ct. 686 (1954).

[7] See, *e.g.*, *Brown v. Board of Education (II)*, 349 U.S. 294, 75 S.Ct. 753 (1955).

1. Student raped by a classmate while in attendance at a regional career-technical student organization conference.

2. Student stumbles over an open desk drawer and falls, causing severe injuries to the head.

3. Two students in a cooperative plan attending an employee recognition banquet "spike" the punch. They leave the dinner intoxicated and are involved in an accident. Six people are killed—one of the students and five passengers in the other car.

4. Two students begin fighting in the classroom. The teacher in attempting to break up the fight, grabs one of the students by the arm and dislocates the student's shoulder.

Obviously, the best protection against a lawsuit is to know some basics on how to avoid being sued. Many suits could be prevented if teachers would become fully aware of the danger points and would maximize their defenses against the problems.

THE FEDERAL ROLE IN EDUCATION

Although the federal role in education is indirect, the U.S. Constitution does confer basic individual rights that must be respected by schools. Under authority of the "general welfare clause," Congress controls the use of federal funds for education.[8] Certain operational functions, such as transportation, safety, and labor regulations, are regulated by the authority given Congress under the "commerce clause."[9]

Federal legislation.—Through legislation, federal support for various educational programs is channeled to the states. There is no doubt that the federal government's involvement in public schools, although indirect, is real. Federal statutes directly affect educational policies. When a state elects to accept federal funds, it is bound to observe the statute's conditions. The interpretation of the conditions included in federal statutes is the province of the federal courts. From career and technical education's perspective, the Morrill Act of 1862, which created land-grant colleges, was the first federal statute giving impetus to career and technical education. The Smith–Hughes Act of 1917 was the first federal legislation directed to support career and technical education in public schools.[10] In addition to the federal legislation providing financial assistance to schools, Congress has enacted several statutes that require or prohibit certain conduct by state and local school boards receiving federal financial assistance. These statutes cover discrimination as it pertains to "race, color or national origin,"[11] "sex,"[12] and "otherwise qualified handicapped individuals."[13] To assist schools in achieving equal access, Congress has enacted federal legislation providing funds for special needs populations. The Bilingual Education Act of 1968[14] and the Education for All Handicapped Children Act of 1975[15] are both examples of legislation directed to serve the needs of "exceptional" children.

The U.S. Constitution.—As mentioned previously, the federal government's powers to be involved in education are derived from (1) the general welfare

---

[8]Under the "general welfare clause," in Article I, Section 8, of the U.S. Constitution, Congress has the power "to lay and collect taxes, duties, imposts and excises, to pay the debts and provide for the common defense and general welfare of the United States. . . ."

[9]Under Article I, Section 8, Clause 3, of the U.S. Constitution, Congress has the power "to regulate commerce with foreign nations, and among the several States, and with Indian tribes." Traditionally, the commerce clause has been broadly interpreted. Through a broad interpretation of "commerce," the courts have expanded the federal government's role in regulating activities impacting commercial functions.

[10]The Carl D. Perkins Vocational and Applied Technology Education Act of 1990 is the federal statute in support of vocational education until new legislation is passed to take its place.

[11]42 U.S.C.A., Sections 2000d–2000d–4, popularly called Title VI of the Civil Rights Act of 1964.

[12]20 U.S.C.A., Section 1681, *et seq.*, popularly called Title IX of the Education Amendments of 1972.

[13]29 U.S.C.A., Section 794, popularly called Section 504 of the Vocational Rehabilitation Act of 1973.

[14]20 U.S.C.A., Section 3221, *et seq.*

[15]20 U.S.C.A., Section 1401, *et seq.*

clause, Article I, Section 8, of the U.S. Constitution, (2) the commerce clause, Article I, Section 8, Clause 3, of the U.S. Constitution, and (3) the protection of the rights and liberties of individual citizens under the U.S. Constitution, particularly the First, Fourth, Fifth, Eighth, and Fourteenth amendments.

The First Amendment protects freedom of religion, speech, and press and the right to petition. It is designed to insure certain basic personal freedoms.[16] Cases involving students' rights of freedom of speech and press and teachers' rights to academic freedom have generated many legal challenges under the First Amendment.[17]

The Fourth Amendment forbids "unreasonable searches and seizures." It guarantees the right of citizens "to be secure in their persons, houses, papers, and effects against unreasonable searches and seizures." Students' protection under this amendment was addressed by the Supreme Court in *New Jersey v. T.L.O.*[18]

Although school functions essentially involve civil, not criminal matters, the **Fifth Amendment** is important in several school situations, such as those in which a state board of education or local school boards acquire property for school purposes and in which teachers are dismissed for activities that occurred outside the school's boundaries. The Fifth Amendment provides "due process" regarding acts of the federal government. Although most school due process issues are initiated under the Fourteenth Amendment, which is related directly to state actions, due process cases in the District of Columbia must be brought under the Fifth Amendment.[19]

The **Eighth Amendment** protects citizens from cruel and unusual punishment by government agents. Although the amendment is used more often in cases challenging the treatment of prisoners, it has been used in corporal punishment cases.[20]

The **Fourteenth Amendment** is the most frequently used in school cases because it specifically relates to state actions. The right to "due process" and "equal protection of the laws," the last two clauses of the Fourteenth Amendment, have the most application to public school issues.

1. **Due process clause**—The Fourteenth Amendment's *due process* clause prohibits states from depriving citizens of life, liberty, or property without due process of law. This clause has possibly been construed more frequently than any other provision of the Constitution. Although the Supreme Court has not defined "due process of law" in clearly stated rules, it has ruled that students facing punishment have the right to due process.[21]

Due process has two aspects: *substantive* and *procedural*. Substantive due process relates to the legislation (the rule) itself. Determining if the rule satisfies the substantive due process aspect includes answering two questions: Is the rule's purpose within the power of the law maker to pursue? Is it rationally related to accomplishing its purpose? Under the second aspect, procedural due process, the question is: Was the rule implemented fairly? In *Carey v. Piphus*, a school suspension case, the Supreme Court ruled that the school was justified in depriving the students of property interest, the right to go to school (the substantive due process test was met), but the rule was procedurally defective.[22] Thus, the Court found the rule to be in violation of the Fourteenth Amendment.

---

[16]The First Amendment reads: "Congress shall make no law respecting an establishment of religions, or prohibiting the free exercise thereof, or abridging the freedom of speech, or of the press; or the right of the people peaceably to assemble, and to petition the Government for a redress of grievances."

[17]See, *e.g., Tinker v. Des Moines School Board*, 393 U.S. 503, 89 S.Ct. 733 (1969); *Pickering v. Board of Education of Township High School District 205, Will County*, 391 U.S. 563 (1968).

[18]105 S.Ct. 733 (1985). See *supra* note 3.

[19]Only federal laws apply in Washington, D.C. The District of Columbia has its own federal district court and circuit court of appeals.

[20]See *supra* note 5.

[21]*Goss v. Lopez*, 419 U.S. 565, 95 S.Ct. 729 (1975). See *infra* note 9.

[22]435 U.S. 247, 98 S.Ct. 1042 (1978).

2. **Equal protection clause**—The Fourteenth Amendment provides that no state shall "deny to any person within its jurisdiction the equal protection of the laws." This clause has frequently been used in school cases involving alleged discrimination.[23] Generally, school rules that discriminate against some students are unconstitutional, since they deny treatment as equals.[24]

Students' Rights in the Classroom

The federal Constitution confers on all people significant substantive rights. These rights are guaranteed and cannot be infringed upon by any state or local government action unless overriding government interest exists.[25]

When students exercise these protected rights, a conflict of interests often results between the school officials and the students. The courts have cautiously guarded the constitutional rights of students against undue encroachment by school districts.

In addition to the guaranteed rights conferred by the U.S. Constitution, students have specific rights defined by federal and state statutes. For example,

the Education for All Handicapped Children Act of 1975 and the Bilingual Education Act of 1968 delineate specific requirements.[26]

U.S. CONSTITUTIONAL RIGHTS

Until rather recently, it was believed that a student's constitutional rights were limited. But over the past four decades, the constitutional rights of students and teachers have been examined and in many situations clearly defined in a rash of state and federal court decisions.[27]

First Amendment rights.—Probably no area of student rights litigation has received as much attention as the protected rights conferred through the First Amendment. The courts have clearly stated that in the area of student press and distribution, students have legal rights protected by the U.S. Constitution. Students may publish and distribute their publications without being disciplined by school officials as long s such materials do not ". . . justify a 'forecast' of substantial disruption or material interference with the school's policies or invade the rights of others."[28] Although the Court recognized in *Scoville v. Board of Education* . . . that school officials have ". . . 'comprehensive authority' to prescribe and control conduct in the schools through reasonable rules," the rules must be "consistent with fundamental constitutional safeguards."[29]

Student publications that merely offend and displease school officials or even materials that raise controversial issues cannot be prohibited. Even when

---

[23]See, *e.g.*, Susanne Martineq, "Sexism in Public Education: Litigation Issues," *Inequality in Education*, No. 18, October 1972, 6–7.

[24]Treatment as an equal is not the same as equal treatment. "Equal treatment" means all persons must be treated alike; *e.g.*, all persons must be free from illegal searches and seizures. "Treatment as an equal" means, *e.g.*, a school may not establish a double standard, as, *e.g.*, one set of rules for young people enrolled in vocational education, another for those not enrolled.

[25]See *Tinker v. Des Moines School Board*, 393 U.S. 503, 89 S.Ct. 733 (1969).

[26]See, *e.g.*, Part E—"Bilingual Vocational Training," Sec. 441, "Program Authorized," of the Carl D. Perkins Vocational and Applied Technology Education Act of 1990, which states that "the Secretary of Education is authorized to make grants to and to enter into contracts with appropriate State Agencies, local education agencies, post-secondary educational institutions, private non-profit vocational training institutions, and other non-profit organizations specifically created to serve individuals who normally use a language other than English, for bilingual vocational education and training for individuals with limited English proficiency to prepare such individuals for jobs in recognized occupations and new emerging occupations. Such training shall include instruction in the English language to ensure that participants in such training will be equipped to pursue such occupations in an English language environment."

[27]See, *e.g.*, infra note 34 (student freedom of speech); *Pickering v. Board of Education of Township High School District 205, Will County*, 391 U.S. 563 (1968) (teacher freedom of speech); in re *Gault, et al.*, 387 U.S. 1 (1967) (due process rights of juveniles charged with being "delinquents"); *Goss v. Lopez*, 419 U.S. 565, 95 S.Ct. 729 (1975) (student due process rights in a suspension situation); *Scoville v. Board of Education of Joliet Township High School District 204*, 425 F.2d 10 (1970) (freedom of expression); *New Jersey v. T.L.O.*, 105 S.Ct. 733 (1985) (student searches by school officials).

[28]*Tinker v. Des Moines School Board*, 393 U.S. 503 at 514, 89 S.Ct. 733 at 740 (1969).

[29]*Scoville v. Board of Education*, 425 F.2d 10 at 13 (1970).

school officials have attempted to control student publications by cutting "the purse strings," the courts have ruled that school districts have the burden of showing justification for their actions. The legal test requires the school officials to show they had a reasonable belief that the publication would have caused substantial disruption of school activities or that it would have invaded the rights of others.[30]

Generally speaking, federal courts oppose the placing of prior restraint on student publications. In *Fujishima* v. *Board of Education*, the Supreme Court found a school regulation requiring a review or approval of a student publication before distribution to be unconstitutional.[31] Even though other federal courts have upheld prior restraint rules, they have not allowed school officials much discretion as to what could be censored from distribution.[32]

Just because court decisions have validated the First Amendment press and distribution rights of students does not mean the courts have concluded school officials have no authority over students exercising their freedom of the press rights. In *Baker* v. *Downey City Board of Education*, the Supreme Court ruled that even though students cannot be disciplined for what they say, school officials had the right to control the use of "vulgar" and "profane" terms used by the students to express their views.[33]

[T]he Court determines that the plaintiffs' First Amendment rights to free speech do not require the suspension of decency in the expression of their views and ideas. The right to criticize and to dissent is protected to high school students but they may be more strictly curtailed in the mode of their expression and in other manners of conduct than college students or adults. The education process must be protected and educational programs properly administered.[34]

In finding the principal's actions in halting the distribution of the school newspaper reasonable, the court in *Williams* v. *Spencer* strongly stressed that prior restraint by school officials of student publications required greater court scrutiny than discipline imposed after the student publications had been distributed.[35] The court also noted that advertisements (commercial speech), although protected, are not protected to the same high standard as other types of speech.[36] The *Williams* v. *Spencer* case appears to stand for the proposition that school policies need to shift from attempts to exercise authority by means of prior restraint efforts to rules that hold students accountable after the fact for their publications.[37] It should be remembered that student publications are intended to provide students with learning experiences; thus, students should be allowed to make mistakes just as professional journalists do. Students need to learn that they are responsible for what they print. Freedom to print is the rule. The few exceptions include: (1) substantial disruption or material interference with the school program; (2) obscene and pornographic publications; (3) vulgarity and/or words that may lead to disruption ("fighting words"); (4) defamation, such as libel; (5) malicious criticism of school officials; and (6) invasion of the privacy of others.

Symbolic speech.—The First Amendment issue in *Tinker* v. *Des Moines School Board* was symbolic speech—the conveying of any message without the use of words.[38] The students wore black armbands to school to protest the hostilities in Vietnam. The principal, fearful that disruption would occur, ordered the students either to take off the armbands or to go home. The Supreme Court concluded that the ". . . problem [in the case] involves direct, primary First Amendment rights akin to 'pure speech'"

---

[30]See *Reineke* v. *Cobb County School District*, 484 F.Supp. 1252 (N.D. Georgia, 1980).

[31]460 F.2d 1355 (1972).

[32]See, *e.g.*, *Shanley* v. *Northeast Independent School District*, 462 F.2d 960 (1972).

[33]307 F.Supp. 517 (C.D. California, 1969).

[34]*Id.* at 527.

---

[35]622 F.2d 1200 (1980).

[36]*Id.*

[37]See Huffman and Trauth, *High School Students' Publication Rights and Prior Restraint*, 10 I.L. and Educ. 485 at 492–493 (1981).

[38]393 U.S. 503 (1969).

and thus the ". . . prohibition cannot be sustained."[39] For the school district to be able to justify the prohibition, it needed to show that to allow the students to exercise the forbidden right would ". . . materially and substantially interfere with the requirements of appropriate discipline in the operation of the school.[40] Generally speaking, as long as there is no substantial disruption, students can wear freedom buttons, black berets, and other insignia—all forms of symbolic speech. The Court in *Burnside* v. *Byars*, mindful of prior decisions, concluded that:

> [T]he liberty of expression guaranteed by the First Amendment can be abridged by state officials if their protection of legitimate state interests necessitates an invasion of free speech. The interest of the state in maintaining an educational system is a compelling one, giving rise to a balancing of First Amendment rights with the duty of the state to further and protect the public school system. The establishment of an educational program requires the formulation of rules and regulations necessary for the maintenance of an orderly program of classroom learning. In formulating regulations, including those pertaining to the discipline of school children, school officials have a wide latitude of discretion. But the school is always bound by the requirement that the rules and regulations must be reasonable. It is not for us to consider whether such rules are wise or expedient but merely whether they are a reasonable exercise of the power and discretion of the school authorities.[41]

Thus, if the situation was such that the school board could ". . . reasonably forecast substantial disruption of or material interference with school activities . . . or intru[sions] into the school affairs or the lives of others . . ." then the board would be justified in "forecasting the harmful consequences . . . "[42] On the other hand, courts have recognized that walkouts or boycotts, sit-ins, and even excessive noise can constitute a material or substantial disruption of the school and thus have ruled that school officials were justified in curtailing the symbolic speech.[43]

Freedom of assembly and associations.—

The First Amendment includes the right to peaceable assembly, including the right to freedom of association and petition for redress of grievances. Generally, students have the right to conduct meetings and to circulate petitions as long as such activities do not "materially and substantially" conflict with the educational process. But, the right of students to organize student clubs has not been given similar protection. Courts have consistently upheld the right of school boards to deny recognition to social and secret student clubs. Typically, courts have ruled that secret organizations ". . . tend to engender an undemocratic spirit of caste to promote cliques, and to foster a contempt for school authority."[44] On the other hand, the right of students to join student clubs that advocate controversial beliefs but that have open membership has been protected by some courts, and school districts have been required to recognize such groups officially.[45] To deny such groups recognition, school officials must present clear evidence that the organizations would produce a substantial threat of disruption to the educational process.[46]

In order to maintain school discipline and to assure "safe schools," school officials may regulate the time and place of student meetings, speakers and assemblies, and the distribution of petitions. Typically, school board policies barring all outside speakers and assemblies have prevailed when chal-

---

[39]*Id.* at 504.

[40]*Id.* at 507.

[41]363 F.2d 744 at 749 (1966).

[42]*Tinker* v. *Des Moines School Board*, 393 U.S. 503 at 514, 89 S.Ct. 733 at 740 (1969).

[43]See, e.g., *Tate* v. *Board of Education*, 453 F.2d 975 (1972) (boycott); *Gebert* v. *Hoffman*, 336 F.Supp. 694 (E.D. Pennsylvania, 1972) (sit-in); *McAlpine* v. *Reese*, 309 F.Supp. 136 (E.D. Michigan, 1970) (excessive noise).

[44]*Burkitt* v. *School District No. 1, Multnomah County*, 246 P.2d 566 at 567 (Oregon, 1952).

[45]See Edward L. Winn, "Legal Control of Student Extracurricular Activities," *School Law Bulletin*, Vol. VII, No. 3, at 4 (July 1976).

[46]See *Dixon* v. *Beresh*, 361 F.Supp. 253 (E.D. Michigan, 1973).

lenged, unless school officials have acted arbitrarily or discriminated against certain groups. In *Wilson v. Chancellor,* a federal district court concluded that the school board's policy barring political speakers from the high school violated the First Amendment.

> The board's only apparent reason for issuing the order which suppressed protected speech was to placate angry residents and taxpayers. The First Amendment forbids this; neither fear of voter reaction nor personal disagreement with views to be expressed justifies a suppression of free expression, at least in the absence of any reasonable fear of material and substantial interference with the educational process.[47]

Although the courts are willing to restrict the personal liberties included in the First Amendment in order to preserve "the educational process," they have repeatedly reiterated that schools must respect the First Amendment rights of students. One court expressed this judicial sentiment when it stated: "We have both compassion and understanding of the difficulties facing school administrators, but we cannot permit those conditions to suppress the First Amendment rights of individual students.[48]

Fourteenth Amendment provisions.—As explained earlier, the Fourteenth Amendment is probably the most widely used amendment in school litigation because it directly pertains to actions by the states. The due process clause of the Fourteenth Amendment prohibits states from depriving citizens of life, liberty, or property without due process of law. Typically, students have a state-created property right to an education because of state compulsory attendance laws.[49] Thus, when a student's right is deprived, challenging the adequacy of the procedures followed in taking away the property right is a due process issue.

Due process challenges are common in school suspension and the expulsion cases and in the placement of students who are disabled.

The equal protection clause of the Fourteenth Amendment provides in part that "no State shall . . . deny to any person within its jurisdiction, the equal protection of the laws." Prior to the Warren Supreme Court, the Supreme Court allowed state policies and laws to provide differential treatment of individuals as long as the basis for the distinctions bore only a rational relationship to a legitimate government objective. Under this standard, only if the challenged act(s) is clearly wrong, or is the use of arbitrary power and not an exercise of judgment, is the act invalid. When this standard is used, the state action usually prevails. In an attempt to afford greater individual protection, the Warren Court established what is commonly called in the legal profession as the *strict scrutiny test.* Under this standard any government act (policy or statute) that racially discriminates against an individual, either for the purpose of providing differential treatment or for effecting a fundamental constitutional right, is strictly scrutinized by the courts, and evidence of a compelling government interest must be proven. Seldom do state actions involving a suspect class or a fundamental right meet the compelling interest standard.

When the Supreme Court in 1973 ruled that education was not among the fundamental rights granted by the U.S. Constitution, the Court's majority announced that challenged state actions regarding education triggered only the rational basis equal protection test, unless a suspect classification (*e.g.,* alienage, race, nationality) was involved.[50] To identify a suspect classification, the Supreme Court has established a set of criteria that considers whether or not the classification is based on an immutable characteristic; whether or not the persons in the class are stigmatized; whether or not there has been a history of discrimination against the class; and whether or not the class is politically powerless.

[47]418 F.Supp. 1358 at 1364 (D. Oregon, 1976).

[48]*Nitzberg v. Parks,* 525 F.2d 378 at 384 (1975).

[49]See *Brown v. Board of Education (I),* 347 U.S. 483, 74 S.Ct. 686 (1954).

[50]*San Antonio Independent School District v. Rodriguez,* 411 U.S. 1, 93 S.Ct. 1278 (1973). See supra note 1.

The courts have recognized what educators have known for a long time to be a fact—that the individual needs of students require differences in "treatment." Valid classification schemes designed to tailor the educational experiences of students to meet individual differences are accepted as legitimate pedagogy. But, courts are sensitive to the bases for certain student classifications and to the procedures used to identify and assign students to particular classifications. Substantial litigation has focused on the extent to which school classifications preclude students from "equal educational opportunities." Most courts have interpreted *"equal educational opportunity"* as requiring school districts to take affirmative action to overcome the deficiencies of certain groups of students.[51] "Separate but equal" practices and procedures that "treat all students neutral" no longer satisfy the requirements of the Fourteenth Amendment.

Controlling Classroom Behavior

If a student behaves disruptively in the classroom, makes little progress in academic work, and usually fails to accomplish anything of significance, who is to blame—the student, the student's parents, the teacher, or the community in general? Certainly the student's home, the life, community surroundings, and the associations with peers contribute to the student's behavior in school. But, most educators would also admit that they are responsible for at least some of the student's behavior. Often the way the teacher responds to the student is the major determinant of the student's success or failure within the classroom. Frequently career-technical education teachers observe that a student who behaves very disruptively in one class may respond well in another class. Obviously, something within the two classroom environments causes the different behavior. By attending to appropriate behavior, by formulating clearly defined

performance objectives, by altering instructional activities, and by initiating contingency teaching–learning experiences, a teacher may be able to modify the behaviors of even the most disruptive students. But, even though good teaching will eliminate most classroom behavior problems, that "kids will be kids," a few may accurately be called school delinquents, and some may be termed incorrigible.[52] How educators respond to inappropriate behavior has become a matter for the courts to review.

THE COURTS' ROLE IN SCHOOL DISCIPLINE

Even though the Supreme Court has clearly stated that students do not shed their constitutional rights at the schoolhouse, including their rights to both substantive and procedural due process, federal and state courts have generally been reluctant in student discipline issues to substitute their judgment for that of the school boards.[53] The Court in *Donaldson v. the Board of Education for Danville School District No. 118* summarized the judicial system's reluctance when it said:

> School discipline is an area which courts enter with great hesitation and reluctance—and rightly so. School officials are trained and paid to determine what form of punishment best addresses a particular student's transgressions. They are in a far better position than is a black-robed judge to decide what to do with a disobedient child at school.[54]

---

[51]See *supra* note 3.

[52]See, *e.g., Brown v. Board of Education (I),* 347 U.S. 483, 74 S.Ct. 686 (1954) (see *supra* note 53); *Della Casa v. Gaffney,* No. 171673 (Cal. Super.Ct. 1973) (exclusion of female students from traditionally male vocational classes [*e.g.,* auto body and auto mechanics] is a denial of equal protection guarantees); *Hickey v. Black River Board of Education,* No. 73–889 (N.D. Ohio, 1973) (home economics and industrial arts classes must be available on equal terms to both male and female students); *Hobson v. Hansen,* 269 F.Supp. 401 (D.D.C., 1967). (Court ruled that ability tracking as used in Washington, D.C., was unconstitutional. But, tracking systems per se were not ruled unconstitutional.)

[53]See *supra* note 7.

[54]98 Ill. App.3d 438 at 439, 424 N.E. 2d 737 at 739, 53 Ill. December 946 at 948.

On the other hand, most courts have not been reluctant to intervene when a student has been abused. Some courts do not believe that students have any right to constitutional protection from punishment and seem to suggest that more discipline in public schools is needed.[55] One circuit court has ruled that the proper forum for changing school rules is not the courtroom but the school board.[56] Courts have refused to hold school officials to the rules governing police actions, primarily because teachers are perceived to be benevolent toward their students, not adversaries as the police are with criminal suspects.[57] While the Court in New Jersey v. T.I.O. rejected the state's position that students have a diminished right of privacy due to the special school–student relationship, the Court favored the position that there is great need to maintain discipline in public schools. In a six to three decision, the Supreme Court ruled: "The special need for an immediate response to behavior that threatens either the safety of school children and teachers or the educational process itself justifies the Court in exempting school searches from the warrant and probable cause requirement."[58]

Many states' statutes protect school personnel, both certificated and non-certificated, from liability in administering as much force as is necessary and reasonable to control a student's behavior that is threatening physical injury to others, to quiet a disturbance that is endangering the safety of other persons, and to defend themselves.

CORPORAL PUNISHMENT

Generally speaking, the courts see teachers standing in *loco parentis* to students under their supervision and as such they have not only the right but also the duty to act on the behalf of students. The broad powers flowing from the *in loco parentis* doctrine often allow educators to get by with practices that would be unconstitutional in other settings, if not a school.[59] But, even though teachers standing *in loco parentis* have some latitude in disciplining students, they can be held liable for assault and battery if they administer corporal punishment to students.[60]

On the one hand, the Supreme Court seems to uphold the right of school personnel to administer corporal punishment, while at the same time ruling that students are persons and that they do not shed their constitutional rights at the schoolhouse gate.[61] In *Ingraham v. Wright*, the Supreme Court ruled that corporally punishing students in school is not cruel and unusual punishment as defined in the Eighth Amendment.[62] The Court also stated that a hearing was not required before reasonable corporal punishment could be administered and that parental consent was not necessary before the corporal punishment could be administered. The Court concluded that elaborate procedures to protect due process rights would dilute the purpose of corporal punishment. It ruled that the threat of civil and perhaps even criminal action should the corporal punishment be extreme and/or result in serious injury was sufficient to protect the student's due process rights.[63] The Court said: ". . . disciplinary corporal punishment does not per se violate the public school child's

---

[55]See, e.g., *Fisher v. Burkburnett Independent School District*, 419 F.Supp. 1200 at 1201 (N.D. Texas, 1976); *Petrey v. Flaugher*, 505 F.Supp. 1087 (E.D. Kentucky, 1981).

[56]*Mitchell v. Board of Trustees of Oxford Municipal Separate School District*, 625 F.2d 660 (5th Cir., 1980).

[57]See, e.g., *Racine Unifield School District v. Thompson*, 107 Wis. App. 657, 321 N.W. 2d 334, 4 E.L.R. 1294 (1982).

[58]*New Jersey v. T.L.O.*, 105 S.Ct. 733 at 748 (1985).

[59]See, e.g., *Stern v. New Haven Community Schools*, 529 F.Supp. 31 (D.C.E.D. Michigan, 1981). The court upheld the school district's power to install two-way mirrors in the school's lavatories so as to allow administrators to observe student drug exchanges. The court in balancing the interests of the community against the privacy rights of the students found the community interests more compelling.

[60]See, e.g., *Johnson v. Horace Mann Mutual Ins. Co.*, 241 So.2d 588 (La. App., 1970). A Louisiana appeals court awarded the student $1,000 in damages for pain, suffering, and humiliation. Generally, before the courts will convict a teacher of criminal assault, the state prosecutor must prove beyond a reasonable doubt that the corporal punishment was excessive and unnecessary to maintain school discipline.

[61]Schools are still the only institutions permitting corporal punishment.

[62]430 U.S. 651 97 S.Ct. 1401 (Florida, 1977).

[63]Damages are sought in civil suits, while criminal charges are prosecuted by the state.

substantive due process rights." But, in *Hall* v. *Towney*, the Court held that corporal punishment that is "so brutal, demeaning, and harmful as literally to shock the conscience of the court" is subject to redress in federal court under the Eighth Amendment.[64] In essence, the Supreme Court concluded per its ruling in *Ingraham* v. *Wright*, and coupled by the lower federal court's holding in *Hall* v. *Tawney*, issues regarding the use of corporal punishment should generally be handled by state courts under the provisions of existing state laws.

Many states have enacted corporal punishment regulations. In Ohio, for example, the state's corporal punishment law has been found to be constitutional by the Sixth Circuit Court of Appeals. Up until 1985, when the state legislature enacted SB 174, local school boards could not establish rules denying school personnel the right to administer corporal punishment. The new law allows local school boards to decide if corporal punishment will be included in the district's disciplinary policy.[65] Some states prohibit corporal punishment by statute.[66] In still other states that permit corporal punishment, rules stipulate how and when corporal punishment may be used.[67] In recent years, especially after *Baker* v. *Owen*, those school boards permitted by state statute to use corporal punishment have established policies regulating its use.[68] Many require that (1) the infliction of corporal punishment must be witnessed by

another certified person; (2) corporal punishment can only be used after other discipline measures have been taken; (3) the student must be told in advance what behaviors will cause a spanking; (4) detailed records of the incidents must be kept; and (5) parents must be informed when their child has been spanked. Ohio's statute requires each school district, at the end of each school year, to submit to the state board of education a "corporal punishment summary." The summary must include (1) the number of times corporal punishment was used in each school within the district, (2) the violations for which corporal punishment was used, (3) "a breakdown of the number of times corporal punishment was inflicted by the age, sex, and grade level . . ."and (4) "a breakdown of the number of pupils who received corporal punishment more than once by the number of times corporal punishment was inflicted."[69]

One court has ruled that whether or not the corporal punishment was reasonable can be ascertained by reviewing the entire situation existing at the time and by examining whether or not the corporal punishment created a substantial risk of serious physical harm.[70] Generally, courts in determining the reasonableness of the punishment consider the student's age and past behavior, the nature of the offense, the instrument used, the gravity of the injury inflicted, and the motivation of the school employee administering the corporal punishment.[71]

SEARCH AND SEIZURE

After many years of debate, the Supreme Court ruled in *New Jersey* v. *T.L.O.* that public school officials may conduct searches of students under standards less demanding than those required in the Fourth Amendment.[72] This case involved a 14-year-old girl, T.L.O., and her friend, whom a math teacher

[64]621 F.2d 607 (4th Cir. 1980).

[65]Am. Sub SB 174, amending Section 3313.20, *Ohio Revised Code.*

[66]See, *e.g.*, Massachusetts and New Jersey prohibit corporal punishment by state statute; Maryland prohibits corporal punishment by a rule promulgated by the State Board of Education. In most states where corporal punishment is permissible, local school boards are allowed to develop policies and procedures that restrict its administration.

[67]See, *e.g.*, *California Education Code* permits the use of corporal punishment but only after prior written approval of the student's parent or guardian. In addition the California statute requires local boards to inform parents that corporal punishment cannot be used without their prior consent.

[68]*Baker* v. *Owen*, 3395 F.Supp. 294 (NC, 1975), affirmed 423 U.S. 907, 96 S.Ct. 210 (1975). This court established procedural requirements for administering corporal punishment. The Supreme Court affirmed the opinion without delivering a written opinion.

[69]Am. Sub. SB 174, amending Section 3319.41 (B), *Ohio Reserved Code.*

[70]*State* v. *Hoover*, 5 Ohio App.3d 207, 450 N.E.2d 710 (1982).

[71]*Hale* v. *Pringle*, 562 F.Supp. 598 (M.D. Alabama, 1983).

[72]105 S.Ct. 733 (1985). See *supra* note 3.

found holding lighted cigarettes in the girls' restroom. Although the school permitted smoking in specific areas, the school's restrooms were off-limits. The teacher "marched" the girls to an assistant principal's office. T.L.O.'s friend admitted she was smoking. She was required to attend a smoking clinic for three days as punishment. T.L.O. denied she was smoking in the restroom and claimed she did not smoke at all. The assistant principal asked T.L.O. to open her purse; when she complied, he found a package of cigarettes in her purse. He also observed cigarette rolling paper. Thus, he continued looking in her purse and found (1) a metal pipe used to smoke marijuana, (2) a plastic bag containing some marijuana, and (3) evidence that T.L.O. was probably selling marijuana. After finding these items, he called T.L.O.'s mother and the police. At the police station, T.L.O. admitted she was involved in selling marijuana to other students. The police charged her with possession of marijuana with the intent to distribute. T.L.O. was suspended from school for 10 days.[73]

In juvenile court, T.L.O. moved to have the criminal complaint dismissed and to suppress the evidence taken by the assistant principal from her purse. Her motions were denied, and she was placed on one year's probation. She appealed the decision and the superior court, appellate division, affirmed the juvenile court decision. The New Jersey Supreme Court reversed the lower court decision and ruled in favor of suppressing all evidence obtained by the assistant principal.[74] The New Jersey Supreme Court stated that although it believed school searches should be under different criteria when used by school officials than when used by law enforcement officials, in this case the assistant principal had no reasonable grounds to believe the purse contained cigarettes. The mere desire on the part of the assistant principal for proof to contradict T.L.O.'s claim that she was not smoking in the restroom and that she did not

smoke did not justify searching her purse.[75] The state of New Jersey appealed the case to the U.S. Supreme Court. Justice White wrote the Court's majority opinion, which focused on the "proper standard for assessing the legality of searches conducted by public school officials and the application of that standard to the facts in this case."[76] The Court concluded that neither a search warrant nor "probable cause" is required before a public school official can search a student. The legality of a search in a school setting depends on the "reasonableness, under all the circumstances of the search."[77] To determine "reasonableness," the Supreme Court held that a two-pronged test must be used: (1) whether or not "reasonable grounds for suspecting the search will turn up evidence that student has violated or is violating either the law or the rules of the school" and (2) whether or not the "measures adopted are intrusive in light of the age and sex of the student and the nature of the infraction."[78] The Court went on to say that school searches are permissible in scope "when the measures adopted are reasonably related to the objectives of the search and are not excessively intrusive."[79]

In ruling the search was legal, the Court clearly did not sanction unjustified or unnecessary intrusive searches.[80] The two-pronged test alerts school personnel to the fact that their authority is not unrestrained. The decision is highly controversial. Most legal writers concur with the conclusion that the circumstances of the school setting establish extraordinary government interests sufficient to justify an exception to the requirement of a warrant before a search is conducted. But, they, as did Justice Brennan in a dissenting opinion, believe that under no circumstances should a standard less than probable cause be allowed. Justice Brennan, joined by Justice Marshall, stated, "the

---

[73]Id. It was not clear if T.L.O. was suspended before or after going to the police station. Three days of the suspension was for smoking in the girls' restroom and seven days was for possession of marijuana.

[74]State in re T.L.O., 94 N.J. 331, 463 A.2d 934 (1983).

[75]Id. at 343, 463 A.2d 934 at 940 (1983).

[76]New Jersey v. T.L.O., 105 S.Ct. 733 at 736 (1985).

[77]Id. at 743–744.

[78]Id. at 744.

[79]Id. at 742.

[80]Id. at 744.

majority opinion sanctions school officials to conduct full-scale searches on a 'reasonableness' standard whose only definite content is that it is not the same test as the 'probable cause' standard found in the text of the Fourth Amendment."[81] Although the decision establishes little "new" law, since it simply affirms the majority of lower federal court decisions, the Supreme Court has spoken on the issue. Several issues were expressly left unresolved. These include: (1)students' right of privacy, (2) standards for searches of lockers and desks, (3) the use of dogs in school searches, (4) school board liability for illegal searches, and (5) the use of the fruits of an unlawful search in expulsion–suspension hearings.[82]

Student Conduct at School-Related/Sponsored Activities

Generally speaking, courts have taken the stand that schools can regulate student behavior at functions that are school-related and/or school-sponsored. In such cases, courts have, for example, upheld a student's (or students') being suspended from school for conduct that interfered with the safe operation of a school bus, for disobeying rules while on a field trip, and for drinking at a state-wide school tournament.[83] In some instances, courts have even said that schools can discipline a student for behavior that did not occur in a school- related activity if such misconduct had a "direct and immediate effect on the discipline or general welfare of the school."[84]

Other Forms of Discipline

DETENTION

Courts have upheld school policies that impose reasonable detention for misbehavior. It appears that no court has ruled that before a detention can be assigned, a student must be given due process of law as required for suspension and expulsion.[85]

ACADEMIC SANCTIONS

Courts do not usually review teachers' grading decisions. However, any "blanket" policy for lowering of grades without giving students a chance to be heard "is suspect" and apt to be court challenged.[86] For example, assigning a zero for each day missed, or lowering a grade by one letter after three class meetings have been missed, is "suspect," since this brings automatic *academic sanction*. Because there may be extenuating circumstances, students must be allowed to explain to the teacher the reasons for the absences. A New Jersey administrative law judge found that a student's rights were prejudiced when he was given a zero for truancy and the zero was weighed against his results on a make-up exam.[87]

Courts have ruled that prohibiting a student who has completed course credit requirements for graduation from receiving the diploma because of improper behavior is not permissible. The issuing of high

[81]*Id.* at 750.

[82]See *supra* note 4.

[83]See, e.g., *Clements* v. *Board of Trustees of Sheridan City School District*, 585 P.2d 197 (Wyoming, 1978). The Court ruled that "It is generally accepted that school authorities may discipline pupils for out-of-school conduct having a direct and immediate effect on the discipline or general welfare of the school." *Id.* at 204–205.

[84]*Caldwell* v. *Cannady*, 340 F.Supp. 835 at 838 (N.D. Texas, 1972). The court ruled that the school district could prohibit students from possessing and using dangerous drugs in school, in a school-related activity, and even in a non-school–related event.

[85]The Court in *Fielder* v. *Board of Education*, 346 F.Supp. 722 (Nebraska, 1972), said there was no constitutional issue at stake in detaining a student in order to make up lost in-class time.

[86]See "Grade Reduction, Academic Dismissal and the Courts," *A Legal Memorandum*, Reston, Virginia: National Association of Secondary School Principals, October, 1977.

[87]*Minorics* v. *Board of Education*, N.J. Comm. of Educ. Decisions (1972). See, e.g., *Hamer* v. *Board of Education of Township High School District No. 113*,383 N.E.2d 231 (Ill. App., 1978). The Court ruled the school's grade-lowering penalty policy was not related to valid discipline objectives. The Court went on to say that the school board had the power to enact disciplinary regulations but that because of the importance of grades to the student's future employment and educational opportunities, the board did not have the authority to use academic sanctions for disciplinary reasons.

school diplomas is considered a ministerial duty over which school officials have no discretionary control. But, if evidence exists that a disruption may occur, a student may be barred from participating in the graduation exercises as punishment.[88]

School Rules and Regulations About Student Conduct

It appears that the language of the school's policies will determine whether or not the courts will uphold the authority of boards of education to control conduct at non-school–related instances. In setting rules and regulations to control student conduct on and off campus, in school-related or non-school–sponsored events, districts need to take steps to (1) acquaint all students with the policy, (2) be sure the language of the policy is clear, and (3) inform students of the penalties that will be imposed if the policy is violated. School rules and regulations that have no valid educational purpose because they are not rationally related to the achievement of the educational purposes of the school and those that infringe on any protected constitutional right are certain to be invalidated by the courts.

WHAT KIND OF BEHAVIOR RULES WILL THE COURTS SUPPORT?

Although courts have held that rules including such language as "willful disobedience" and "intentional disruption" were sufficiently clear as to be enforceable, "insubordination" and "disorderly" have been upheld only if due process had been given the student when the punishment was determined.[89]

Some courts have declared a rule invalid for vagueness. In the case of *People v. Barksdale*, a California court expressed vagueness to exist when "[a] statute which either forbids or requires doing of an act in terms so vague that men of common intelligence must necessarily guess at its meaning and differ as to its application violates first essentials of due process."[90] Also, courts have held that school rules that are overly broad are not enforceable.[91] In addition, any rule that "automatically" penalizes a student is per se suspect (in and of itself), because it is possible that at an impartial hearing the student might be able to show cause why the actions were taken and why the behavior was as it was.

Generally, courts will uphold school board rules and policies that are found to be reasonable, not arbitrary. In addition, to support a district's behavior rules, courts require that there be a connection between the behavior and the sanction. Simply stated, the punishment must fit the crime. For example, the Court in *Dorsey v. Bale* found the school's rule that school work could not be made up for unexcused absences, including disciplinary suspension days, and the deduction of 5 percentage points from the nine-week grade for each day missed, to be *ultra vires*—outside the authority of the school district. The Court determined the sanction to be *ultra vires* in that it was additional punishment for the offense that originally resulted in a suspension.[92]

But, a Connecticut court rejected the *ultra vires* and unconstitutional claims of a student that a school policy that withheld course credit in a year-long course in which the student was absent from more than 24 class periods. In addition to the absence limit, the student's course grade was subject to a fine point reduction for each unapproved

---

[88]A New York court said that because there was no evidence that a disruption would occur, the student could not be barred from graduation exercises "without due process." The Court said that "...such a means of punishment may not be an appropriate regulatory act by the board of education." *Ladson v. Board of Education of Union Free School District No. 9*, 323 N.Y.S.2d 545 at 546 (Sup.Ct., Nassau County, 1971).

[89]See, e.g., *Murray v. West Baton Rouge Parish School Board*, 472 F.2d 438 (Louisiana, 1973); *Reid v. Nyquist*, 319 N.Y.S.2d 53 (1971).

[90]503 P.2d 257 at _____ (1972).

[91]*Dunn v. Tyler Independent School District*, 327 F.Supp. 528 (Texas, 1971).

[92]521 S.W.2d 76 (Kentucky, 1975). In the case of *Katzman v. Cumberland Valley School District*, 479 A.2d 671 (Pa. Commw.Ct., 1984), the court held that the action of the board in reducing the student's grades would result in a clear misrepresentation of the student's scholastic achievement. Misrepresentation of achievement was held to be both improper and illegal.

absence after the first. Because the sanction only applied to unexcused absences, it was not determined by the court to be "double punishment" as in *Dorsey v. Bale.*[93]

Labor Laws and Regulations

In March and June of 1990, the Department of Labor conducted a nationwide sweep of 6,600 fast-food stores, groceries, bakeries, restaurants, manufacturers, laundries, and candy businesses. Of those inspected, 2,060 were found to be in violation of the nation's child labor laws. "The department found more than 20,000 children were working under illegal conditions. They included 14- and 15-year-olds working too long and too late, children under 14 working at all, and 16- and 17-year-olds operating dangerous equipment (meat-cutting machines, paper-box bailers, etc.)."[94] As the number of teens working (over 50 percent) has increased over the past 10 years, U.S. government statistics show a marked rise in child labor violations. In 1992, the Department of Labor logged 19,443 such offenses.[95] Burger King was fined $500,000, the largest child labor penalty in history, for letting 14- and 15-year-olds work late on school nights.

Former Secretary of Labor Elizabeth Dole stated that "Protecting our children—America's future—from exploitation in the workplace is a fundamental duty of the Labor Department." The teacher-coordinator must protect the youth employed through the cooperative educative plan. While strict adherence to training station selection and training plan and training agreement guidelines is extremely important, violations (to a lesser degree) will occur and the teacher-coordinator must be prepared to report the

violation and to immediately pull the student from the training station. Anything less opens the teacher-coordinator, employer, and school corporation to claims of negligence and subsequent law suits. Furthermore, failure to comply with legislation may risk the student's health, safety, or welfare; damage the image of the program; and depose the coordinator to sharp criticism.

In a 1993 study of Marketing Education Association members, Stewart W. Husted and Frank Whitehouse found that most teacher-coordinators reported at least "sometimes" observing students working "excessive hours" or working late hours past the legal limits.[96] Most violations are probably due to individual managers rather than company policies; however, company policies should clearly state that violations will not be tolerated.

While these violations are serious, none are as serious as the operation of dangerous equipment by minors. The National Safe Workplace Institute estimates that in 1990, 139 minors died on the job and 71,660 were injured. Of those, "the reports says, about 20,000 were hurt in the restaurant industry—primarily working with fast-food—as a result of slips and falls, cuts, burns, electrical shock, vehicle shock, heavy lifting, chemical exposure and sleep loss."[97] These figures are hard to verify, because the Department of Labor (Wage and Hour Division) prior to 1994 did not have a reporting system that gave age or circumstances for injury or death. In 1991, 31 states participated in a Census of Fatal Occupational Injuries (CFOI). "A total of 3,465 fatal injuries were identified that year in the 31 states for the year. Ten of these involved children under the age of 16, another 19 involved teenagers aged 16 or 17. Another 120 young adults (ages 18–20) suffered fatal injuries while at work."[98]

[93]*The Fair Labor Standards Act* and the *1966 Amendments,* U.S. Department of Labor.

[94]Lee Saulk, "After-School Jobs: Are They Good for Kids?" *McCall's,* October 1990, pp. 102–106.

[95]Brian Dumaine, "Illegal Child Labor Comes Back," *Fortune,* April 5, 1993, pp. 86–96.

[96]Stewart W. Husted, "Most Admired Training Stations," Unpublished research study, 1993.

[97]"Report: Thousands of Teens Hurt Working at Fast Food Eateries," *News & Advance,* September 6, 1992, p. A-7.

[98]Private letter to Stewart W. Husted from James W. Knight, Regional Administrator for Wage and Hour Administration, March 4, 1993.

COVERAGE OF CHILD LABOR PROVISIONS[99]

All employees of certain enterprises (annual gross volume is not less than $500,000) having workers engaged in interstate commerce, producing goods for interstate commerce, or handling, selling, or otherwise working on goods or materials that have been moved in or produced for such commerce by any person are covered by the Federal Labor Standards Act (FLSA). Teacher-coordinators should assume that the profit and non-profit businesses employing students are subject to the provisions set forth by the FLSA. The only non-agriculture exemptions are children under 16 years of age employed by their parents in occupations other than manufacturing or mining, or occupations declared hazardous by the Secretary of Labor; children employed as actors or performers in motion pictures, theatrical, radio, or television; children engaged in the delivery of newspapers to the consumer; and homemakers engaged in the delivery of wreaths composed principally of natural materials.

Employment of 14- and 15-year old minors is limited to certain occupations under conditions which do not interfere with schooling, health, or well-being. Fourteen- and 15-year-old minors may not be employed during school hours except as provided for in Work Experience and Career Exploration Programs (WECEP); more than 3 hours per day—on school days; before 7 a.m. or after 7 p.m. except 9 p.m. from June 1 through Labor Day (time depends on local standards); more than 18 hours a week—in school weeks; more than 8 hours a day—on non-school days; and more than 40 hours a week—in non-school weeks.

Fourteen- and 15-year-olds may be employed in office and clerical work; cashiering, selling, modeling, artwork, work in advertising departments, window trimming, comparative shopping; price marking and tagging, assembling orders, packaging and shelving; bagging and carrying out customer's orders, errand and delivery work (by foot, bike, or public transportation), cleanup work, maintenance of grounds; kitchen work; work connected with cars and trucks (gas dispensing, car washing and polishing); and cleaning fruits and vegetables when it is done in areas that are physically separated from meat preparation.

The Federal Labor Standards Act establishes minimum ages for covered employment in agriculture unless a specific exemption applies. Covered employment in agriculture includes employees whose occupations involve growing crops or raising livestock that will leave the state directly or indirectly through a buyer who will either ship them across state lines or process them as ingredients of other goods that will leave the state. The child labor provisions may apply to employment in any hazardous occupation (e.g., operating 20 PTO horsepower tractors, corn and cotton pickers, hay bailers, grain combines, potato diggers, posthole diggers, fork lifts, power saws, or working around bulls, boars, or stud horses or sows with suckling pigs or cows with newborn calves) regardless of farm size or the number of days of farm labor used on that farm.

Sixteen is the minimum age for students to be employed during school hours and in any agricultural occupation declared to be hazardous. A student must be at least 14 years old to be able to be employed outside school hours in any agricultural occupation not declared hazardous by the Secretary of Labor; however, 12- and 13-year-olds may be employed with written parental consent or on farms where their parents or persons standing in place of the parents are also employed; minors under 12 may also be employed with written parental consent on farms where employees are exempt from the federal minimum wage provisions.

Minors of any age may be employed by their parents or persons standing in place of their parents at any time in any occupations on farms owned or operated by their parents or persons standing in place of them. Other exemptions include student-learners in a bona fide CTE program (specific provisions apply) and 14- and 15-year-old minors that hold 4-H certificates of completion of either the tractor operation or

---

[99]This section is excerpted from the *Child Labor Requirements in Nonagriculture Occupations Under the Fair Labor Standards Act*," U.S. Department of Labor, Employment Standards Administration, Wage and Hour Division, WH–1330, rev. August 1990, and *Child Labor Requirements in Agriculture Under the Fair Labor Standards Act* (Child Labor Bull. No. 102), 1984.

machine operation programs (special provisions apply). Farmers who employ minors who have completed these programs must keep a copy of the certificates on file with the minors' records. Every employer of minors (except parents or persons standing in place of parents) must maintain records containing the following data about each minor: (1) name in full, (2) place where the minor lives while employed, (3) date of birth, (4) written evidence of consent from a parent or a person standing in place of that parent.

It should be noted that while the Federal Labor Standards Act provides minimum guidelines for the employment of minors 14 and 15 years old, the states can require stricter standards for any age minor. For example, Indiana restricts the number of days and hours that 16-year-olds who are still enrolled in school may work. In Indiana, 16-year-olds are limited when school is in session to working 8 hours per day, 40 hours per week, and no later than 10 p.m. on any night that is followed by a school day. With written permission from parents, they may work until midnight on any night that is not followed by a school day. When school is not in session, they may work with their parents' written permission up to 9 hours per day, 48 hours per week, and until midnight.

In addition, many states require work permits or *employment certificates* to be on file in a place of business before a student 16-years-old or younger may be employed. The certificate is often issued by a school official and signed by the parents and employer. The minor must submit a birth certificate to the issuing officer, as a means of verifying his/her age. The certificate is returned to the issuing officer after the student's employment ends. *Age certificates* can also be issued for minors 17 to 21 to protect an employer during a Department of Labor inspection.

MINIMUM WAGE

Teacher-coordinators need to stay abreast of laws and provisions related to student pay. Profit and non-profit businesses subject to FLSA regulations must pay the federal minimum wage. As of July 2001, the minimum wage remained at $5.15 per hour. Persons working more than 40 hours per week are entitled to no less than $1\frac{1}{2}$ times their regular rate of pay for each additional hour worked.

According to the U.S. Department of Labor, a sub-minimum wage may be paid to trainees and student learners age 16 or older in the following three categories:

1. Student-learners in a vocational training program as defined in Title 29, Section 520 of the U.S. Code of Federal Regulations.

2. Full-time students working in retail establishments, in service establishments, or in institutions of higher learning where they are enrolled.

3. Students with disabilities participating in cooperative vocational education.

Students being paid a sub-minimum wage must have a written employment agreement that establishes provisions regarding a schedule of organized work processes, safety instructions, and supervision. Some states go beyond the sub-minimum wage guidelines established by the federal government. Therefore, teacher-coordinators should research regulations set forth by their own state governments.

HAZARDOUS OCCUPATIONS

The FLSA established guidelines regarding hazardous occupations. The following are occupations prohibited to minors ages 14 and 15:

1. Cooking, other than at lunch counters and snack bars and within the view of the customer

2. Manufacturing, mining, and processing

3. Most transportation jobs

4. Work in warehouses and workrooms

5. Work for construction firms, except in the office

6. Any job involving hoists, conveyor belts, power-driven lawnmowers, and other power-driven machinery

Except as noted below, minors under the age of 18 may not participate in the following 17 hazardous occupations:

HO 1: The manufacturing or storing of explosives

HO 2: Work as a driver of a motor vehicle or work as an outside helper on a motor vehicle, either as an occupation or as part of an occupation

HO 3: The mining of coal

HO 4: Work involving logging or saw milling

HO 5: Work using power-driven woodworking machines, including saws on construction sites

HO 6: Work involving exposure to radioactive substances

HO 7: Work involving the operation of power-driven hoisting devices, including the use of forklifts, cranes, or non-automatic elevators

HO 8: Work using power-driven metal forming, punching, or shearing machines

HO 9: All mining other than coal mining, including work at gravel pits

HO 10: Work involving meat harvesting, packing, processing, or rendering, including the operation of power-driven meat slicers in retail stores

HO 11: Work involving the operation of power-driven bakery machines

HO 12: Work involving the use of power-driven paper-products machines, including the operation or loading of paper balers in grocery stores

HO 13: Work in the manufacturing of brick, tile, or kindred products

HO 14: Work involving the use of circular saws, band saws, or guillotine shears

HO 15: All work involving wrecking, demolition, or ship-breaking

HO 16: All work in roofing operations

HO 17: All work in excavating, including work in a trench as a plumber

A student enrolled in a career and technical education program may perform seven of the hazardous occupations (HO 5, HO 8, HO 10, HO 12, HO 14, HO 16, and HO 17) under established guidelines documented in a written agreement with the employer. These guidelines include the following:

1. All hazardous work will be performed under the direct and close supervision of a qualified and experienced person.

2. Safety instructions are given by the school and reinforced by the employer with on-the-job training.

3. The job training follows a schedule that reflects organized and progressive skills development.

Questions and Activities

1. What legal topics have direct impact on career and technical education teachers and their students?

2. Explain the importance of case decisions in establishing law.

3. Enumerate basic student rights that must be recognized in the classroom.

4. Explain how teachers act in loco parentis when they are working with students.

5. Explain how teachers can legally administer corporal punishment to students.

6. Explain how schools and teacher-coordinators can legally regulate student behavior at school functions. How does this apply to student organizations?

7. Explain why it is necessary to monitor minors who are employed. What is the role of the teacher-coordinator?

8. Recall your study of training agreements and training plans. Are these legally binding documents? Discuss. If you need to do so, refer to Chapter 11 on training agreements and training plans.

9. Discuss labor laws that apply to the employment of teen workers.

For references pertinent to the subject matter of this chapter, see Reading Resources.

SECTION FOUR

Application of the Systems Approach

INTRODUCTION TO SECTION FOUR

Sections One, Two, and Three dealt with the various components of the systems approach to improving instruction. Section One described the theory of the **strategy** (goals), Section Two described the **structure** (curriculum) of using the work environment as a learning experience, and Section Three described the **system** of improving instruction.

Section Four focuses on the various fields of cooperative education that utilize the systems approach: (1) agricultural occupations, (2) business occupations, (3) health occupations, (4) family and consumer sciences occupations, (5) marketing occupations, and (6) trade and industrial occupations. Organized and operated according to the underlying concepts and principles outlined in Sections Two and Three, the cooperative plan of instruction can be used within a program of instruction in each of these areas at the secondary and post-secondary levels. However, the general techniques described in the previous chapters may need to be modified because each occupational field has certain unique characteristics. These characteristics are derived from various factors, among them, (1) the number of direct training personnel found in the occupations, (2) the degree to which the occupations demand cognitive versus manipulative or affective skills, (3) the degree to which learning is best accomplished in school rather than on the job, and (4) the extent to which the mores of the occupations are adopted.

17

The Plan in Agricultural Occupations

In a training station where the student-learner will be dealing with farmers, it is wise for him/her to have an agricultural background and/or some 4-H experience.

KEY CONCEPT: A training station in agricultural cooperative education can be on a farm, in an agribusiness, in a nursery, in a food processing plant, at a state park, etc.

GOALS

After successfully completing the study of this chapter, answering the questions, and carrying out the activities, the student should be able to:

- Identify and list appropriate training station and career objectives available in agriculture.

- Develop an appropriate curriculum for a local plan in cooperative occupational education for agricultural occupations.

- List the major sources of instructional materials for agricultural occupations education.

- Devise appropriate training plans for students at various training stations.

- Evaluate other student-learners in the classroom and on the job.

Research and investigation have revealed that in addition to farming, there are many other occupations in which agricultural competencies are required. Such occupations may be classified broadly into two categories: (1) nonfarm agricultural occupations in which the agricultural competencies are of primary importance and other competencies are of secondary importance and (2) non-agricultural occupations in which some agricultural competencies are needed. Much research is needed to distinguish between the two so that the proper combination of instructional activities may be devised.

Agribusiness Occupations

Agribusiness occupations are those nonfarm agricultural occupations in business and industry in which agricultural competencies are primary. Some are allied to marketing, such as those found in feed and fertilizer sales, elevators, farm produce and livestock brokerage, farm machinery sales, and garden and horticultural sales. Others are allied to industrial occupations, such as those found in farm machinery repair.

Through preparatory courses the student receives a background in the various areas of agricultural science, horticulture, and agricultural technology. In the twelfth grade, the student is offered supervised experience in a nonfarm agricultural occupation. In some occupations, the experience might be approximately half-days for the school year, as is common in the other five cooperative education plans. The student-learner is enrolled in an agricultural class, taught by the agricultural teacher-coordinator, in which directly related instruction is given. Electives might be courses in agriculture, business, marketing, or industrial education, depending on the nature of the student's occupational choice.

Agricultural Occupations

Agricultural occupations clusters are composed of groups of related courses or units of subject matter that are organized for carrying on learning experiences concerned with preparation for or upgrading in occupations requiring knowledge of and skills in agricultural subjects. The functions of agricultural production, agricultural supplies, agricultural mechanization, agricultural products (processing), ornamental horticulture, forestry, agricultural resources, and the services related thereto are emphasized in the instruction designed to provide opportunities for students to prepare for or to improve their competencies in agricultural occupations. An agricultural occupation may include one or any combination of these functions.

The agriculture CIP Code 01 includes the following clusters of occupations.[1]

01.00   Agriculture, General
01.01   Agricultural Business and Management
01.02   Agricultural Mechanization
01.03   Agricultural Production Operations
01.04   Agricultural and Food Products Processing
01.05   Agricultural and Domestic Animal Services
01.06   Applied Horticulture / Horticultural Business Services
01.07   International Agriculture
01.08   Agricultural Public Services
01.09   Animal Sciences
01.10   Food Science and Technology
01.11   Plant Sciences
01.12   Soil Sciences
01.99   Agricultural Business and Production, Other

[1]Robert L. Morgan and E. Stephen Hunt, *A Classification of Instructional Programs—2000*, National Center for Education Statistics, Presidential Building, 400 Maryland Avenue, S.W., Washington, DC 20202, 2000, pp. 3–4. Consult your state department of vocational education for a more detailed list.

COOPERATIVE EDUCATION PLANS

An agricultural cooperative education plan may occur in one of two ways: (1) in a small school having only one agriculture teacher, only one option area in agriculture (*i.e.*, horticulture, agribusiness, agriculture production) would be offered at the junior and senior levels, with cooperative education offered during the senior year or (2) in a larger school having two or more agriculture teachers, several options might be offered, and seniors in cooperative education could pursue employment in basically any agricultural field.

The first situation is unlikely, since cooperative education for agricultural students only is not feasible in a one-teacher program. In this case, the small number of agricultural students who want cooperative education enroll in interrelated (diversified) cooperative plan.

An example of a schedule of a senior-level student enrolled in the cooperative agricultural education plan at a medium-sized secondary school, in Paris, Illinois, is as follows.[2]

Class Schedule

1st Hour	Agricultural Business Management
2nd Hour	English IV
3rd Hour	Government/Contemporary Problems
4th Hour	Released time to go to the training station on the farm as farm worker, livestock laborer, a paid worker as a student-learner. (See the training plan, Figure 17–3.)

An example of how a large school system has organized the career-technical agriculture program is that of Omaha, Nebraska. In the Omaha high schools, a four-year CTE program in agribusiness is offered in the following sequence of courses and experiences.

---

[2]Courtesy, Michael Gray, Coordinator of Agricultural Cooperative Education, Paris High School, Paris, Illinois.

Grade 9: Agribusiness I and II

Grade 10: Agribusiness III and IV or V and VI or VII and VIII or IX and X

Grade 11: Agribusiness III and IV or V and VI or VII and VIII or IX and X

Grade 12: Agribusiness XI and XII and Agribusiness Laboratory I and II or any agribusiness course offered during the 10th and 11th grades

Agribusiness III and IV (grades 10, 11, and 12)—offers instruction in the areas of soil development, fertility, and conservation. Also included are activities such as land judging, surveying, and environmental control.

Agribusiness V and VI (grades 10, 11, and 12)—offers instruction in the selection, feeding, breeding, grading, identification, and evaluation of animals.

Agribusiness VII and VIII (grades 10, 11, and 12)—offers instruction in the study of crops, lawn and turf management, greenhouse and nursery operations.

Agribusiness IX and X (grades 10, 11, and 12)—offers instruction in arc and oxyacetylene welding and cutting. Other learning activities include tractor and machinery operations, safety, and maintenance.

Agribusiness XI and XII (grade 12)—taken concurrently with Agribusiness Laboratory I and II. Offers instruction in areas such as job application, employer–employee relations, sales, taxes, insurance, and business operations.

Agribusiness Laboratory I and II (grade 12)—offered under the cooperative plan and taken concurrently with Agribusiness XI and XII. Students spend a minimum of 15 hours a week in supervised on-the-job training stations in cooperative local businesses. Agribusiness XI and XII is the related instruction under the cooperative plan.

Under the cooperative plan, students are typically placed in on-the-job training stations in grade 12, as illustrated in the Omaha example. Studies show that most agricultural occupations of the future will be in the area of products and services, not in the area of production agriculture. Thus, appropriate training stations for cooperative agricultural education include florists, fertilizer dealers, farms, implement dealers, nurseries, processing companies, and agricultural marketing organizations.[3]

Bona-fide cooperative education plans in agriculture will include paid employment in approved on-the-job training stations under qualified job supervisors and certified teachers.

Related Instruction

THE NATURE OF INSTRUCTION

In cooperative education the workplace is the place where the technical skills, merchandise knowledge, and human relations particular to the organization are to be taught and learned. The classroom instruction must teach the technical information related to the organization as well as the general information required of all workers. For this purpose the local board of education purchases and places necessary technical information in the related classroom; the coordinator organizes the technical materials; and the student-learners meet daily for one period or two periods (according to the plan adopted) for the express purpose of assimilating the information. It is at this point that the coordinator assumes the task of guiding and directing the individualized study of each student-learner.

When planning related instruction for student-learners, the teacher-coordinator should consider teacher–learner objectives and instructional methods, materials, and classroom facilities.

OBJECTIVES OF INSTRUCTION

In cooperative agricultural education, there are three areas of instructional objectives. The first area of objectives emphasizes *the development of information, concepts, and attitudes that are basic to all agricultural occupations* and that the student-learners use to advance successfully toward their career objectives. These instructional objectives are taught primarily in group meetings of the class devoted to what is known as general related instruction.

The second area of objectives stresses the *student-learner's building of occupational skills and knowledges that are applicable to the particular initial job in which he/she is placed.* Each student's individual learning outcomes are stated in the training plan outline, which identifies the job tasks, functions, and processes in which the student-learner will be trained. Some of these objectives are accomplished by in-school instruction through group study by student-learners who have common needs by being in the same type of agricultural business or service establishment. Some of these learning outcomes are emphasized in the informal individual conferences of student-learner and teacher-coordinator and/or training station sponsor.

The third area of objectives deals with improving *the student's general education.* The teacher-coordinator should certainly try to help the student to read, write, compute, and communicate better.

INSTRUCTIONAL METHODS, MATERIALS, AND FACILITIES

The instructional methods in cooperative agricultural education plans evolve from the two types of instruction necessary for developing vocational competence. These are *general related instruction* and *specific related instruction*—general and technical.

General related instruction.—The term "general related instruction" refers to instruction in topics considered of common importance to all workers in agricultural occupations.[4] In some cooperative agricultural education plans, one period per day from

---

[3]Lloyd J. Phipps and Edward W. Osborne, *Handbook on Agricultural Education in the Public Schools,* Fifth Edition, Danville, Illinois: Interstate Publishers, Inc., 1988.

[4]Many state departments have developed guides and course outlines for general related instruction.

the two-hour block is devoted to such instruction. Group instruction methods, with emphasis upon readings, lectures, and discussions, are appropriate for general related instruction, which is primarily conceptual and attitudinal in nature.

Although the topics in the general related class may vary from program to program, the following topics may be considered common.

Units of Study

School and Community Relations

The Cooperative Agricultural Training Plan
What Makes a Good Trainee?
School Rules and Regulations
Responsibilities of the Student-Learner
Grooming and Dressing for the Job
Choosing a Means of Earning a Living
Getting a Job and Growing In It
Job Versus Training Station
Employer–Employee Relations
Paying for Services Through Taxes
Private Insurance Plans
Social Security
Legal Regulations for Young Workers
Labor Unions
Wage and Hour Records
Filing Your Income Tax Report

Career Opportunities in Agriculture

Agriculture Mechanics
Agriculture Sales and Service
Agribusiness
Forestry
Animal Science
Crop Science
Landscape and Ground Management
Veterinary Technology

Recurring Topics

Personal Development

Understanding Ourselves
Keeping Out of Trouble
Character and Personality
Successful Job Relations
Practice of Manners
Effective Study Habits
Budgeting Your Income
Saving and Investing
Credit and Money Management
Understanding Contracts
Consumer Legal Problems
Using Negotiable Instruments
Planning Your Career

Human Relations

Understanding Others
Courtesy to Others
Cooperation

Organization Activities

Aims and Purposes
Robert's Rules of Order
Making a Speech
Learning Opportunities
Recreational Opportunities

Contact your state department or a teacher-educator in your state for some suggestions.

Specific related instruction.—The term "specific related instruction" is used to denote instruction given to the students in the related class in the technical information requisite to their jobs. At least half of the time available for related instruction should be devoted to these needs. Whereas general topics of value to all class members may be handled through group instruction, the technical requirements of the numerous agricultural occupations must be handled through individual instruction. The technical requirements of the various jobs and occupations differ. The rate and extent of learning vary considerably among student-learners. To be effective, technical instruction in cooperative agricultural education must be individualized, and student-learners must accept the responsibility for applying themselves with the guidance of their teachers.

Training plan.—Because each agricultural student-learner needs instruction in the specific job activities performed on the job, the teacher-coordinator must determine what those specific learnings are. The technique that accomplishes this is the development of a *training plan*. There **must** be a training plan for every individual student-learner, showing what is to be learned on the job. This plan should be worked out by the teacher-coordinator in cooperation with the employer. This plan will be in the form of a job analysis of the phases of the occupation that the student-learner is expected to learn.

Training will be given in many occupations with which the coordinator is not familiar and about which assistance will have to be sought. In practically all instances, a craft committee or an advisory committee can help. This committee may consist of the employer or supervisor and an employee in the establishment, or it may consist of people engaged in the occupation outside the educational agency. The employer should be encouraged to have a hand in working out the training plan.

This training plan may be worked out in any one of a number of forms, but the three-column variety will probably be the most usable and convenient. The first, or left-hand, column contains a listing by units of the manipulative skills to be learned on the job, while the related information that correlates with these job units appears in the second column. In the third column are references to specific books, magazines, etc., that indicate exactly where the desired related information can be found.

When the training plan is completed, it becomes a part of the training agreement. An example of a training plan is shown in Figure 17–3.

When a schedule of processes or other job analyses has been worked out for each student-learner, the teacher-coordinator still faces the problems of learning to guide each student's individual study, making

necessary instructional materials available, and developing a classroom suitable for individual study.[5]

SPECIAL LABORATORIES AND CLASSROOMS

If the school already has a career-technical agricultural education classroom and laboratory, and these facilities are available to be used for related instruction, then new facilities may not be necessary to accommodate the cooperative agricultural education instruction. If a laboratory is not available, then a classroom suited to both group and individual instruction is desirable.

Movable tables might well be arranged in a "U" shape, with chairs outside the "U." Writing board space, bulletin board space, reference book shelves, magazine racks, and filing cabinets are needed. Technological advancements are significantly impacting all career and technical education, including agricultural education. Computers, a digital camera, an audiotape recorder, a TV, a video camcorder, a VCR, a DVD player, an overhead projector, and an LCD multimedia projector can play significant roles in the related instruction classroom.

A conference room available for individual student-coordinator conferences is an important feature of the cooperative education classroom. It will provide privacy for the discussion of personal problems or problems encountered at training stations. Glass partitions or large glass windows in the conference room will allow the activities going on in the classroom to be in full view of the teacher-coordinator while he or she is participating in an individual conference. If the school has other cooperative education plans, such as health occupations or family and consumer sciences occupations, then this specially designed classroom could conceivably be shared with teacher-coordinators in those areas. This will be

---

[5]Training stations and training plans may be developed for a wide range of student-learners, from slow achievers to fast achievers. Competency-based SSLO, described in Chapter 11, may be selected each evaluation period, suitable to the learning abilities of the individual student-learner.

New technology is important in the related instruction classroom. This instructor is preparing a laptop computer and an LCD panel for use in a presentation. (Courtesy, Jasper S. Lee)

dependent on when related classes are scheduled, since all the cooperative students should be available to go to training stations during the afternoon hours. A classroom of this design is also ideal for adult training and development.

SOURCES OF INSTRUCTIONAL MATERIALS

By contacting the following sources and identifying the needs specific for instructional materials for related class instruction in cooperative agricultural education, teacher-coordinators can obtain titles, descriptions, and price lists of textbooks, manuals, audio-visual aids.

AAVIM
220 Smithsonia Road
Winterville, GA 30683
Phone: (800) 228-4689
Fax: (706) 742-7005
Email: sales@aavim.com
Web site: www.aavim.com

Association for Career and Technical Education
1410 King Street
Alexandria, VA 22314
Phone: (800) 826-9972
Email: acte@acteonline.org
Web site: www.acteonline.org

Curriculum and Instructional Materials Center
Oklahoma Department of Career and
    Technology Education
1500 West Seventh Avenue
Stillwater, OK 74074-4364
Phone: (800) 654-4502
Fax: (405) 743-5154
Web site: www.okcareertech.org/cimc

Delmar / Thomson Learning
P.O. Box 15015
Albany, NY 12212-5015
Phone: (800) 477-3692
Fax: (518) 464-7000
Web site: www.delmar.com

Instructional Materials Laboratory
University of Missouri–Columbia,
      College of Education
2316 Industrial Drive
Columbia, MO 65211-8120
Phone: (800) 669-2465
Fax: (573) 882-1992
Email: info_iml@coe.missouri.edu
Web site: www.iml.coe.missouri.edu

Interstate Publishers, Inc.
P.O. Box 50
Danville, IL 61834-0050
Phone: (800) 843-4774
Fax: (217) 446-9706
Email: info-ipp@ippinc.com
Web site: interstatepublishers.com

ITCS Instructional Materials
1401 South Maryland Drive
Urbana, IL 61801
Phone: (217) 244-3906
Fax: (217) 333-0005
Web site: www.aces.uiuc.edu/ITCS/IM

Multistate Academic and Vocational
      Curriculum Consortium
1500 West Seventh Avenue
Stillwater, OK 74074-4364
Phone: (800) 654-4502
Fax: (405) 743-5154
Web site: www.mavcc.org

Ohio Agricultural Education
Curriculum Materials Service
The Ohio State University
254 Agricultural Administration Building
2120 Fyffe Road
Columbus, OH 43210-1067
Phone: (614) 292-4848
Fax: (800) 292-4919
Email: cms@osu.edu
Web site: www-cms.ag.ohio-state.edu

For a bibliography of agriculture school-to-work standards, visit www.mcrel.org/careerstandards/agri-biblio.asp.

The FFA organization is the third dimension of the cooperative education plan. It, along with the classroom and the training station, is an integral part of instruction. (Refer to Chapter 14 for suggestions on how students can benefit from participation in an FFA chapter.)

FFA has long been noted for its stress on parliamentary procedure, neat appearance and appropriate dress, proper etiquette and good manners in public activities, and successful completion of individual and group projects. All of these are important to student-learners. In fact, if the number of student-learners warrant it, a special chapter of FFA for these trainees might be appropriate.

SECURING DESIRED LEARNING ON THE JOB

The experiences in the training station are vital to student-learners to give them opportunities to put into practice under supervision those skills and knowledges learned in school and on the job. Consult Chapters 10, 11, and 12 again for suggestions on how to integrate effectively in-school instruction with on-the-job experiences.

Teacher-coordinators are reminded that the proper establishment of training stations is very important. Training stations should provide more than just jobs for students. Approved training stations must be agriculturally related to the career objectives of the students. Training stations should provide adequate supervision and appropriate hours per week and for the school year so that a quality educational experience is available.

A training profile for each potential training station, as shown in Figure 17–1, is a useful tool in deciding whether adequate learning experiences are available at a given business. Some coordinators use this profile as a review of what has transpired at a training station during the school year. The student and job supervisor help develop the profile at the end of the school year. Some coordinators use this form as a training station analysis when they are selecting or approving training stations. In those cases, the degree of progress items are changed to:

TRAINING PROFILE — AGRICULTURAL MECHANICS HELPER

A combination of subject matter and activities designed to develop abilities in the student that are necessary for his/her assisting with and/or performing the common and important operations or processes concerned with the selection, operation, maintenance, and use of agricultural power, agricultural machinery and equipment, structures and utilities, soil and water management, and sales and services.

Student's Name_____School Year _____

Training Station _____

Recommended Level of Training _____

Degree of Progress:

1. **No** training
2. **Very little** training
3. **Adequate** training

4. **Great deal** of training
5. Not applicable to program

Dates of Training Progress Evaluations: A _____ B _____ C _____ D _____

Task	A	B	C	D
Orientation				
1. Understands nature and purpose of firm.				
2. Knows regulations concerning working hours, tardinesses, absences, coffee breaks, holidays.				
3. Recognizes the job in relationship to the work flow at the firm.				
*Duties and responsibilities**				
1. Cleans the shop.				
2. Identifies and locates tools.				
3. Washes service rack.				
4. Cleans oil pump.				
5. Operates battery charger.				
6. Operates power grinder.				

FIGURE 17–1

TRAINING PROFILE (Continued)

Task	A	B	C	D
Duties and responsibilities* (cont.)				
7. Operates steam cleaner.				
10. Adjusts tractor wheel width.				
12. Tests and services batteries.				
13. Installs truck bumpers and trailer hitches.				
18. Services cooling system.				
20. Operates power paint sprayer.				
21. Sharpens disks.				
26. Selects and changes oil and filter.				
28. Reassembles parts.				
29. Assembles new equipment according to instructions.				
30. Has mechanic check repair.				
35. Repairs body parts of machinery with only acetylene and arc welders.				
38. Identifies iron and steel parts and fasteners.				
40. Adjusts and calibrates machinery.				
41. Tightens and adjusts bolts, pulleys, sprockets, and chain.				
Miscellaneous duties				
1. Uses telephone.				
2. Conducts public relations.				
3. Carries out special assignments by management.				

*Note that because of space limitations, only selected tasks are shown from the many others that would be practiced; therefore, the numbers skip.

FIGURE 17–1 (Continued)

1. No training available.
2. Very little training available.
3. Adequate training available.
4. A great deal of training available.
5. Does not apply.

STUDENT EVALUATION

A weekly student-learner report from each student is advisable. This report (see Figure 12-1) may be prepared in class each Monday morning as a summary of the previous week's experiences on the job. Ask each student to prepare one and keep the reports filed in the student's folder.

An appropriate rating sheet should be used at the end of each grading period, filled out preferably by the coordinator in consultation with the job supervisor, and in some cases, with the student present. Figure 17–1 is an example of one form. Other examples of various forms that can be adapted to specific needs are presented throughout the text. Each coordinator should seriously consider preparing appropriate forms in consultation with advisory committee members.

STUDENT APPLICATION

With a knowledge of the demands of the cooperative plan upon students, the teacher-coordinator accepts applications from prospective student-learners. An appropriate application form should be developed to serve the needs of the specific school system and its students. (Refer to Chapter 7 for information regarding interviewing, counseling, and admitting student-learners.)

TRAINING STATION OPPORTUNITIES

After studying the capabilities and needs of the student-learners accepted into the cooperative agricultural education plan, the teacher-coordinator needs to prepare a training agreement for each training station and for each student-learner placed there. A training agreement (see Figure 17–2) which has

---

CASE-IN-POINT
Kenney Shows Promise

The Southern States farm cooperative store in Lexington, Kentucky, deals in seed grains, hardware supplies, paint, small engine products, fence, feeds, feed mill services, etc. The cooperative agricultural education coordinator, Harvey Gibbs, feels that this organization would make a perfect training station for one of his students who has grown up on a farm. The student, Kenney Dobbs, has a pleasant personality, a good record in vocational agriculture, and an overall B grade average in school. He has shown leadership in the FFA and in football. Everything points to his potential for an eventual supervisory or management career.

Mr. Gibbs secures an appointment to consult with the manager, Everett Toll, regarding the cooperative as a training station for the coming school year. Mr. Gibbs takes along to the appointment a profile of the types of duties and responsibilities that a student-learner with Kenney's potential could experience in a retail store operation during the senior year in cooperative education. Mr. Toll has room on his payroll for a part-time employee. The Southern States organization likes to bring promising employees up the career ladder into its management ranks. Its many stores across the Midwest are always searching for good assistant managers and managers.

The opportunity to employ a student-learner appeals to Mr. Toll. He and Mr. Gibbs outline a training plan, adapting the profile to what is possible in the cooperative business. They agree on a beginning wage with possible raises during the school year and to set up an interview for Kenney, as well as several other possible trainees. If all goes well, and Kenney is chosen, he will start at his new training station the day after school begins in the fall.

---

been developed in consultation with the training station owner or the on-the-job supervisor. The appropriate matching of student-learners with training stations providing experiences suitable to each student's interests, abilities, and career objectives will contribute to the success of cooperative education in the school system and in the business and agricultural communities.

COOPERATIVE EDUCATION TRAINING AGREEMENT

Agriculture Placement Program

This training agreement is for the purpose of (1) establishing the general conditions of the student-learner's employment and (2) indicating the on-the-job areas of experience in which the student-learner will be working.

AGREEMENT: The employer, _____(Name)_____ , agrees to employ _____(Name of Student)_____ , who is enrolled in cooperative education at Paris High School for the purpose of receiving training and experience in _____(Occupation)_____. The employment period will begin on ____(Date)____ and will be for a minimum of _____ hours per week during the school day. This training program is designed to extend over one school year. The beginning wage will be _____ per hour, with future consideration of increases directly proportional to the increased productivity of the student.

While in the process of work experience, the student will have the status of a student worker. The wages, employment, and work experience of the student shall be in accordance with the federal, state, and local laws and regulations.

All complaints will be made to and adjusted by the coordinator. The employer and cooperating instructor have the right to discuss the student's progress and to share records kept on the student that are relevant to the program.

1. The EMPLOYER agrees to:

    a. Supervise the student-trainee on the job.

    b. Rotate the student through the various areas of the job in accordance with the jointly developed training plan.

    c. Notify the teacher-coordinator at once of any unsatisfactory developments.

    d. Provide the instructor time for a short conference regarding the progress of the student's work and to prepare a written report or evaluation with the instructor at the end of each nine weeks' school period.

2. The COORDINATOR agrees to:

    a. Periodically visit the training station to assess the progress and the needs of the student-learner.

FIGURE 17–2

COOPERATIVE EDUCATION TRAINING AGREEMENT
(Continued)

    b. Withdraw or transfer the student when it is mutually agreed to be in the best interest of everyone concerned.

    c. Work with the student, parents, and employer to insure that the student gets the most value from the cooperative education plan.

3. The PARENTS agree to:

    a. Be responsible for the student's transportation to and from work and back to school.

    b. Be responsible for the student's personal conduct while he/she is in the cooperative education plan.

4. The STUDENT agrees to:

    a. Consult with the teacher-coordinator when any difficulties arise at the training station.

    b. Conform to all rules and regulations of the training station.

    c. Exhibit a cooperative attitude, proper attire, and good work habits, and try to work in a manner that will reflect credit upon himself/herself, the school, and the work program.

    d. Be punctual and dependable in working the designated hours.

    e. Be in school and at work every day, unless confined at home with an illness or excused by the teacher-coordinator and the employer from regular school activities. An unexcused absence will lower the student's quarter grade one whole letter.

    f. Inform the employer by 8:00 A.M. if he/she has to miss work. Absence from work is treated the same as absence from classes.

    g. Inform the teacher-coordinator and the school by 8:00 A.M. if he/she is unable to attend class and/or work.

    h. Have a work permit on file in the school office.

FIGURE 17–2 (Continued)

School Attendance

Attendance at school is a must. The student is expected to attend school if he/she is able to go to work. If the student works on the job on any day, attendance in school for all classes and activities on that day is expected. In the event the student is unable to attend classes and/or other school activities, then he/she cannot go to work (the exception being if the student has an excused absence approved by a school official). Non-compliance with these rules will result in a deduction of 40 percent from the quarter grade. Repeated non-compliance will result in the student being dropped from the cooperative plan.

Failures

Failing any class for the first quarter places the student on probation for the second quarter.

APPROVALS

Employer	Date	Student	Date
Address	Phone	Address	Phone
Coordinator	Date	Parent	Date
Address	Phone	Address	Phone

FIGURE 17–2 (Continued)

STEP-BY-STEP TRAINING PLAN

for

Student-Learner	Birthdate	Soc. Sec. No.

Katelain Jessup 06/10/86 123–45–6789

Training Station: Ron Martin
Address: R.R. #4, Paris, IL 61944
Phone: (217) 463–5630

Job Definition: 410.664-010 Farmworker, Livestock Laborer
Dictionary of Occupational Titles

Performs any of the following tasks to tend swine: Mixes feed and additives, fills feed troughs with feed, and waters livestock. Examines animals to detect diseases and injuries. Vaccinates animals by placing vaccine in drinking water or feed or by using syringes and hypodermic needles. Applies medications to cuts and bruises, sprays livestock with insecticides, and herds them into insecticide bath. Binds or clips testes or surgically removes testes to castrate livestock, clips identifying notches in ears to indicate litter. Clamps metal rings into nostrils of livestock to prevent rooting. Using disinfectants, shovels, and brushes, cleans stalls and sheds. Grooms, clips, and trims animals for exhibition; may maintain buildings and equipment. May plant, cultivate, and harvest feed grain for stock. Maintains breeding, feeding, and cost records.

Description of Training Station

The Martin Farm breeds both purebred and commercial animals. It is a farrow-to-finish operation, which has both confinement and outdoor growing space. This training station grows its own feed on owned land and with owned equipment.

Courtesy: Paris High School, Paris, Illinois.

FIGURE 17–3

STEP-BY-STEP TRAINING PLAN (Continued)

Learning Experiences	Training Station	School Instruction Group	Individual	Time Schedule
A. Performing Administrative and Record-keeping Functions				
1. Applies for registration with purebred swine association.	X		X	
2. Appraises farm properties to determine their total value.	X	X	X	
3. Assigns jobs to workers.	X			
4. Calculates cost of gain on feeder pig operation.	X		X	
5. Calculates cost of rations and feed mixture.	X			
6. Dismisses employees.		X		
7. Hires employees.		X		
8. Keeps breeding and farrowing records.	X		X	
9. Keeps labor records.	X		X	
10. Files swine operations documents.	X			
11. Maintains record system.	X	X	X	
12. Makes inventory of swine and equipment.	X			
13. Orders fuel and petroleum supplies.	X			
14. Orders replacement parts for equipment.	X			
15. Plans for increased land requirements by leasing.		X	X	
16. Plans for increased land requirements by purchasing.		X	X	
17. Plans goals for swine farm.	X		X	
18. Plans pasture rotation system.	X		X	
19. Plans public relations program for the farm.	X	X	X	

FIGURE 17–3 (Continued)

STEP-BY-STEP TRAINING PLAN (Continued)

Learning Experiences	Training Station	School Instruction Group	School Instruction Individual	Time Schedule
20. Plans daily work schedule.	X		X	
21. Prepares cash flow statement.		X	X	
22. Prepares farm income tax return.		X	X	
23. Prepares marketing records.	X	X	X	
24. Prepares production performance records.	X	X	X	
25. Prepares purchase orders.		X	X	
26. Prepares swine feeding records.	X		X	
27. Purchases new or used equipment.		X	X	
28. Records fuel, oil, and hydraulic fluid consumed.	X			
29. Records swine health information.	X		X	
30. Records weaning weight of pigs.	X		X	
31. Sets priorities for purchasing equipment.	X	X	X	
B. Maintaining a Safe Environment				
1. Disposes of unused chemicals.	X	X	X	
2. Extinguishes fires.	X	X	X	
3. Keeps a record of accidents.	X		X	
4. Keeps work areas ventilated.	X			
5. Makes inspections for safety hazards.	X		X	
6. Stores chemicals, flammable materials, and medications.	X	X	X	
C. Maintaining Hog Herd Health				
1. Administers medication by injection.	X		X	
2. Administers oral medication.	X			
3. Applies insecticides in buildings.	X			
4. Applies medication to cuts and bruises.	X			

FIGURE 17–3 (Continued)

STEP-BY-STEP TRAINING PLAN (Continued)

Learning Experiences	Training Station	School Instruction Group	School Instruction Individual	Time Schedule
5. Cleans swine pens and housing.	X			
6. Disinfects boots and clothing.	X			
7. Disinfects buildings and equipment.	X			
8. Dusts swine.	X		X	
9. Examines hogs for external parasites.	X		X	
10. Examines hogs for internal parasites.	X		X	
11. Gives iron shots.	X		X	
12. Paints sow's underline with iron compound.	X		X	
13. Plans disease control program.	X	X	X	
14. Plans parasite control program.	X	X	X	
15. Separates sick, weak, and injured animals.	X		X	
16. Sprays swine.	X			
17. Takes temperature of animals.	X	X	X	
18. Treats hogs for contagious diseases.	X		X	
19. Treats hogs for scours.	X		X	
20. Vaccinates animals.	X	X	X	

(Because of space limitations, the remainder of this training plan shows only the main headings of the learning experiences and omits the detailed breakdown.)

D. Formulating Feeds and Feeding

E. Constructing and Maintaining Swine Operations Buildings, Structures, and Equipment

F. Marketing and Shipping Hogs

G. Breeding Swine

H. Fitting and Showing Swine

I. Handling and Caring for Animals

FIGURE 17–3 (Continued)

Questions and Activities

1. Prepare a 20-minute talk that you could give before a civic club, trade group, or union explaining the cooperative agricultural education plan. Prepare charts or other visual aids to supplement your talk. (Add the notes of this talk to your files.)

2. What are the minimum qualifications for becoming certified as a cooperative education coordinator in a reimbursable program in your state?

3. How adequate is the list of topics for related instruction suggested in this chapter? If you were the teacher-coordinator, which areas would you emphasize the most? Use a specific-type program as a case for your analysis.

4. If you were preparing a long-range (three-to five-year) plan for purchasing equipment and supplies for a cooperative agricultural training classroom, what essentials would you request the first year? What would be some desirable items you would put in your plan for purchase later on? (Add the notes of these ideas to your files.)

5. Analyze the learning outcomes that could be expected in a specific prospective training station in your local community. Determine the possibility of a student-learner's progress toward his/her career objective in this training station by the end of the school year. (Refer to Chapter 11 for suggestions on how to proceed.)

6. Discuss the merits of the step-by-step training plan (or schedule of processes) in organizing a training station and the merits of the profile approach to evaluating a training station experience.

7. How do FFA organization activities provide educational experiences for student-learners? How do they differ from other occupational areas?

8. From examples you have studied, prepare for each of the following a form that you could use in a local cooperative agricultural training program: a rating sheet on the student-learner at the training station and a weekly report form on which the student-learner could report his/her training station activities. (Add a copy of each of these to your files.)

9. How might you, as a teacher-coordinator, participate in a local agricultural adult training and development program?

10. State the learning outcomes of a training plan in behavioral terms. (Refer to Chapter 11 for suggestions.)

For references pertinent to the subject matter of this chapter, see Reading Resources.

18

The Plan in
Business Occupations

Recent technological changes have made computer instruction for word processing and desktop
publishing prominent in business occupations.

KEY CONCEPT: Although those competencies needed for business occupations are necessary in many types of businesses, only those occupations that require competencies primarily in recording, storing, reproducing, and communicating information can be called business occupations.

GOALS

After successfully completing the study of this chapter, answering the questions, and carrying out the projects, the student should be able to:

- Describe recent changes in business education curricula and curricula content.

- Identify appropriate training stations that are available in business occupations and their career objectives.

- Choose a proper curriculum for a local plan in cooperative business education.

- Devise appropriate training plans for training stations.

- Evaluate other student-learners in the classroom and on the job.

- Describe student-learner membership benefits in BPA and FBLA.

- Prepare a list of topics to be included in a cooperative business education student handbook.

For many years, secondary education for business was called "commercial education." In the 1950's, the term "business education" came into vogue. It was used to describe an educational program that provided educational experience both for and about business. Simply put, it indicated that business educators were to assist all persons in becoming knowledgeable, economic citizens and wise consumers who would be equipped with basic but personal skills of business. In addition, business educators were to furnish students with exploratory experiences related to careers in business and to instruct them in ways to gain entry employment and later upgrading and advancement.

Within the framework of business education is the office occupations education plan. Although the term "business and office occupations" continues to be used to define the target occupational areas in which the cooperative plan provides career and technical instruction, many state departments of education and the Association for Career and Technical Education call the discipline area "business education." In this chapter, the terms "business occupations" and "business education" broadly refer to all the occupations included in *A Classification of Instructional Programs—2000*[1] as being career-technical business and office related programs. The families of business occupations for reporting purposes are:

11.00	Computer and Information Services
11.01	Computer Science
11.02	Computer Programming
11.03	Data Processing
11.04	Information Sciences/Studies
11.05	Computer Systems Analysis
11.06	Data Entry/Microcomputer Applications
11.08	Computer Software and Media Applications
11.09	Computer Systems Network and Telecommunications

11.10	Computer Information Technology Administration and Management
11.99	Computer and Information Sciences, Other
22.03	Legal Support Services
51.07	Health and Medical Administrative Services
52.04	Administrative and Secretarial Services
52.0401	Administrative Assistant / Secretarial Science, General
52.0402	Executive Assistant / Executive Secretary
52.0406	Receptionist
52.0407	Business / Office Automation / Technology / Data Entry
52.0408	General Office Occupations and Clerical Services
52.05	Business Communications
52.06	Business / Managerial Economics

The Business Education Curriculum

Business education curricula are being challenged as never before to be responsive to the new office occupations and the new technologies impacting all business occupations. Although some of the curriculum changes needed have been minor, most have required that students develop new skills so that they can be productive workers in today's information age. In many cases, these changes have required remodeling facilities, purchasing new equipment, developing competency-based curricula, training teachers in the new technologies, and even hiring additional staff members to accommodate the changes.

BUSINESS OCCUPATIONS IN THE INFORMATION AGE

The U.S. Bureau of Labor Statistics projects that administrative support occupations, including clerical, are expected to grow by 14 percent during the next decade. Technological advances are projected

___
[1]Robert L. Morgan and E. Stephen Hunt, *A Classification of Instructional Programs—2000*, National Center for Education Statistics, Presidential Building, 400 Maryland Avenue, S.W., Washington, DC 20202, 2000, pp. 8–9, 20, 36–37, 41–42.

to slow employment growth for stenographers and word processors. Jobs for receptionists and information clerks will grow faster than average, spurred by rapidly expanding industries such as business services. Because of their large size and substantial turnover, clerical occupations will offer abundant opportunities for qualified job seekers in the years ahead. Thus, cooperative education programs for office occupations will continue to be faced with many challenges.

NEW CURRICULUM CONTENT

The business education curriculum has undergone major changes in recent years. An effort has been made to update the business education program content and teaching methodologies required to prepare students for emerging employment needs. This has occurred partly in response to technological advances impacting business occupations and partly in response to national educational thrusts. Increased productivity from new office technologies will change the demand for secretaries and stenographers as well as the preparation needed for them to secure employment. Not only must business education prepare students to use electronic mail, facsimile machines, and voice message retrieval systems, but it must also provide them with extensive training in the use of computer software and in the development of programming skills.

In 1998, there were 3,194,000 secretaries: 285,000 legal, 219,000 medical, and 2,690,000 other. Employment of secretaries is expected to grow more slowly than the average for all occupations through the year 2008. However, employment opportunities should be quite plentiful, especially for well-qualified and experienced secretaries, who, according to many employers, are in short supply. General office clerk occupations are expected to experience rapid growth, increasing 17 percent by 2008.

In 1985, the U.S. Department of Education, Office of Vocational and Adult Education, published the results of a project conducted at East Carolina University, Greenville, North Carolina. Entitled *National Standards of Excellence for Business Education*,[2] the project developed standards for excellence in business education programs and instructional standards for information processing curricula at both the secondary and post-secondary/adult levels. Program standards were developed in nine topical areas: Philosophy and Purpose, Organization and Administration, Curriculum and Instruction, Instructional Staff, Financial Resources, Instructional Support Systems, Public Relations, Student Development Services, and Evaluation. The six areas in the information processing standards were Organization, Content, Related Content, Methods and Resources, Instructional Support Systems, and Evaluation. The primary intent of the study was to provide business educators with evaluative criteria so they could identify and correct program deficiencies. The information gathered in the self-evaluation established a program profile from which program improvement plans and implementation strategies could be formulated.

In 1995, the National Business Education Association published the *National Standards for Business Education: What America's Students Should Know and Be Able to Do in Business*. This document was established using a curriculum model that included 12 business education content areas: Accounting, Business Law, Career Development, Communications, Computation, Economics and Personal Finance, Entrepreneurship Education, Information Systems, International Business, Management, Marketing, and Interrelationships of Business Functions. It contains all the subjects taught in business education and identifies specific standards, levels of attainment, and performance expectations. The document is based on the research conducted by a nationally recognized task force that included business-related organizations, business education teachers from all levels of

---

[2]*National Standards of Excellence for Business Education*, a project at East Carolina University's School of Technology, Greenville, NC. Funded and published by the U.S. Department of Education, Office of Vocational and Adult Education, Washington, DC, 1985. Printed and disseminated by the National Business Education Association, 1914 Association Drive, Reston, VA 22091-1596.

education, and business and managerial personnel throughout all regions of the United States. Task force members directed the standards project toward the development of a quality-based continuous-curriculum structure that could regularly be updated and refined to meet the ever-changing needs of students and business environments.[3]

Both of these studies reinforce the premise that if business curricula are to prepare employees to be productive in new business environments in which technology is impacting all business operations, then curricula must change. Those changes have caused an evolution, if not a revolution, in business education. Many changes are apparent; others are being planned. The following are some examples of the directions business education curricula are taking.

1. Although accounting principles have not changed, the methods of performing accounting tasks have changed. Accounting firms now use computers and accounting software to do their work. Business education students must have hands-on time on computers to complete accounting assignments.

2. The business education curriculum should offer, in the middle school, a short-term keyboarding class, preferably for nine weeks, to teach the keyboard and the 10-key number pad by touch.

3. In most schools, typing classes have been replaced by word processing courses. Advanced word processing and software instruction, procedures, and applications are now being emphasized.

4. Electronic records management is essential to contemporary business operations. In addition to learning basic alphabetic filing rules, business students need to know about the media that can be used to scan and store records, such as microfilm, microfiche, computer disk, and CD-ROM. They need to learn how to set up electronic files and to attach labels and descriptors to documents so that they can be located quickly.

The Cooperative Plan in Secondary Business Education Curriculum Patterns

Those teacher-coordinators who are part of the team of business educators need to consider how cooperative plans of instruction become an integral part of contemporary business education curriculum patterns. There are four basic patterns that are used to provide business occupations education at the secondary level. In brief, these are:

1. **Separate course pattern**—A number of courses offered in various areas of business content, with some, such as word processing, accounting, and machine transcription, planned as both a beginning and an advanced sequence. The cooperative plan is often offered as an advanced senior experience for students who have completed several of these basic preparatory courses. Depending on the school, specialized sequences may be used with the cooperative plan for each specialization—clerical, secretarial, bookkeeping–accounting, introduction to business, economics, and business law.

2. **Cooperative education sequence**—The sequence of courses for grades 9–12 is the following:

Grades 9–11: Prerequisite business subjects, depending on the aptitudes and interests of the students, such as keyboarding, information– word processing, bookkeeping–accounting, and introduction to business.

[3]*National Standards for Business Education: What America's Students Should Know and Be Able to Do in Business.* Published by the National Business Education Association, 1995. Printed and disseminated by the National Business Education Association, 1914 Association Drive, Reston, VA 22091-1596.

Grade 12: Cooperative office experience—approximately one-half day; related class for cooperative student-learners—either one or two periods, entitled "Cooperative Office Practice," "Cooperative Business Occupations," "Cooperative Related Instruction," "Cooperative Office Education," or "Cooperative Business Education"; electives—one or more periods in general education, such as English, or in related business education courses, such as advanced bookkeeping–accounting, office procedures, data processing.

3. **Skill sequence with a capstone laboratory**—A program of instruction consisting of one or more basic and core skill courses, plus a culminating senior laboratory experience on a block-time schedule, which integrates prior learnings with a defined entry and career objective. The cooperative plan may be offered to students in this plan as part of their experiences, especially in the method of integrating their single-skill experiences into a definable behavioral pattern.

4. **Intensified senior program**—An intensified senior program that crams, for those with few prior business courses, a complete entry-job curriculum into a block-time schedule. This approach uses the cooperative plan as an instructional device when it is needed during the year, especially at the end when the student is both employable and ready to undertake the task-oriented experiences of the real job.

These approaches are basic because they attempt to structure offerings to student needs and commitment to various time schedules. Each approach uses the cooperative plan of instruction. What varies in each is the dependence of the institution upon the cooperative plan to provide for a whole range of instructional objectives.

Guidelines for Using the Cooperative Plan

When the cooperative plan is used as part of the business education curriculum structure, several guidelines should be observed. They are:

1. The nature of the required prerequisite courses is determined by the career objectives of the students and the placements that can be arranged. For example, while taking a course in word processing, accounting, or general office procedures in his/her senior year, a student with only one year of keyboarding could be placed as a general clerk. Or, while taking a course in advanced word–information processing, a potential secretary could be placed as an office clerk assistant or first-level receptionist.

2. If the cooperative related class is a two-hour time block, many skills that would normally be developed in other individual classes can be developed in that class. For example, these include operation of accounting and reproduction equipment, word–information processing, and records management.

3. Because the cooperative related class is a laboratory, a maximum of 18 to 25 students can be enrolled. If more than that number is enrolled, then two or more sections must be scheduled. When this occurs, there are several alternatives: (a) The sections can be split with, for example, clerical or word processing student-learners in one and accounting student-learners in the other, thus allowing for specialized instruction without waste. (b) If a six-period day is used with three periods for coordination, then the related class must be reduced to one hour with the teacher-coordinator or two teacher-coordinators must be employed.

4. The two-year period of occupational experience is used infrequently because it has some undesirable features. For example, the eleventh-grade student usually has not had sufficient time to develop skills necessary for placement. The time devoted to occupational experience in the eleventh grade can be better used for enrollment of students in skills courses and for the development of more mature personal/social skills that students need for employment.

5. A one-semester occupational experience may be desirable for some students.[4] However, one-semester students should be placed in a related instruction class by themselves so that the related instruction can be tailored to their needs.

Using Short-Term Employment Experiences

If students are in a career-technical business laboratory course, several short-term directed experiences in the business world can serve purposes that regular classroom instruction is unlikely to serve. These experiences, if directed and geared to the students' career goals, as well as correlated with classroom learnings, are a form of cooperative experience. They are, in essence, a series of "mini-cooperative" programs.

There are several reasons why students should have actual job experiences. These are:

1. **Job exploration**—To find out what the adult world demands, what a small office is like compared to a large one, how the office position fits into the total operation. In short, students are asked to relate to real situations—they must come to grips with reality in this self-actualization process.

2. **Skills application**—To try out what has been learned according to classroom standards, to test theory (principles) against actual practice. In effect, the students are asked to see if they can "cut it," to bring back to school and share with others the methods by which various firms apply basic business skills and knowledges.

3. **Skills development**—To learn skills involving equipment and systems not available in the school, to acquire knowledges of a given industry or a line of business, to practice and to achieve better performance under actual conditions.

4. **Integration and polished performance**—To "put it all together" when performing a set of tasks, for example, those associated with a real position in a real office, and to solve problems by applying what has been learned under conditions of stress and interruption.

Short-term experience projects may be a day, a week, or several weeks in duration. They are different for every student, not only in length but in timing over the year. A project is defined as a learning activity that is constructed to meet an individual student's career goal, learning style, and state of readiness.

1. Experience projects are most easily accomplished when the business occupations program is at least a two-hour block. A block of the last two periods of the day has the obvious advantage of continuing the experience after school hours. But, Saturdays, vacations, and nights also should be considered.

2. In-firm experiences are scheduled and controlled by a human resource manager who makes sure that all the students are met, assigned, supervised, and evaluated.

3. The competencies needed for students to achieve each level of the career ladder are shown in each student's career-goal plan.

4. All experiences are treated as valuable, regardless of whether they are observational or work-producing or whether or not they result in pay.

---

[4]Some state plans currently do not permit reimbursement of less than a full-year program. Consult your own state plan on this matter.

5. The range of practices in business firms, or the reality of the work world, is seldom represented by experiences in a school office, even though such experiences may be useful.

6. Experiences are reinforced when they are brought back to the classroom and shared with other students who enjoy them, enlarge their scope, and learn vicariously from the ones who are sharing them.

7. The time for planning, managing, and evaluating the experience projects on an individual basis must be allocated to the teachers who use them.

8. Not only should short-term experience projects be viewed by local supervisors and the state department as being essential to CTE instruction, but they also should be equated with cooperative plans in terms of reimbursement for time and energy inputs by the laboratory teachers, who thus become teacher-coordinators.

The following is a list of examples of some situations that portray what creative and community-wise teachers can do.

1. The entire class assists in inventory and gets out a mass mailing for a major firm for two weeks in the afternoons.

2. A student observes an office assistant for three days after school.

3. A firm employs two students as temporary workers, at nights and on Saturdays, to solve a labor shortage problem.

4. A group of students plan and do the clerical work for the campaign of a charitable organization.

5. A student works full time as a data entry operator over the Christmas holiday.

6. Several students plan and carry out the office functions of the Booster Club.

7. One student prepares computer records for a legal firm every afternoon after school for a semester.

8. Several students work half-days for a month with selected employers in various business occupations.

9. Several students observe the same job in four dissimilar firms one day a week for a month and then each writes a report.

10. The entire class operates an office for a month, taking in contract work for local firms, school officials, and community agencies.

Basic Competencies and Instructional Objectives

Generally, the business education program at the secondary level uses the cooperative plan primarily as a capstone experience that "tops off" a series of instructional experiences through separate courses. Some business education curricula are intensive senior or pre-graduation programs, with the cooperative plan being used as an integral part of the block-time arrangement. Regardless of the curriculum approach used, the instructional objectives fall into similar areas and are based on the major competency areas demanded by the job cluster and the career goals.

The basic objectives of the plan of instruction are: (1) common occupational competence, (2) business and office job cluster competence, (3) initial job and career objective competence, and (4) general learnings for competence as adults.

Common occupational competence.— Acquisition by student-learners of those competencies and attitudes needed by all workers regardless of their occupations. These learnings, previously described as *occupational related learnings* or *general related learnings*, should be taught to the entire group. These areas of learning include the following examples: (1) employer–co-worker relationships, (2) job safety, (3) laws affecting workers, (4) personal development, (5) union–management relations, and (6) preparation for future education.

Job cluster competence.—Acquisition by student-learners of occupational skills and job intelligence basic to all business occupations so that the student-learners may advance successfully toward their career objectives. These learnings are primarily sought in the formal group classwork in which specific areas are studied. These often include:

School and office relationships (general)

Good grooming

Communications media (personal—telephone, telegraph, letters, fax machines, e-mail, voice mail system)

Personality analysis and development

Human relations and public relations

Filing, billing, recording

Office systems

Fundamentals of word–information processing

Records management and maintenance

Bookkeeping fundamentals

Business machines operation

Business mathematics

Business English (proofreading and grammar)

Advanced word processing and related skills

Clerical workers, and secretaries' vocabulary (both general and specific)

Initial job and career objective competence.—Acquisition by student-learners of those occupational skills and knowledges applicable to the particular initial jobs in which they are placed. Each student's individual learning outcomes are stated in the step-by-step training plan created by the teacher-coordinator and the training station sponsor. These learning outcomes involve office relationships, office information, office techniques, and acquaintance with resource materials on the job. They are emphasized in the informal individual conference of the student-learners and teacher-coordinator and/or training station sponsors. The areas of experience and training providing the core elements of a training plan are illustrated by a list of learning outcomes for a student-learner assigned to a general office/clerical position. The training plan provides information as to whether these learning outcomes will be emphasized in the classroom or on the job or both. The training plan also can identify which learnings will be integrated into the career-technical student organization's activities, as illustrated in Figure 18–1. Frequently, a complete training plan includes both a training agreement and a step-by-step training plan (Figure 18–2).

General learnings for competence.—Learnings that bring about improvement in the students' ability to read, write, and compute. The plan should also contribute to their economic understanding of the business world.

RELATED INSTRUCTION—CONTENT AND SEQUENCE

The related instruction content of the cooperative business education plan is derived from the four basic objectives outlined under the preceding heading. Disregarding the learnings that occur on the job and in other business classes, the learnings in the directly related class taught by the teacher-coordinator deal with three areas: (1) general learnings applicable to all occupations, (2) learnings common to all business occupations, and (3) learnings common only to the student-learners on the job and those needed to meet the student-learners' career objectives. The proportionate amount of time spent in the related class on each area of learning is best shown by Figure 18–3, representing one school year, with the greater proportion of class time (per week or month) being spent on general learnings, at the beginning. This is desirable because the students' skills are usually suffi-

TRAINING PLAN*

Student _____   Training Station _____

Career Objective _____   Training Supervisor _____

_____   Address of Training Station _____

Current Job Title _____   _____

Taxonomy Code _____   Telephone _____

Daily Duties	Weekly Duties	Office Equipment Used (Be Specific)
1. _____	1. _____	1. _____
2. _____	2. _____	2. _____
3. _____	3. _____	3. _____
4. _____	4. _____	4. _____
		5. _____

Monthly Duties	Occasional Duties	
1. _____	1. _____	6. _____
2. _____	2. _____	7. _____
3. _____	3. _____	8. _____

FIGURE 18–1

Competencies	Job Descriptions — Frequency of Use				Instruction Will Occur			Training Station Evaluation				
To Be Completed by the Student-Trainee								To Be Completed by the Supervisor				
					Check / Date in Appropriate Column When Task First Performed			5 – Outstanding 4 – Excellent 3 – Good 2 – Satisfactory 1 – Unsatisfactory				
Skills, Knowledges, and Attitudes	Daily	Monthly	Weekly	Occasionally	Related Class	Student Organization	Training Station	5	4	3	2	1
Keyboarding												
Letters												
Envelopes												
Interoffice memoranda												
Reports												
Rough draft												
Statistical copy												
Legal documents												
Business forms												
Word Processing / Data Processing												
Access keys												
Data entry												
Editing												
Information analysis												
Composing Documents												
Letters												

FIGURE 18-1 (Continued)

	To Be Completed by the Student-Trainee								To Be Completed by the Supervisor				
Competencies	Job Descriptions — Frequency of Use				Instruction Will Occur				Training Station Evaluation				
					Check / Date in Appropriate Column When Task First Performed				5 – Outstanding 4 – Excellent 3 – Good 2 – Satisfactory 1 – Unsatisfactory				
Skills, Knowledges, and Attitudes	Daily	Monthly	Weekly	Occasionally	Related Class	Student Organization	Training Station		5	4	3	2	1
Reprographics													
Printing / duplicating													
Recordkeeping / Bookkeeping													
Dispersing money													
Receiving money													
Making journal entries													
Posting													
Banking													
Check writing													
Depositing													
Reconciliation													
Banking services													

FIGURE 18–1 (Continued)

	To Be Completed by the Student-Trainee							To Be Completed by the Supervisor				
Competencies	Job Descriptions — Frequency of Use				Instruction Will Occur			Training Station Evaluation				
					Check / Date in Appropriate Column When Task First Performed			5 – Outstanding 4 – Excellent 3 – Good 2 – Satisfactory 1 – Unsatisfactory				
Skills, Knowledges, and Attitudes	Daily	Monthly	Weekly	Occasionally	Related Class	Student Organization	Training Station	5	4	3	2	1
Processing Mail												
Incoming mail												
Outgoing mail												
Interoffice mail												
Meter mail												
Calculating												
Keyboarding												
Whole numbers												
Decimals												
Adding												
Subtracting												
Multiplying												
Dividing												
Desktop Publishing												
Designing												
Typesetting												
Merging Files												

FIGURE 18–1 (Continued)

Competencies	Job Descriptions — Frequency of Use				Instruction Will Occur			Training Station Evaluation				
					Check / Date in Appropriate Column When Task First Performed			5 – Outstanding 4 – Excellent 3 – Good 2 – Satisfactory 1 – Unsatisfactory				
Skills, Knowledges, and Attitudes	Daily	Monthly	Weekly	Occasionally	Related Class	Student Organization	Training Station	5	4	3	2	1
Proofreading												
Correcting												
Records Management												
Sorting records												
Indexing												
Alphabetic filing												
Numeric filing												
Retrieving records												
Telephoning												
Incoming calls												
Outgoing calls												
Recording messages												
Screening calls												
Handling appointments												

To Be Completed by the Student-Trainee / To Be Completed by the Supervisor

FIGURE 18–1 (Continued)

Competencies	Job Descriptions — Frequency of Use				Instruction Will Occur			Training Station Evaluation				
					Check / Date in Appropriate Column When Task First Performed			5 – Outstanding 4 – Excellent 3 – Good 2 – Satisfactory 1 – Unsatisfactory				
Skills, Knowledges, and Attitudes	Daily	Monthly	Weekly	Occasionally	Related Class	Student Organization	Training Station	5	4	3	2	1
Taxes												
Sales												
FICA												
Income												
Federal												
W–2 form												
W–4 form												
State												
City												
Employability Skills												
Employment sources												
Application letter												
Résumé												
Application form												
Preparing for the interview												
Handling the interview												
Employment												
Testing												
Follow-up letter												

Header spanning note: "To Be Completed by the Student-Trainee" covers Competencies, Job Descriptions, and Instruction Will Occur; "To Be Completed by the Supervisor" covers Training Station Evaluation.

FIGURE 18–1 (Continued)

	To Be Completed by the Student-Trainee									To Be Completed by the Supervisor				
Competencies	Job Descriptions — Frequency of Use				Instruction Will Occur					Training Station Evaluation				
					Check / Date in Appropriate Column When Task First Performed					5 – Outstanding 4 – Excellent 3 – Good 2 – Satisfactory 1 – Unsatisfactory				
Skills, Knowledges, and Attitudes	Daily	Monthly	Weekly	Occasionally	Related Class	Student Organization	Training Station			5	4	3	2	1
People-to-People Skills														
Courteous														
Coping ability														
Able to set priorities														
Considerate														
Flexible														
Able to be entrusted with confidential matters														
Self-disciplined														
Able to listen attentively														
Cooperative														
Positive attitude														
Proper office attire														
Performance														
Quality of work														
Quantity of work														
Knowledge of work														
Planning and organization of work														
Initiative														
Teamwork														
Attendance														
Punctuality														
Compliance with instructions														

FIGURE 18–1 (Continued)

The Plan in Business Occupations / 351

To Be Completed by the Student-Trainee								To Be Completed by the Supervisor				
Competencies	Job Descriptions — Frequency of Use				Instruction Will Occur			Training Station Evaluation				
Skills, Knowledges, and Attitudes	Daily	Monthly	Weekly	Occasionally	Check / Date in Appropriate Column When Task First Performed			5 – Outstanding 4 – Excellent 3 – Good 2 – Satisfactory 1 – Unsatisfactory				
					Related Class	Student Organization	Training Station	5	4	3	2	1
Job Specialties												
Other												

Employer _____

Instructor _____

Student-Trainee _____

*Developed by the Northcentral COE 1982–83 Task Force.

FIGURE 18–1 (Continued)

BUSINESS / OFFICE OCCUPATIONS TRAINING AGREEMENT

Student _____   School _____

Firm _____   Kind of Business _____

This training agreement briefly describes the responsibilities of the student and the coordinator. It also confirms the consent of parent or guardian for the student to work and to study business / office occupations. The student is expected to observe the same regulations that apply to other employees of the firm. Furthermore, the student agrees to try to improve in skills, knowledges, and concepts, in order to be regarded as being an efficient worker, and in being industrious both in studying at school and in working for the employer.

The training the student will receive in the business / office occupations classroom will be correlated with the experience received on the job, with the ultimate objective of preparing for the occupation of

_____.

The employer recognizes that the program prepares students for careers in business / office work. The employer, therefore, will see that close supervision and instruction are given in each of the following areas of training in order to aid the student to reach the chosen occupational objectives.

_____   _____   _____

_____   _____   _____

Schedule of Experiences
First Semester

First 9 Weeks	Second 9 Weeks
(State competencies, goals, objectives in behavioral terms.)	

FIGURE 18–2

BUSINESS / OFFICE OCCUPATIONS TRAINING AGREEMENT (Continued)

Schedule of Experiences
First Semester (Continued)

First 9 Weeks	Second 9 Weeks
(State competencies, goals, objectives in behavioral terms.)	

Schedule of Experiences
Second Semester

Third 9 Weeks	Fourth 9 Weeks
(State competencies, goals, objectives in behavioral terms.)	

FIGURE 18–2 (Continued)

BUSINESS / OFFICE OCCUPATIONS TRAINING AGREEMENT (Continued)

The beginning wage will be $ _____ per _____ for _____ per week. Progress and advancement made by the student will be evaluated by the employer each _____. The coordinator will cooperate with the training sponsor in an effort to plan classroom instruction in accordance with the schedule of experience.

Training will cover the period from _____, ——, to _____, ——.

_____
(Parent or Guardian)

_____
(Employer)

_____
(Student)

_____
(Coordinator)

FIGURE 18–2 (Continued)

Proportionate Allocation of Instruction for a Cooperative Program

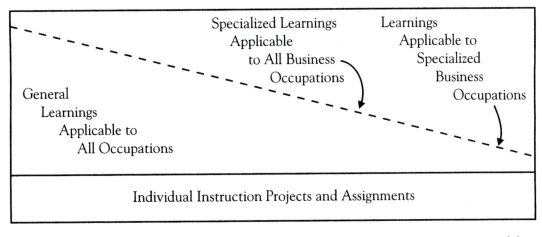

FIGURE 18–3. Sequence of instruction in the related class.

cient then to meet the demands of their employers. As the year progresses, the greater amount of time can be spent on learning new office procedures and improving the performance of those already learned. Notice that individual projects are scheduled throughout the year to help each student-learner meet the demands of the job. Assignments from group learning topics also should be made so that each student applies such learning to his/her job.

The general office procedures area is described well by the widely used office practices textbooks available from major publishers in business education. However, the textbook order of the topics will not be suitable for most related instruction topics. Topics should be sequenced according to the needs of the majority of the student-learners, as they arise from the students' present state of competency and their immediate job needs. For example, filing is typically an early assignment for most student-learners and should be taught near the beginning of the year in the related class.[5]

---

[5]Outlines of instructional content for the related class in business education may be found in the manuals issued by career and technical divisions of many state departments of education.

"Filling out" the related instruction class.—School administrators, faced with demands for maintaining the teacher–student ratio, often wonder if the related class can be "filled out" with students who are not cooperative student-learners. Practically, this procedure may be carried on as long as it involves only a few students. However, the practice is not desirable, because non-cooperative students do not have the motivation of the work responsibility, nor is it possible for them to apply their learnings to a work situation. Their presence usually dampens the instruction for the cooperative student-learners.

Equipping the related instruction classroom.—The directly related instruction for the cooperative business education plan should take place in a classroom especially designed and equipped to facilitate instruction in business occupations. A good deal of flexibility needs to be built into the classroom in order for related instruction to take care of both group and individual requirements and to provide practice in the multitude of procedures that the student-learners need. For example, if one or more of the student-learners are learning desktop publishing, then a computer, printer, scanner, and desktop software must be available for individual practice.

The classrooms used for preparing students with the business skills needed for today's information age must include state-of-the-art equipment. Considering the cost of information–word processing equipment and software, teachers must become very selective of the materials they purchase. For example, equipment purchases should be directed toward multi-use machines, such as personal computers, rather than single-function, stand-alone word processors. Purchasing rarely used machines or machines that will soon be out-of-date is probably not a sound investment. Often, leasing machines, such as copiers, is the best solution. Modern, up-to-date equipment is critical if students are to be prepared with the employment skills needed for today's and tomorrow's jobs.

The related instruction classroom may also be used for non-cooperative office practice classes to fill out the daily schedule and to secure maximum room utilization. The classroom is also useful for adult training and development classes and for training classes after regular school hours.

An office for the teacher-coordinator should be provided in an adjacent location. The necessary facilities should be planned with consultative help from state department and teacher education personnel and with advice and assistance from the advisory committee. Often a three-to five-year plan for acquisition of equipment and supplies will enable the administration to budget the necessary funds.

Coordination Techniques

General principles of coordination for all cooperative plans were discussed in Chapters 6 through 9. This section concentrates on special problems of coordinating the business occupations education plan.

CASE-IN-POINT
Framatome Technologies Student Apprenticeship Program

They keep offices running. In truth, they are the driving force behind American business. They are secretaries.

Secretaries make it (business) tick. Framatome Technologies of Lynchburg, Virginia, is in partnership with local schools and surrounding counties to rekindle interest among students in the secretarial sciences.

There are a number of reasons students have lost interest in the secretarial profession. First, society says that all graduates should go to college. Not necessarily. Many good students are not inclined to go to college, for a variety of reasons, and should seriously consider today's secretarial field.

Secondly, those working in the profession often put it down. This leads to their younger counterparts avoiding the field at all costs.

Finally, TV projects a poor image of secretaries by making them appear to be flighty and not very intelligent.

Many people don't realize the responsibilities involved with being a secretary. Secretaries keep their bosses' calendars, make travel arrangements, and do accounting, in addition to the typing, dictation, and other office duties. Today's secretary wears many hats. And, the secretary of today has to be skilled in using the latest business machines including computers, word processors, printers, and fax machines.

Framatome is extremely excited about the latest endeavor through Campbell County Schools with the Office Systems Apprenticeship Program. Jean Yates, manager of Training and Development, spent the summer months working with Mr. Richard Edwards, Mr. David Moseley, and Mr. Calon Burrus to understand how the program works and to administratively get things in order to begin with two students when school started in September. Currently, four young ladies are working as student apprentices in the company. Things seem to be going well, especially for the students.

In the Apprenticeship Program, the students, the company, and the school all have specific responsibilities to fulfill in order to be successful.

(Continued)

SELECTING STUDENT-LEARNERS

Business skills developed in school under controlled practice are vital to placement. Therefore, it is particularly important in admitting students into the program to assess their performance in business classes to determine their business competency levels. In a course such as keyboarding, it is important to know what the course grade was. Even more significant would be a reference from the keyboarding teacher regarding which aspects of the skill are strong and which are weak. For example, a B student in beginning keyboarding might be only average in speed but very accurate, thus boosting the number of words per minute on which the grade might have been primarily based. On the other hand, that student might also be particularly strong, or weak, on tabulation problems. Thus, in order to make a proper placement, the teacher-coordinator needs to know not only the gross estimates of skills but also the strengths of the component competencies.

The teacher-coordinator should also acquire information about the students' attendance and punctuality, general academic aptitude, and career interests. Much of this can be obtained from the application/admission form (see Figure 18–4) the stu-

Potential training station supervisors enjoy lunch together after an orientation session.

BUSINESS OCCUPATIONS
STUDENT-LEARNER APPLICATION / ADMISSION FORM

Name _____ Year of Graduation _____

Address _____ Telephone _____

Birthdate _____ Parent or Guardian _____

Parent / Guardian's Telephone Number: Home _____ Work _____

Extracurricular activities

Music _____ Publications _____

Athletics _____ Organizations _____

_____ _____

_____

Classes and grades	Semester I	Semester II
Bookkeeping / accounting	_____	_____
Keyboarding	_____	_____
General business	_____	_____
Economics	_____	_____
Word – data processing	_____	_____
Introduction to computers	_____	_____
Spreadsheets	_____	_____
Data management	_____	_____
Desktop Publishing	_____	_____

Achievement record

Keyboarding: copying speed of _____ words a minute _____ errors (gross words)

Transcribing speed grade _____ Grade for mailability of work _____

Knowledge of _____ calculator, _____ computer, _____ electronic typewriter, _____ word processor

Knowledge of filing _____

FIGURE 18–4

dents complete and from interviews and personal data sheets (Figure 18–5). Other teachers who have previously had the students in class should be asked to comment on their work habits and attitudes, relationships with others, and career interests.

Particularly important for good placement in business occupations plans are the image the students present—an image of composure, self-confidence, inner tranquility—and the demeanor they display.

While many factors influence whether or not a student should be enrolled in the cooperative plan, the most important ones are: (1) the student must be genuinely interested in business employment and (2) the student must be motivated to learn the skills necessary to be successful in business employment.

FINDING APPROPRIATE TRAINING LABORATORY PLACEMENTS

Finding appropriate occupational laboratory placements requires the careful matching of two sets of factors: (1) the current abilities of the student-learners, their capacity to grow, and their career objectives and (2) the variety of experiences available at the training stations, the beginning job standards, and the potential full-time employment possibilities. Placement in the cooperative business occupations education plan is somewhat different from that of other occupational fields because of the skills the cooperative business student-learners are expected to have upon initial job entry.

Consideration of the following factors will help facilitate placement.

1. Personality traits are very important in most office positions where employees work, often in close proximity to one another, for long periods of time and where work assignments are often closely interdependent. Thus, compatibility of the personalities of the student-workers and the prospective job sponsors and co-workers needs careful consideration. The experienced teacher-coordinator is aware that some office workers may be jealous and may view the students as a threat to themselves.

2. Very small offices frequently offer a more extensive variety of learning experiences and more responsibility than some larger offices, unless they adhere to a careful rotation schedule. On the other hand, in small offices, training and supervision may be less effective and performance standards not as highly organized. The office in which the student-learner is the only employee offers the student a chance to exercise independence and responsibility, but supervision may be almost entirely lacking.

3. Very large offices usually give the student-learners the opportunity to try out, or at least to observe, operations that may not be present in smaller offices. For example, the large firms usually have a data processing unit or a word processing center. Very large offices may offer the advantage of more rapid advancement because of more personnel turnover.

4. Teacher-coordinators should not rule out placements in unusual situations—offices that are not well decorated or those that are in out-of-the-way locations in buildings; for example, good placements may exist in the offices in trucking terminals, warehouses, and small factories.

5. Teacher-coordinators should consider the possibility of building the business program around two sets of trainees, a morning group and an afternoon group. Thus, the employers have the advantage of keeping their work situations fully occupied. Even in some small schools where training stations are limited, teacher-coordinators may utilize this procedure if the non-business courses that are needed and the related class can be scheduled.

In making placements, teacher-coordinators should be fully cognizant of the nature of business occupations and should avoid placements where the duties are primarily non-office, such as marketing or industrial. Although business occupations exist in many types of businesses, only those jobs that require primarily competencies in recording, storing, reproducing, and communicating information should be termed "business occupations."

(For Advisement Use)
BUSINESS OCCUPATIONS
PERSONAL DATA SHEET

Date _____

Phone _____

Name _____ Address _____
　　　Last　　　　　First　　　　Middle

Do you live with your parents? _____ If not, with whom? _____

Age _____ Birthday _____ Height _____ Weight _____ Male _____ Female _____

How long have you lived in this city? _____ How many brothers? _____ (names, ages) _____

_____

How many sisters? _____ (names, ages) _____

Your responsibilities at home _____

Father's name _____ Birthplace _____ Occupation _____

Mother's name _____ Birthplace _____ Occupation _____

Is it necessary for you to work? _____ If so, why? _____

_____

Give details of any illnesses, accidents, or operations within the past five years.

_____

_____

_____

Describe briefly any defects in sight, hearing, speech, or other physical impairment.

_____

_____

To what school organizations do you belong? _____

_____

What are your favorite sports, and how do you spend your "off-work" hours? _____

_____

_____

_____

FIGURE 18–5

School Training

List the courses you have taken in school and the grade you received in each.

Freshman Year	Sophomore Year	Junior Year
1. _____ ____	1. _____ ____	1. _____ ____
2. _____ ____	2. _____ ____	2. _____ ____
3. _____ ____	3. _____ ____	3. _____ ____
4. _____ ____	4. _____ ____	4. _____ ____
5. _____ ____	5. _____ ____	5. _____ ____

What has been your major field of study? _____

What has been your favorite subject(s)? _____

Past Work Experience

List below the places where you have worked in the past.

	Type of Work	Name of Firm	Dates Employed	Employer
1.				
2.				
3.				
4.				

Names of Schools Attended

1. _____ 2. _____

3. _____ 4. _____

Attendance Record

Has your attendance been: Excellent _____ Good _____ Fair _____ Poor _____?

Do you wish to work this summer? _____ If so, when would you like to begin? _____

What means of transportation will you have? _____

Names of two teachers who will recommend you. _____

_____

FIGURE 18–5 (Continued)

Future Plans

College _____ Marriage _____ Career plans _____

Military _____

STUDENT: DO NOT FILL IN THE BLANKS BELOW.

SCHOOL RECORDS: Absences _____ Tardinesses _____

Grade average _____ Best grade in _____

Remarks: _____

_____

_____

_____

FIGURE 18–5 (Continued)

USING THE STEP-BY-STEP TRAINING PLAN

The step-by-step training plan is derived jointly by the teacher-coordinator and the training station sponsor from a realistic analysis of the tasks, duties, and responsibilities of the student-learner on the part-time job. The development of specific training plans, together with follow-up activities, is one of the most important responsibilities of teacher-coordinators. The teacher-coordinator must carefully analyze the job skills needed by the student-worker on the job daily, weekly, monthly, or occasionally. Thus, the teacher-coordinator should be able to (1) eliminate the teaching of unnecessary skills, (2) spend more time assisting the students to acquire those skills essential for success on the job, and (3) sequence the related instruction so that it truly prepares and reinforces the on-the-job experiences.

If effectively carried out, the training plan makes the cooperative business education approach sharply distinctive from an unsupervised work experience program. The training plan makes a specific attempt to relate classroom instruction to on-the-job training and to allow students to gain all the potential learning advantages therefrom. It serves as the basis for guiding student-learners through worthwhile educational experiences on the job. It indicates the specific objective to be achieved and whether it is to be stressed in the related classroom or at the training station, or both, and whether or not the student organization can further enhance the skill development (see Figure 18–1). By identifying those competencies in which the student organizations can provide the opportunity for development, the plan emphasizes the importance of the student organization as an integral part of the instructional program.

GUIDING THE BUSINESS OCCUPATIONS STUDENT ORGANIZATION

The general principles of establishing student organizations are presented in Chapter 14, wherein both national business student organizations, FBLA and BPA, are discussed. Included in this section are some specific activities for the business occupations student organization. BPA members with a variety of business skills will find numerous opportunities for the application of their special career interests, ranging from management of the chapter's records to word processing and office administrative skills to accounting and financial management. The FBLA provides its members with countless opportunities to increase their understanding of the content of the business classroom. Whether the organization is a local program or one affiliated with a national association, the student organization is a vital part of the total educational process. The vast majority of the goals and objectives of business education are compatible to those of any student organization and the student organization's activities reinforce and expand the important concepts taught in the classroom.

The student organization's activity program should involve community resources to further instructional outcomes. For example, speakers may be invited to discuss topics such as (1) color and design in the modern office, (2) reasons for joining the American Business Club or the Business and Professional Women's Club, (3) requirements for becoming a certified professional secretary or a CPA (certified professional accountant), (4) ways to develop composure and self-assurance, and (5) effects of electronic communications on office jobs. Businesspersons can demonstrate proper business etiquette (grooming and attire). The student organization can undertake surveys of the types of jobs available in businesses in the community, or it can offer consultative advice on office accounting procedures to the marketing education and industrial education student organizations.

Activities that could provide financing for student organization functions might include: (1) preparing programs, for a fee, on a copier for civic or professional clubs; (2) holding a bake sale in the school cafeteria or in the lobby of a cooperating business; and (3) carrying out a cooperative effort with the marketing education or the industrial education student

A training supervisor checks out a computer procedure with a trainee in an office station.

organization, in which the special talents of the student-workers could be utilized.

Evaluation Procedures

When evaluating student-learners, teacher-coordinators of the cooperative business education plan have two different tasks. First, they must evaluate the student-learners' performance, and second, they must evaluate the total plan. Evaluation of the total plan is discussed in Chapter 15.

EVALUATING THE PERFORMANCE OF THE STUDENT-LEARNERS

Evaluation of the student-learners' related classroom instruction is based upon their total class performance, including (1) written subject matter assignments, (2) work completed in assigned manuals, (3) oral presentations, and (4) other individual and group projects designed for learning various business functions and for operating specialized equipment. The evaluation of the student-learners' on-the-job learning and classroom performance is based on observations of the teacher-coordinator and the student-learner evaluation completed by the training station sponsor.

The evaluation can be a part of the training plan, as it is in Figure 18–1. The evaluation to be completed by the training sponsor allows the job supervisor to rate the student's performance with regard to each competence identified in the training plan. Some schools have developed a separate evaluation form and ask the training sponsor to rate the student-worker. Figures 18–6 and 18–7 are examples of such evaluation forms.

In the cooperative business occupations education plan, particular attention in evaluation should be paid to both the quantity and the quality of performance. Many business skills can be measured quite accurately, and any deficiencies described well. Therefore, teacher-coordinators should tell student-learners in which skill(s) they are performing below standard. For example, if a student keyboards letters well, except for sub-standard corrections, the teacher-coordinator should tell the student that his/her correction techniques need to be improved, rather than the letter needs improvement. If a student's telephone technique is faulty, the teacher-coordinator should tell the student that his/her voice needs to be louder, or more pleasant, or that he/she should be more tactful. In addition to evaluating skill competency, the teacher-coordinator also should evaluate the student-workers on their ability to solve problems. Are they able to put skills together to accomplish a task?

Evaluation as part of coordination visit.— Generally speaking, during each coordination visit, the teacher-coordinator should prepare an evaluation. Since it is customary for the teacher-coordinator to calculate coordination time on the basis of one-half hour per student per week, the evaluation report should be short and precise.

The frequency of coordination visits to the job training site depends on the needs of the student, the

BUSINESS OCCUPATIONS
STUDENT-LEARNER EVALUATION FORM

Student-Learner _____ Office _____ Date _____

Note to job supervisor: Please rate the above-named student-learner in each of the areas listed below. Your assessment will assist the coordinator in designing instruction that will help the student become more effective on the job.

In the right-hand columns, please check your impression of the following, omitting points that do not apply to the activities performed by the student-learner in the training station.

Skills	Superior	Above Average	Average	Poor	Unsatis-factory
Word Processing	___	___	___	___	___
Spreadsheets	___	___	___	___	___
Recordkeeping	___	___	___	___	___
Filing	___	___	___	___	___
Machine duplication	___	___	___	___	___
Proofreading	___	___	___	___	___
Desktop publishing	___	___	___	___	___

Production

Volume of work	___	___	___	___	___
Quality of work	___	___	___	___	___
Steadiness	___	___	___	___	___

Fundamentals

Handwriting	___	___	___	___	___
Spelling	___	___	___	___	___
Arithmetic	___	___	___	___	___

Business Techniques

Ability to:

Meet people	___	___	___	___	___
Work harmoniously with others	___	___	___	___	___
Use office telephone	___	___	___	___	___
Follow instructions	___	___	___	___	___
Handle supplies efficiently	___	___	___	___	___

FIGURE 18–6

Personal Traits	Superior	Above Average	Average	Poor	Unsatisfactory
Appearance	_____	_____	_____	____	_____
Manners	_____	_____	_____	____	_____
Health	_____	_____	_____	____	_____
Personal hygiene	_____	_____	_____	____	_____
Speech	_____	_____	_____	____	_____
Initiative (keeping busy)	_____	_____	_____	____	_____
Tact	_____	_____	_____	____	_____
Accuracy	_____	_____	_____	____	_____
Judgment	_____	_____	_____	____	_____
Patience	_____	_____	_____	____	_____
Self-confidence	_____	_____	_____	____	_____

Other Comments (unusual strengths and weaknesses): _____

_____

_____

_____

_____

_____

(Job Supervisor)

FIGURE 18–6 (Continued)

EMPLOYER'S REPORT
COOPERATIVE BUSINESS EDUCATION
_____ High School

Student _____   Supervisor _____

Firm _____   Coordinator _____

Skills and Traits	Excellent	Good	Satisfactory	Fair	Unsatisfactory	Comments
Reliability						
Attendance, promptness						
Personal appearance						
Personality						
Cooperation						
Attitude toward work						
Work habits						
Production rate						

Quality of Work Performed	Excellent	Good	Satisfactory	Fair	Unsatisfactory	Comments
Word processing						
Duplicating						
Filing						
Alphabetic ☐						
Numeric ☐						
Bookkeeping						
Computation						
Telephoning/receptioning						
Office housekeeping						
Other						

What improvement has the trainee shown? _____
_____
_____

Date _____   Employer's Signature _____

FIGURE 18–7

characteristics of the training station and the training position, the time of the year (more frequent visits are made at the beginning of the school year than later in the year when the student-worker is well established on the training job). An example of an evaluation form to be completed on a coordination visit is presented in Figure 18–8. This form is designed to be used on each coordination visit during the semester. The teacher-coordinator lists the job tasks observed and/or discussed and rates the student's performance in doing each job task. At the end of the school year and/or semester, the student's

on-the-job performance can quickly be determined according to the teacher-coordinator's coordination assessment. The teacher-coordinator's rating can then be compared to the employer's rating. The student's grade for the on-the-job phase of the cooperative education plan should be based upon both the teacher-coordinator's and the employer's assessment of the student's performance and achievement.

Student's self-evaluation.—Some teacher-coordinators have their student-learners complete a weekly performance report, which can be signed by

---

Coordinator _____   Date _____

EVALUATION — COORDINATION VISIT

Student _____   Job Title _____

Firm _____   Training Sponsor _____

Rating scale
5 – Outstanding
4 – Good
3 – Meets expectations
2 – Needs improvement
1 – Very poor

Job Tasks Observed / Discussed	Performance					Comments
	5	4	3	2	1	
1. _____						_____
_____						_____
2. _____						_____
_____						_____
3. _____						_____
_____						_____
4. _____						_____
_____						_____

Copy to student, employer, coordinator's file.

FIGURE 18–8

STUDENT-EMPLOYEE WEEKLY PERFORMANCE REPORT

Trainee _____   Supervisor _____

Job Station _____   Telephone _____

What new skills did you learn through experience or instruction on the job this week?

_____

_____

_____

_____

Was there any change in your salary or in your job tasks during the past week? If so, explain. _____

_____

Are there any particular problems you are having on the job that you need and want instructional help to resolve? If so, explain. _____

_____

_____

_____

_____

FIGURE 18–9

the on-the-job training supervisor. These reports can assist the teacher-coordinator in identifying the student-worker's strengths and weaknesses. Such information, being prepared by the student, is especially important in modifying and refining the teaching–learning sequence. This information can help the teacher-coordinator in tailoring the related instruction to better correlate to the student-worker's on-the-job tasks. Figure 18–9 is an example of a student-employee weekly performance report.

Questions and Activities

1. Prepare a 10-minute talk that you could give before a business group, such as the local chapter of the Administrative Management Society, explaining the cooperative business education plan. Prepare charts, transparencies, a Microsoft PowerPoint presentation, or other visual aids that you could use.

2. From your state plan for career and technical education, determine the educational and occupational experience requirements necessary for certification as a business education coordinator.

3. Using census data or information from your state employment service, prepare a brief forecast report for business occupations in your county for the next decade.

4. Some authorities have claimed that office practice classes generally have been too skill-oriented or too machine operations–oriented and that more time

SOURCES OF INSTRUCTIONAL MATERIALS

Glencoe/McGraw-Hill
Customer Service Department
P.O. Box 543
Blacklick, OH 43004-0544
Phone: (800) 334-7344
Fax: (614) 755-5682
Web site: www.glencoe.com

Harcourt School Publishers
6277 Sea Harbor Drive
Orlando, FL 32887-4420
Phone: (800) 225-5425
Fax: (800) 269-5232
Email: hbspcs@harcourtbrace.com
Web site: www.harcourtschool.com

Houghton Mifflin Company
One Beacon Street
Boston, MA 02108
Phone: (800) 257-9137
Fax: (617) 227-5409
Web site: www.hmco.com

Interstate Publishers, Inc.
P.O. Box 50
Danville, IL 61834-0050
Phone: (800) 843-4774
Fax: (217) 446-9706
Email: info–ipp@ippinc.com
Web site: www.interstatepublishers.com

John Wiley & Sons, Inc.
605 Third Avenue
New York, NY 10158
Phone:(212) 850-6000
(908) 302-2300 (Somerset, NJ)
Web site: www.wiley.com

Simon & Schuster, Inc.
Educational Books Division
113 Sylvan Avenue
Englewood Cliffs, NJ 07632
Phone: (201) 592-2000 or (212) 698-7000
Fax: (800) 445-6991
Web site: www.simonandschuster.com

South-Western Publishing company
5101 Madison Road
Cincinnati, OH 45227
Phone: (800) 543-0487
Fax: (513) 527-6940
Email: swinformation@thomsonlearning.com
Web site: www.swep.com

Wadsworth Publishing company
10 Davis Drive
Belmont, CA 94002
Phone: (800) 354-9706
Web site: www.wadsworth.com

should be devoted to human relations skills, office concepts, office systems, etc. Do you agree?

5. If two sections of the related class are taught, would homogeneous grouping, such as clerical and word processing, provide more effective instruction? Explain your answer.

6. Prepare a sample set of forms that you feel you could use in a cooperative education plan job: performance rating sheet, weekly report form, student application/admission form, etc.

7. The text suggests that in some instances the cooperative related class may be substituted for other courses that the student-learner might have taken. Show by specific learning experiences what topics and practice activities might be shifted from other courses to the related class.

8. Prepare a list of job tasks and behaviors that a teacher-coordinator could use in evaluating a student's on-the-job performance.

9. Assume that you are the chairperson of a district-wide task force of teacher-coordinators responsible for developing a cooperative business education student handbook. Prepare a list of topics you would include in such a handbook and then prepare in final draft form two of the handouts that would be a part of the handbook.

10. "A competency-based cooperative business education plan must be the basis for delivering education in the information age." Do you agree? Explain your answer.

For references pertinent to the subject matter of this chapter, see Reading Resources.

19

The Plan in
Health Occupations

Health occupations trainees enroll in science courses as part of the background they will need to pursue their chosen career objectives.

KEY CONCEPT: Health occupations encompass those career areas that provide services to individuals in need of health care. They include direct patient-care and supportive services.

GOALS

After successfully completing the study of this chapter, answering the questions, and carrying out the activities, the student should be able to:

- Identify and list appropriate training station and career objectives available in health occupations.

- Understand how to develop an appropriate curriculum for a local plan in cooperative occupational education for health occupations.

- List the major sources of instructional materials for health occupations education.

- Devise appropriate training plans for students at various training stations.

- Evaluate other student-learners in the classroom and on the job.

Health Occupations Education

Health occupations encompass those career areas that provide services to individuals in need of health care. They include direct patient-care and supportive services. Employment in this discipline is available in a wide variety of settings: health-care facilities (medical offices, hospitals, nursing homes, and extended-care facilities), community health service agencies, and professional offices. Preparation for these careers may be offered in a number of educational settings. Some are located in high schools, career technical schools, community–junior colleges, four-year colleges, and private career schools. Hospitals also offer specific technical programs in addition to on-the-job training.

When educational administrators are determining the kinds of programs to be offered, they need to consider the following factors: student needs, community needs, and employer needs.

1. **Student needs**—Opportunities should be available for students to enroll in programs that develop concepts and related skills to prepare them for health careers. This preparation must be realistic to the student's abilities, interests, job opportunities, career mobility, continued education, and advancement possibilities.

2. **Community needs**—Competent health workers are needed to assist in providing quality health care for the local citizens.

3. **Employer needs**—Health occupations programs that prepare individuals to be competent employees are needed.

INSTRUCTIONAL PROGRAM CLUSTERS

The health occupations education CIP Code 51 includes the following clusters:[1]

---

[1]Robert L. Morgan and E. Stephen Hunt, *A Classification of Instructional Programs—2000*, National Center for Education Statistics, Presidential Building, 400 Maryland Avenue, S.W., Washington, DC 20202, 2000, pp. 36–41.

51.00 Health Services/Allied Health, General

51.02 Communication Disorders Science and Services

51.04 Dentistry

51.05 Advanced/Graduate Dentistry and Oral Sciences

51.06 Dental Support Services and Allied Services

51.07 Health and Medical Administrative Services

51.08 Allied Health and Medical Assisting Services

51.09 Allied Health Diagnostic, Intervention, and Treatment Professions

51.10 Clinical/Medical Laboratory Science and Allied Professions

51.11 Health and Medical Preparatory Programs

51.12 Medicine

51.15 Mental and Social Health Services and Allied Professions

51.16 Nursing

51.17 Optometry

51.18 Opthalmalmic/Optometric Support Services and Allied Professions

51.19 Osteopathic Medicine / Osteopathy (DO)

51.20 Pharmacy, Pharmaceutical Sciences, and Administration

51.21 Podiatric Medicine / Podiatry (DPM)

51.22 Public Health

51.23 Rehabilitation and Therapeutic Professions

51.24 Veterinary Medicine (DVM)

51.25 Veterinary Biomedical and Clinical Sciences (Cert, MS, PhD)

51.26 Health Aides / Attendants / Orderlies

51.27 Medical Illustration and Informatics

51.40 Dietetics and Clinical Nutrition Services

51.41 Bioethics / Medical Ethics

51.42 Alternative and Complementary Medicine and Medical Systems

51.43 Alternative, Complementary, and Somatic Practice Health and Therapeutic Services

51.99 Health Professions and Related Sciences, Other

JOB OPPORTUNITIES

Job opportunities in health occupations are many and varied in nature. The following list indicates the breadth of the health occupations field and illustrates by descriptive title that preparation for many of the positions requires knowledge and background in disciplines other than health.

Patient-Care Services

Nursing Services

Registered nurses
LPN's/LVN's
Nurses' aides/attendants
Orderlies
Burn unit technicians
Pediatric technicians
Homemaker/home health aides
Emergency medical technicians
Obstetrical technicians

Mental Health Services

Mental health technicians
Mental health assistants
Mental deficiencies technicians
Psychiatric technicians
Psychiatric aides/assistants
Occupational therapy aides
Recreational therapy aides

Diagnostic Services

Certified laboratory assistants
Medical laboratory technicians
Laboratory aides
Cytotechnologists
Histological technicians
Electro-cardiagraphic technicians
Electro-encephlegraphic technicians
Pulmonary function technicians
Radiological technologists

Physical Rehabilitation Services

Physical therapy assistants
Physical therapy aides/orderlies
Ortho-prosthetic technicians

Podiatric assistants
Certified respiratory therapy technicians
Respiratory therapy aides
Respiratory therapist

Ancillary Services

Urologic physicians' assistants
Operating room technicians
Surgical aides/assistants
Central service aides/assistants
Housekeeping workers

Dietary Services

Dietetic aides
Dietary technicians

Dental-Care Services

Direct Patient Contact Services

Dental assistants
Oral surgery dental assistants
Pedodontic dental assistants
Endodontic dental assistants
Expanded duties dental assistants

Indirect Patient Services

Dental receptionists
Dental laboratory technicians
Dental assistant

Medical and Clerical Services

Medical Records Services

Medical transcribers
Medical stenographers
Medical dental secretaries
Medical librarian assistants
Medical records clerks
Medical records technicians
Insurance clerks

Clinical Services

Ward unit clerks
Ward secretaries
Medical assistants

Animal-Care Services

Nursing Services

  Veterinary assistants
  Groomers
  Cage-care attendants

Diagnostic Services

  Veterinary technicians
  Veterinary X-ray assistants

Auxiliary Services

  Receptionists
  Record keepers

CURRICULUM

Units of related subject matter, or groups of related courses, and planned experiences designed to impart knowledge and to develop understandings and skills required to support the health professions make up the curriculum for health occupations education. Instruction is organized to prepare students for meeting those occupational objectives concerned with assisting qualified personnel in providing diagnostic, therapeutic, preventive, restorative, and rehabilitative services, including those understandings and skills essential to health care and health services to patients.

The curriculum in health occupations is usually arranged according to one of these three options—single occupation, cluster program, sequential program.

Single occupation option.—Preparation for entry into a specific single occupation is the major purpose of the single occupation option. Educational programs may range from short-term ones to long-term ones leading to an associate degree. The program for secondary students may extend from one class period per day to one-half of the school day. Some post-secondary programs may lead to certification and/or licensure. Examples of single occupation positions are

respiratory therapy aide, dental assistant, and dental laboratory technician.

Cluster program option.—The cluster program option at the secondary level is structured to present a core of knowledges and skills common to several health occupations. Upon completion of the core curriculum, students continue the rest of their studies in the single occupation for which they have shown both aptitude and interest. The goal of the cluster program is to provide students with an orientation within the total health field, an exploration of several related health occupations, and marketable skills upon graduation. For administrators it affords flexibility in planning and the opportunity to adapt to student, school, and community needs. The cluster curriculum is based on an "open-entry, open-exit" semester concept. An example follows.

Level I

Health Occupations I

  Overview and Orientation to the Health Field
  (1–2 periods for one semester, grades 9–12)

Health Occupations II

  Career Planning
  (1–2 periods for one semester, grades 9–12)

Level II

Health Occupations III

  Sub-cluster Exploration
  (2 periods for two semesters, grades 10–12)

Level III

Health Occupations IV

  Job Preparation or Category Exploration
  (3 periods for two semesters, grades 11 and 12, in-school and extended laboratory)

Level IV

Job Preparation

> (3 periods for two semesters, grade 12, in-school, extended, or cooperative placement)

In the cluster program, students should enter Level I in the ninth grade and complete Level IV in the twelfth grade. It is possible, however, for twelfth-grade students to select only one semester of Level I and still acquire competencies for specific job titles.

In the Dade County Public Schools, in Miami, Florida, cooperative health occupations is a one-year program for seniors designed to provide the students with the knowledges and skills necessary to function as nonprofessional employees in the field of health occupations and services.

Training facilities, including hospitals, nursing homes, veterinary clinics, and doctors' and dentists' offices, cooperate in providing training stations. Typical areas of training include nurse's aide, physician's aide, veterinary assistant, and radiological aide. Some programs organize instruction under the single occupation option and/or the cluster program option.

A typical schedule for a student in cooperative health occupations is as follows:

Period	Course	Credit
1	General Related Study	1
2	CHO Specific Related Study	1
3	U.S. Government	1
4	English IV	1
5	On-the-Job Training	1
6	On-the-Job Training	1

Sequential program option.—The sequential program begins in the ninth grade with a general exploratory plan of one class period per day to acquaint students with the wide variety of opportunities in health occupations. This continues in the tenth grade on a one-class-period basis. Foundation courses, such as "Elementary Body Structure and Function," "Nutrition," and "Understanding Human Behavior," are coupled with exposure to health-care facilities on an observational basis. In the eleventh grade, students enroll in the single occupation or the cluster program. The twelfth grade is designed as a cooperative education plan. This sequence may be adapted to meet the specific needs of students, as well as those of the school and the community. The goal, obviously, is to prepare students for saleable skills upon graduation.

INSTRUCTIONAL PROCEDURES/PATTERNS

An essential part of each option—single occupation, cluster program, or sequential program—is planned supervised experience in the clinical setting, as well as in-school learning. Health occupations education utilize in-school laboratories, extended laboratories, and the cooperative plan.

In-school laboratories are used to simulate the actual working conditions found in the various health occupations. The facilities and equipment are representative of clinical equipment. Students work in the in-school laboratory just as they would in a real job situation. Some in-school laboratories even provide patient services. In these laboratories the career and technical education students assist the professionals.

The use of extended laboratories is a unique instructional procedure of health occupations education. The school uses actual health facilities as a laboratory, usually for a short specified time. The career-technical health instructor and the students travel from the school to the extended laboratory facility in the community for a preplanned educational experience. Then, the students return to the classroom for additional study. These activities are coordinated and evaluated by the school instructor. During extended laboratory experiences, instruction and supervision may be done on a group or one-to-one basis. Specific predetermined learning objectives provide structure to the extended laboratory experiences.

In the cooperative plan, students acquire on-the-job training through employment in local health agency training stations. Students are placed in train-

CASE-IN-POINT

Carolyn Hollar, a high school sophomore, volunteered her services at Union Hospital on weekends to deliver flowers, gifts, water, and reading material to patients. She enjoyed making sick people happy. The activities of the orderlies, laboratory technicians, and nurses appealed to her. Carolyn chatted with the school nurse about the nurse's background and training. The nurse suggested a contact with the guidance counselor about the health occupations program in the high school

A meeting with the counselor provided an opportunity for Carolyn to look over publicity materials about health occupations. She found that her interest in general science and biology would serve her well in subject matter studied during the program. Why not make an appointment with Mrs. Jane Sponder, the health occupations teacher-coordinator, she thought? After getting her parent's approval, she decided to do so.

Mrs. Sponder was pleased to give her aptitude tests and to review her academic background. Her test scores and course grades proved to be adequate. Therefore, Mrs. Sponder, in consultation with Carolyn and the guidance counselor, planned Carolyn's eleventh and twelfth grade schedules of studies. In the planning, Mrs. Sponder allowed for the cooperative plan whereby Carolyn would interview at appropriate health organizations in the area to be placed in a training station to gain on-the-job experience toward a health occupations career.

ing stations where they can receive supervised experience in the occupations they have chosen. The students are employed for a minimum average of 15 hours a week, under an on-the-job supervisor, for the school year. This is paid employment. A training plan is developed to accompany the training agreement to meet the career objectives of each student. The cooperative plan usually covers the twelfth year as a capstone to a planned sequence of courses, but it may be extended to two years if the training objective warrants it. A certified teacher-coordinator teaches related instruction and coordinates the activities of the training stations and the HOSA student organization.

The organization or pattern of classroom and employment experiences in the cooperative plan should be geared to the needs of the students and the requirements of the occupations being pursued. There are several patterns that may be followed and still guarantee that the student will meet the minimum requirement of an average of 200 minutes of related class instruction each week. The following are examples of scheduling patterns of employment and related class instruction.

1. One 45-minute period a day for job-related class (225 minutes total) plus one-half day of employment.

2. One week of employment, one week of class.

3. One day of employment, one day of class.

4. Four 45-minute periods (180 minutes) of related class on one day plus one period (45 minutes) of related class on another day (225 minutes total). Employment for four days.

5. One day of related class instruction, in addition to daily clinical conferences with the on-the-job supervisor and in-service programs provided by the health-care facility.

RELATED CLASS INSTRUCTION

Related class instruction in health occupations education generally includes *orientation units*, which are instructional units containing the basic information and technical knowledge needed by student-learners to be occupationally competent in all health occupations, and *job-related units*, which can be studied on an individual or small-group basis.

Orientation units could include:

1. Applying for a Job
2. Employee Responsibilities
3. Developing Good Work Habits
4. Importance of Proper Grooming

5. Laws and Regulations Affecting Workers

   a. Social Security

   b. Workers' Compensation

   c. Income Tax

   d. Unemployment Insurance

   e. Disability Insurance

6. Elements of Effective Communication

7. Desirable Attitudes and Personal Habits

The content for job-related units is planned by the teacher-coordinator and the supervising employers. The teacher-coordinator and the employer decide which units can best be taught by the instructor in the classroom and which ones can best be taught by the supervisor on the job. Skills and knowledges to be taught will be based on both the job analysis and the level at which the student is expected to perform or is capable of performing.

Teacher-coordinators must develop specific related safety instructional units for each health occupations education program. These units and the student's test scores indicating 100 percent comprehension should be kept on file while the student-learners are in the cooperative plan.

ADEQUATE FACILITIES

The facilities for health occupations education programs will depend upon the occupation(s) being offered. Since cooperative arrangements are made with outside agencies and businesses for on-the-job training, classroom equipment can be kept to a minimum. If necessary, equipment for classroom instruction may sometimes be borrowed from cooperating agencies or other community sources. In addition, some equipment at the training station may be used for further instructional purposes (e.g., suction machine).

The following should be provided in a health occupations laboratory.

1. The instructional area must be adequate to allow for both group and individual instruction. The room should have adequate sink facilities. The instructor should be able to darken the area to show slides, videotapes, etc.

2. Basic equipment and supplies to demonstrate and practice skills and techniques to meet the requirements of the specific primary program should be available. Adequate storage area for all such equipment and supplies is essential.

3. Audio–visual equipment should be readily accessible.

4. A private conference area adjacent to the classroom, where the coordinator may confer with students, is necessary. It is imperative that this office area contain a computer, an unencumbered phone, and a filing cabinet(s) with a lock(s) to store records and reports.

5. Clerical assistance, as needed, should be available.

6. Adequate texts and resource materials pertinent to the program should be located within the instructional area.

The equipment needed for any health occupations education program should be determined with the help of the local advisory committee. In most states, suggested equipment lists and other forms of technical assistance are available from the health occupations consultant in the division of career and technical education.

LICENSURE, REGISTRATION, OR CERTIFICATION

A credentialing system found only in the health field affects certain health occupations education training programs. Whenever a training program leads to licensure, registration, or certification of the graduates, it is necessary for the educational personnel and the specific health occupation regulatory organizations to collaborate in designing and developing the program. For example, to establish an approved practical nursing program, the school administrators, the health occupations education teachers, and the state board of nurses, registration and nursing education work together in its development and continued operation. Students completing the program are required to take the state license examination administered by the state board of nurses. Students should check their own state requirements.

Most entry- or aide-level health occupations are not regulated by licensure, registration, or certification, with the training time being less than one year. Students may enter the job market directly after high school, or they may continue in a more advanced health occupations education training program.

SOURCES OF INSTRUCTIONAL MATERIALS

Suggestions for instructional materials may be obtained from publishers, from state supervisory personnel, from employers, from individuals already employed in occupations for which students are training, and from individuals contacted during field trips.

By writing to the following sources, among others and by identifying their specific needs, teacher-coordinators can obtain titles, descriptions, and price lists of textbooks, manuals, and visual aids.

AAVIM
220 Smithsonia Road
Winterville, GA 30683
Phone: (800) 228-4689
Fax: (706) 742-7005
Email: sales@aavim.com
Web site: www.aavim.com

Center on Education and Work
The University of Wisconsin–Madison
964 Educational Science Building
1025 West Johnson Street
Madison, WI 53706-1796
Phone: (800) 446-0399
Fax: (608) 262-9197
Email: cewmail@education.wisc.edu
Web site: www.cew.wisc.edu

Delmar / Thomson Learning
P.O. Box 15015
Albany, NY 12212-5015
Phone: (800) 477-3692
Fax: (518) 464-7000
Web site: www.delmar.com

Instructional Materials Laboratory
University of Missouri–Columbia,
   College of Education
2316 Industrial Drive
Columbia, MO 65211-8120
Phone: (800) 669-2465
Fax: (573) 882-1992
Email: info_iml@coe.missouri.edu
Web site: www.iml.coe.missouri.edu

Teacher-coordinators should always be looking for materials that they can get free, such as those donated by local employers and associations. If school policy permits, it may be wise to ask student-learners to purchase some manuals of their health specialty, just as they would buy uniforms and professional shoes. This is a prudent investment of some of the money they earn.

STUDENT ORGANIZATION

The third dimension of cooperative health occupations education, in addition to related class instruction and the training station, is the health occupations student organization—Health Occupations Students of America (HOSA). As indicated in Chapter 14, HOSA includes events in parliamentary procedures, public speaking, interviewing, writing, and medical technology at its annual conventions on the local level if the coordinator plans these events in the student organization programs. (Refer to Chapter 14, which discusses how to organize and operate a student organization.)

SECURING DESIRED LEARNING ON THE JOB

Training agreements are mandatory for those programs funded under vocational legislation. This practice of establishing training agreements is deemed highly desirable for any cooperative plan, regardless of its source of funding.

On-the-job training in a cooperative education plan should contribute directly to the development of the student's occupational competency. The teacher-coordinator and the training station supervisor develop each student's training plan (see Figure 19–1). This plan outlines the kinds of experiences that will be offered during the training period. It is most desirable for the related class instruction to parallel as closely as possible the needs of the student on the job. Both the school and the training station supervisor should understand which areas they are responsible for in the training. Through cooperative planning, the educational plan is developed in a logical sequence.

The employer signs a written agreement with the school indicating the specific nature of the training to be offered and the conditions of the training plan. A sample training agreement form is presented in Chapter 11. Also, Chapter 11 discusses how to develop training stations and training plans for a wide range of student-learners, from slow achievers to fast achievers. A competency-based Statement of Student-Trainee Learning Objectives (SSLO) (see Chapter 11), suitable to the learning abilities of the individual student-learners, may be selected each evaluation period.

The teacher-coordinator must be free to supervise students adequately when they are employed at the training stations. Time devoted to each student for on-the-job visitation and supervision should equal an average of at least one-half hour per week.

Illustrative of appropriate community health-care facilities is the following list. When searching for training stations in a local community, the teacher-coordinator will find this checklist useful.

Appropriate Community Health-Care Facilities

Hospitals

Admissions Office
Business Office
Central Supply
Clinics
Dietary Department
Electrocardiogram Unit
Electroencephalogram Unit
Emergency Room
Escort Service
Housekeeping Department
Inhalation Therapy Unit
Laboratory
Maternity Unit
Medical Records Department
Medical–Surgical Units
Occupational Therapy Department
Operating Room
Pediatric Unit
Physical Therapy Unit
Recreational Therapy Unit
Speech Therapy Unit
Social Service Office
Ward Clerk Units

Community Health-Care and Service Facilities

Baby-keep-well stations
Blood banks
Child day-care centers
Chiropractor's offices
Community health centers
Dental laboratories
Dentists' offices
Doctors' offices
Drug addiction centers
Drugstores
Extended-care centers
Health department
Homes for individuals who are mentally retarded
Industrial health offices
Intermediate-care centers
Medical–surgical supply stations
Mortuaries
Neighborhood health centers
Nursing homes
Orphanages
Pharmaceutical companies
Podiatrists' offices
Private medical laboratories
Rehabilitation centers
School nurses' offices

Veterinarian offices
Visiting homemaker services
Voluntary health agencies (heart, cancer, etc.

STUDENT EVALUATION

Evaluating student progress is a joint responsibility of the teacher-coordinator and the employers. Although some school districts require a separate grade for the related class instruction and the student's on-the-job performance, the authors believe the grade given for the course should include both related class instruction and on-the-job performance. Furthermore, it is strongly recommended that the teacher-coordinator and the employer evaluate the student's performance at least once during the grading period.

The teacher-coordinator should develop an evaluation form to meet local needs. Figure 19–2 illustrates a general purpose form with graduated scale for each element of evaluation. A more specific evaluation form is illustrated in Figure 19–3, which is a performance checklist for learning activities in the classroom and/or on the job.

POST-SECONDARY PROGRAMS

Post-secondary health occupations education programs provide training for those individuals seeking occupational preparation in the health field. Students may be graduates of secondary health occupations education programs who desire to continue their occupational training, or they may be individuals just beginning their training in the health occupations field. Students in some post-secondary health occupations education programs may work toward licensure, registration, or certification.

ADULT PROGRAMS

In order to meet the changing needs of adults in health occupations, non-credit health occupations education programs are conducted in comprehensive high schools and area career-technical schools. Individuals may be preparing for entry into gainful employment or upgrading skills in their current positions. Students in some adult programs may work toward licensure, registration, or certification.

Questions and Activities

1. Prepare a 20-minute talk that you could give before a civic club or professional group explaining the cooperative health occupations plan. Prepare charts or other visual aids to supplement your talk. (Add the notes of this talk to your files.)

2. What are the minimum qualifications for becoming certified as a coordinator of the cooperative plan for health occupations? (Refer to your state's certification plan.)

3. How adequate is the list of topics for related instruction suggested in this chapter? If you were the teacher-coordinator, which areas would you emphasize the most?

4. If you were preparing a long-range (three-to five-year) plan for purchasing equipment and supplies for a cooperative health occupations classroom, what essentials would you request the first year? What would be some desirable items you would put in your plan for purchase later on? Referring to current periodicals and state department handbooks should give you some ideas. (Add the notes of these ideas to your files.)

5. Analyze the learning outcomes that could be expected in a specific prospective training station in your local community. Determine the possibility of a student-learner's progress toward his/her career objective in this training station by the end of the school year. (Refer to Chapter 11 for suggestions on how to proceed.)

6. How do the activities provided by a health occupations education student organization provide educational experiences for student-learners?

7. Study current references on classroom equipment and floor plans. Draw a room layout to scale 1/4 inch equals 1 foot) and prepare an equipment list.

8. State the learning outcomes of a training plan in behavioral terms. (Refer to Chapter 11 for suggestions.)

For references pertinent to the subject matter of this chapter, see Reading Resources.

STEP-BY-STEP TRAINING PLAN
CARE OF A SENIOR CITIZEN

Student's Name _____

Student's Address _____

Supervisor's Name _____

Employer's Address _____

A variety of learning experiences are needed for the student-learner to develop the kinds of abilities needed in health and community service occupations. The responsibilities of the job include: (1) assisting an older person in meeting his/her own psychological and physical needs; (2) helping that individual with personal, social, and routine business matters; and (3) securing assistance in case of emergencies. The teacher-coordinator checks the training activities scheduled in the classroom, and the job supervisor checks those scheduled on the job. Students, in turn, benefit from the experiences, which help them toward achieving their career objectives.

Suggested areas of training and experience include the following:

	Training in Class	Experience on the Job
1. Working with senior citizens		
a. Helps person to learn to accept his/her state in the life cycle.	_____	_____
b. Helps with his/her grooming and dressing.	_____	_____
c. Prepares and serves suitable meals.	_____	_____
d. Helps him/her to maintain dignity and a sense of worth.	_____	_____
e. Respects religious and cultural values, patterns, and differences.	_____	_____
f. Takes proper safety measures to prevent accidents.	_____	_____
g. In an emergency situation, administers first-aid before the physician arrives.	_____	_____
2. Attaining desirable personal physical appearance		
a. Dresses appropriately.	_____	_____

FIGURE 19–1

STEP-BY-STEP TRAINING PLAN (Continued)

	Training in Class	Experience on the Job
2. Attaining desirable personal physical appearance (cont.)		
b. Maintains good posture.	_____	_____
c. Improves personal appearance through proper hair care and style changes.	_____	_____
d. Cares for nails, face, and teeth.	_____	_____
e. Learns the importance of personal hygiene.	_____	_____
f. Learns proper way to care for shoes, clothing, and accessories.	_____	_____
g. Wears shoes that are comfortable and that cover the feet.	_____	_____
h. Applies make-up tastefully.	_____	_____
3. Cultivating desirable personal qualities		
a. Is interested in people, especially older persons.	_____	_____
b. Is able to express feelings of tenderness, warmth, and affection.	_____	_____
c. Accepts responsibility.	_____	_____
d. Is responsive to the needs of older persons.	_____	_____
e. Is willing to accept constructive criticism.	_____	_____
f. Controls negative emotions at all times.	_____	_____
g. Is able to make independent decisions when necessary.	_____	_____
h. Adjusts to difficult situations.	_____	_____
i. Is honest.	_____	_____
j. Is patient; understands that older persons usually have slow movements.	_____	_____
k. Is discreet.	_____	_____
l. Respects other people's beliefs and values.	_____	_____
m. Is able to maintain harmonious personal relationships.	_____	_____
n. Is able to speak clearly, write legibly, and read aloud effectively.	_____	_____
o. Communicates easily with others.	_____	_____
4. Understanding employment policies		
a. Knows proper ways to apply for work.	_____	_____
b. Understands policies regarding wages, pay periods, transportation, health examinations.	_____	_____

FIGURE 19–1 (Continued)

STEP-BY-STEP TRAINING PLAN (Continued)

	Training in Class	Experience on the Job
4. Understanding employment policies (cont.)		
c. Understands the special concerns if an individual family or person is employer.	____	____
d. Understands Social Security and other laws affecting employment.	____	____
e. Understands training agreement concerning hours, wages, job to be done.	____	____
f. Understands the mutual responsibilities of both employer and employee.	____	____
5. Understanding physical and mental health needs of senior citizens		
a. Understands the disabilities common in older persons: frailty, loss of hearing, sight, memory.	____	____
b. Understands the needs of older persons who have any of these disabilities.	____	____
c. Understands how to help those with disabilities.	____	____
d. Knows how to use available health services.	____	____
e. Understands the reasons for inadequate nutrition — economics, food habits, food fads, frailty, lack of interest, aversion to eating alone.	____	____
f. Encourages person to help himself/herself as much as possible in daily activities.	____	____
6. Assisting with personal, social, and business matters; acquiring social amenities		
a. Accompanies person to hair styling salon, to doctor's office, to a movie, shopping, on an extended tour.	____	____
b. Addresses greeting cards and writes letters.	____	____
c. Attends to business transactions — going to bank, paying bills, writing checks, making telephone calls.	____	____
d. Accompanies person to church, to social gatherings, to call on friends.	____	____
e. Arranges to entertain person's friends and relatives.	____	____
f. Reads aloud, plays games, is an interested listener.	____	____

FIGURE 19–1 (Continued)

STEP-BY-STEP TRAINING PLAN (Continued)

	Training in Class	Experience on the Job
6. Assisting with personal, social, and business matters; acquiring social amenities (cont.)		
g. Can be entrusted with private or secret matters.		
7. Doing simple household tasks		
a. Keeps the living area clean, orderly, and attractive.		
b. Plans sequence of cleaning jobs to avoid upsetting daily routine.		
c. Uses equipment and methods for each job that cause least amount of confusion and noise.		
8. Helping with personal tasks		
a. Helps the person take a bath.		
b. Helps the person change clothes.		
c. Respects the person's wish for privacy.		
9. Caring for clothing and shoes		
a. Hand washes drip-dry fabrics, hose, fine lingerie, gloves.		
b. Presses and repairs articles of clothing.		
c. Stores items used often in accustomed places.		
d. Takes or sends clothes to commercial cleaners.		
e. Polishes shoes; takes them to be repaired.		
10. Understanding principles of nutrition and meal planning		
a. Understands the principles of food preparation.		
b. Knows what the basic food groups are and the functions of each.		
c. Knows how to plan and prepare special meals that are nutritionally sound.		
d. Understands special nutritional needs and food problems.		
e. Considers cost, flavor, texture, and color in planning attractive, appetizing meals.		
f. Knows how to prevent spoilage and contamination of food.		
g. Knows how to arrange table attractively.		
h. Serves food in appetizing manner and in appropriate quantities.		

FIGURE 19–1 (Continued)

STEP-BY-STEP TRAINING PLAN (Continued)

	Training in Class	Experience on the Job
10. Understanding principles of nutrition and meal planning (cont.)		
i. Understands the importance of using sanitary methods of dishwashing and cleaning up the kitchen.		
11. Understanding how to prevent accidents in the home		
a. Understands causes of home accidents.	_____	_____
b. Encourages elimination of common household hazards, such as small rugs and waxed floors.	_____	_____
c. Encourages installation of safety devices, such as stair railings, wall handles at tub and toilet, carpeting floors.	_____	_____
d. Anticipates danger zones and gives extra precautionary help.	_____	_____
12. Knowing what actions to take in an emergency		
a. Keeps telephone numbers to use in emergencies — numbers of person(s) legally responsible for the injured person, neighbors or friends, doctor, hospital, ambulance, police, fire department, plumber, electrician.	_____	_____
b. Notifies doctor and/or family.	_____	_____
c. Applies appropriate first-aid measures for cuts, burns, or fainting.	_____	_____
13. Completing individual projects		
a. Preparing a job manual based on activities at the training station.	_____	_____
b. Develops career manuals.	_____	_____

Employer _____

Parent _____

Teacher-Coordinator _____

School _____

(Copies to employer, teacher-coordinator, student)

FIGURE 19–1 (Continued)

STUDENT-LEARNER EVALUATION

Date _____ to Date _____

Student's Name _____ Firm _____

Instructions to the Employer: Read each description carefully. For each characteristic listed on the left, place a checkmark on the line over the phrase that describes this worker most accurately. If you think this individual is about halfway between two descriptions, make your checkmark about halfway between them on that line. Any additional comments you wish to make will be appreciated.

ABILITY TO FOLLOW INSTRUCTIONS

Seems unable to follow instructions.	Needs repeated detailed instructions.	Follows most instructions with little difficulty.	Follows most instructions with no difficulty.	Uses initiative in interpreting and following instructions.

ABILITY TO GET ALONG WITH PEOPLE

Frequently rude and uncooperative.	Sometimes lacks poise and understanding; seems indifferent.	Usually gets along well with people.	Usually poised, courteous, tactful in working with people.	Usually tactful and understanding in dealing with all types of people.

ATTITUDE TOWARD APPEARANCE OF WORK STATION

Maintains careless, sloven work station.	Allows work station to become disorganized.	Follows good housekeeping rules.	Takes pride in appearance and arrangement of work station.	Keeps work place outstandingly neat and efficiently organized.

COOPERATION

Uncooperative, antagonistic — hard to get along with.	Cooperates reluctantly.	Cooperates willingly when asked.	Usually cooperates eagerly and cheerfully.	Always cooperates eagerly and cheerfully without being asked.

INDUSTRY

Always attempts to avoid work.	Sometimes attempts to avoid work.	Does assigned work willingly.	Does more than assigned work willingly if given directions.	Shows originality and resourcefulness in going beyond assigned job without continual direction.

FIGURE 19–2. (Courtesy, South Bend Community School Corporation, South Bend, Indiana)

STUDENT-LEARNER EVALUATION (Continued)

QUALITY OF WORK

Does almost no acceptable work.	Does less than required amount of satisfactory work.	Does normal amount of acceptable work.	Does more than required amount of neat, accurate work.	Shows special aptitude for doing neat, accurate work beyond the required amount.

DEPENDABILITY

Unreliable, even under careful supervision.	Sometimes fails in obligations, even under careful supervision.	Meets obligations under careful supervision.	Meets obligations with very little supervision.	Meets all obligations unfailingly without supervision.

PERSONAL APPEARANCE

Slovenly and inappropriately groomed.	Sometimes neglectful of appearance.	Satisfactory appearance.	Neat and appropriately groomed.	Exceptionally neat and appropriately groomed.

PROGRESS

Fails to do an adequate job.	Lets down on the job somewhat.	Maintains a constant level of performance	Shows considerable progress.	Shows outstanding progress.

OVER-ALL ESTIMATE OF WORK

Poor	Below average	Average	Above average	Outstanding

Comments: _____

_____

_____

_____

Employer–Trainer _____ Date _____

FIGURE 19–2 (Continued)

PERFORMANCE CHECKLIST

Student _____ Date _____

	Class Instruct.	Demo	Return Demo	Domain Test
1. Using medical asepsis and cleaning				
a. Washes hands aseptically for nursing tasks.				
b. Cleans tubs and showers used by patients.				
c. Handles clean and dirty linen and equipment.				
2. Positioning and moving patients				
a. Adjusts a Gatch-frame bed for a patient.				
b. Utilizes a supportive aid to support a patient's body parts.				
c. Moves a patient to the head of the bed.				
d. Puts a patient in the supine position.				
e. Puts a patient in the lateral or Sim's position.				
f. Puts a patient in the prone position.				
g. Puts a patient in the Fowler position.				
h. Records position changes.				
i. Applies restraints and records.				
3. Making beds for patients				
a. Makes an unoccupied closed bed for a patient.				
b. Makes an unoccupied open bed for a patient.				
c. Makes an occupied bed for a patient.				
d. Makes a bed with supportive devices (heat cradle, footboard, etc.) for a patient.				

FIGURE 19–3. (Courtesy, South Bend Community School Corporation, South Bend, Indiana)

PERFORMANCE CHECKLIST (Continued)

	Class Instruct.	Demo	Return Demo	Domain Test
4. Dressing and undressing patients				
a. Changes a patient's gown.				
b. Assists a patient with slippers and robe.				
c. Assists a patient with pant-type garments.				
d. Assists a patient who is partially paralyzed.				
e. Puts incontinent briefs on a patient.				
f. Drapes a cotton blanket or a sheet on a patient.				
5. Ambulating and transferring patients				
a. Helps a patient to dangle legs.				
b. Helps a patient from bed to chair and back.				
c. Helps a patient to walk (with a person).				
d. Helps a patient to walk with a walker.				
e. Helps a patient to a wheelchair.				
f. With the help of two other persons, transfers a patient to a chair.				
g. With the help of three other persons, transfers a patient to a chair.				
h. Transports a patient in a wheelchair.				
i. Records assistance given.				

FIGURE 19–3 (Continued)

PERFORMANCE CHECKLIST (Continued)

	Class Instruct.	Demo	Return Demo	Domain Test
6. Assisting patients with oral hygiene				
a. Helps a patient to brush teeth.				
b. Brushes a patient's teeth.				
c. Cleans a patient's dentures.				
d. Administers a mouthwash.				
e. Records care given.				
7. Assisting patients with hair and nail care				
a. Combs a patient's hair.				
b. Shampoos and dries a patient's hair.				
c. Shaves a male patient's face with an electric razor.				
d. Shaves a male patient's face with a safety razor.				
e. Cleans a patient's fingernails and toenails.				
f. Trims a patient's fingernails and toenails.				
g. Records care given.				
8. Assisting patients with elimination				
a. Helps a patient to use the bedpan.				
b. Helps a patient to use the fracture pan.				
c. Helps a male patient to use the urinal.				
d. Helps a patient to use the commode.				
e. Helps a patient to get to the bathroom.				
f. Gives a patient perineal care.				

FIGURE 19–3 (Continued)

PERFORMANCE CHECKLIST (Continued)

	Class Instruct.	Demo	Return Demo	Domain Test
8. Assisting patients with elimination (continued)				
g. Cleans an incontinent patient.				
h. Empties a patient's urinary drainage bag.				
i. Measures a patient's urinary output.				
j. Records care given.				
9. Bathing patients				
a. Washes a patient's face and hands.				
b. Washes a patient's back.				
c. Assists a patient with A.M. care.				
d. Assists a patient with P.M. care.				
e. Assists a patient with a partial bath.				
f. Assists a patient with a tub bath.				
g. Assists a patient with a shower.				
h. Encourages or assists a patient with range of motion.				
i. Gives patient a complete bed bath.				
j. Records care given.				
10. Determining patient's vital signs				
a. Using a glass or an electronic thermometer, determines a patient's oral temperature.				
b. Determines a patient's axillary temperature.				
c. Determines a patient's rectal temperature.				
d. Records results.				

FIGURE 19–3 (Continued)

PERFORMANCE CHECKLIST (Continued)

	Class Instruct.	Demo	Return Demo	Domain Test
11. Assisting patients with nutrition and hydration				
a. Checks the food on a food tray to be served to a patient.				
b. Serves food on a food tray to a patient.				
c. Secures a geriatric feeding chair or a high chair to feed a patient.				
d. Helps a patient to eat.				
e. Feeds a helpless patient.				
f. Measures and/or computes a patient's oral intake.				
g. Records food/fluids taken by a patient.				
h. Brings a patient drinking water.				
12. Weighing and measuring patients				
a. Determines a patient's weight when patient is standing on scales.				
b. Determines a patient's weight when patient is sitting in chair.				
c. Records patient's weight/measurement.				
13. Admitting, transferring, and discharging patients				
a. Prepares a unit for a new patient.				
b. Inventories a patient's personal belongings.				
c. Marks and/or stores a patient's personal belongings.				

FIGURE 19–3 (Continued)

PERFORMANCE CHECKLIST (Continued)

	Class Instruct.	Demo	Return Demo	Domain Test
14. Applying protective devices and administering special skin care to patients				
a. Puts heel and elbow protectors on a patient.				
b. Puts anti-decubitus pads under a patient.				
c. Checks a patient's skin for pressure areas.				
d. Supports a patient's body parts to prevent pressure points.				
15. Understanding and using first-aid practices				
a. Observes and reports a patient who has fallen.				
b. Observes and reports a patient who has fainted.				
c. Assists nurse with appropriate first-aid measures.				
d. Recognizes and reports potential sites for accidents.				
e. Recognizes and reports impending signs of a patient in distress.				
16. Understanding the aging process				
a. Begins to understand the developmental stages of aging (children included).				
b. Recognizes normal structure of body systems.				
c. Recognizes changing functions of body system.				
d. Understands psychosocial aspects of institutionalization.				
e. Understands dependency and independency.				
f. Understands the losses encountered in old age, including the process of grieving.				

FIGURE 19-3 (Continued)

PERFORMANCE CHECKLIST (Continued)

	Class Instruct.	Demo	Return Demo	Domain Test
17. Accountability				
a. Recognizes invasion of privacy, privileged communications and consent, and resident's safety and rights.				
b. Identifies and traces lines of authority in the facility.				
18. Using interpersonal skills/communication skills				
a. Identifies factors that promote effective communication.				
b. Defines verbal and nonverbal communication/ active listening.				
c. Identifies essential attitudes for establishing working relationships with residents, families, and staff.				
d. Responds to a call light or a resident's request.				
e. Communicates and interacts with those who have impaired mental and psycho-social abilities.				

FIGURE 19–3 (Continued)

20

The Plan in Family and Consumer Sciences Occupations

Food preparation is an important training station activity for student-learners in family and consumer sciences occupations.

KEY CONCEPT   The broad scope of occupations related to family and consumer sciences concepts and principles has made it clear that the cooperative plan of instruction is a valuable curriculum structure in this area.

GOALS

After successfully completing the study of this chapter, answering the questions, and carrying out the activities, the student should be able to:

- Identify and list appropriate training station and career objectives in family and consumer sciences (FCS) occupations.

- Understand how to develop an appropriate curriculum for a local plan in cooperative occupational education for FCS occupations.

- List the major sources of instructional materials for FCS occupations education.

- Devise appropriate training plans for students at various training stations.

- Evaluate other student-learners in the classroom and on the job.

Family and Consumer Sciences Wage-earning Occupations

Family and consumer sciences (FCS) education provides career-technical preparation for family life and work life. This is a relatively new role for many FCS educators who have adopted the cooperative plan[1] of instruction in their curricula. Today's family and consumer sciences education is very different from the home economics education of the past. Its new conceptual framework is designed to empower individuals, strengthen families, and enable communities by improving individual family and community well-being; impacting the development, delivery, and evaluation of consumer goods and services; influencing the creation of policy; and shaping social change to enhance the quality of life.

The broad scope of wage-earning occupations outside the home has made clear that the cooperative plan of instruction is a valuable curriculum structure in FCS education. Career-technical educators should refrain from what might be a normal temptation to classify as family and consumer sciences those occupations in which women are primarily engaged. The criterion for curriculum development is the learning source of the competencies needed by a worker. That is, those competencies best learned on the job might be covered in training stations through a cooperative education plan. Those competencies best learned in the classroom should be built into the curriculum to be taught in school classrooms and laboratories.

Family and Consumer Sciences Occupations

FCS occupations plans are composed of groups of related courses or units of instruction organized for the purpose of enabling students to acquire knowledge and to develop understandings, attitudes, and skills relevant to occupational preparation by using the knowledges and skills of FCS. The subject matter of family and consumer sciences applies concepts drawn from the natural and social sciences and the humanities in unique ways. Occupational preparation in FCS areas consists of courses or units of instruction emphasizing the acquisition of competencies needed for getting and holding a job and/or preparing for advancement in an occupational area by using family and consumer sciences knowledges and skills. Instructional content is selected from the unique requirements in a specific occupation related to FCS and is coordinated with appropriate field, laboratory, and work experiences. Occupations include those that provide (1) services to families in the home and similar services to others in group situations; (2) assistance to professional family and consumer scientists and professionals in fields related to FCS; and (3) other services and/or assistance directly related to one or more FCS subject matter areas, such as occupations in food service and child care and service occupations in the textile and clothing industries.

The Family and Consumer Sciences / Human Sciences CIP codes include the following occupational clusters.[2]

19.00 Work and Family Studies

19.02 Family and Consumer Sciences / Human Sciences, General

19.04 Family and Consumer Economics and Related Studies

19.05 Foods, Nutrition, and Related Services

19.06 Housing and Human Environments

19.07 Human Development, Family Studies, and Related Services

19.09 Apparel and Textiles

19.99 Family and Consumer Sciences / Human Sciences, Other

---

[1]The authors choose to use the term "cooperative plan" rather than "cooperative method." The rationale for this practice is explained in Chapters 1 and 4.

[2]Robert L. Morgan and E. Stephen Hunt, *A Classification of Instructional Programs—2000*, National Center for Education Statistics, Presidential Building, 400 Maryland Avenue, S.W., Washington, DC 20202, 2000, pp. 19–20.

The cooperative plan in FCS is used in both one- and two-year programs. Basic instruction is combined with supervised on-the-job experiences in caring for children in nursery schools, day-care centers, and kindergarten programs; in clothing management, alterations, industrial sewing, clothing maintenance; and in health centers, restaurants, institutional kitchens, and cafeterias. Some of the many occupations include bridal consultant, fabric coordinator, child-care aide, household employee, home furnishings aide, hotel/motel aide, alterations worker, dietary aide, tailor, millinery aide, food service worker, and foods, equipment, or textiles tester.

Students who graduate from a FCS occupations plan are prepared to assume positions in the occupations for which they have been trained. Those entry-level through mid-management positions may lead to job advancement requiring additional education and work experience. This is known as the *vertical ladder concept*. Utilizing the lattice concept, the students might also move horizontally to other jobs requiring comparable skills. A listing of job opportunities in FCS occupations follows. A variety of jobs that utilize students with career-technical training or advanced college and university preparation are available in FCS.

Types of Family and Consumer Sciences Occupations Training Stations

Restaurants	Hotels/motels	Cafeterias
Day-care centers	Convalescent centers	Bakeries
Nursing homes	Baking laboratories	Specialty shops
Hospitals	Clothing departments	Alterations departments

THE COOPERATIVE PLAN

When organized as a cooperative plan, FCS occupations operate partly in the schools and partly in businesses and industries in the community. Cooperative students attend a one-hour related class per day at school and then work approximately 15 hours per week in businesses or industries, usually being paid the minimum wage at the start of the training. The selection of training stations is based upon the students' career goals in their chosen occupations. The cooperative education teacher-coordinators are responsible for teaching basic skills and world-of-work information in the related classroom, placing the students in appropriate training stations, and supervising students on the job. The employers provide the technical skills training and practice. Cooperatively, the teachers, students, and employers develop individual training agreements and training plans. These specify the competencies each student should learn in school and on the job.

In some instances, FCS programs are a combination of both in-school laboratory and cooperative experiences. The junior year may be structured as an in-school laboratory program, with the senior year used as a cooperative plan. This dual combination provides the structure of laboratory training in the first year with actual work experience the following year. Also, the combination program has advantages when the in-school facilities are only large enough to handle one beginning class, or if the number of community training stations is limited.

FACILITIES AND EQUIPMENT

In-school laboratory students receive all their training within the school before they apply for a job. Because of this, the most realistic and up-to-date facilities and equipment are necessary to aid the students in making an easy transition from school to work. In addition to the laboratory facility, a fully equipped classroom is a vital component for each program. The classroom needs to be conveniently located near the laboratory, with facilities to accommodate large-group, small-group, and individualized instruction.

Movable tables might well be arranged in a "U" shape, with chairs outside the "U." Writing board space, bulletin board space, reference book shelves, magazine racks, and filing cabinets are needed.

Computers and other technological advances can play important roles in the related instruction classroom (Courtesy, Jasper S. Lee)

Technological advancements are significantly impacting all career and technical education, including family and consumer science education. Computers, a digital camera, an audiotape recorder, a TV, a video camcorder, a VCR, a DVD player, an overhead projector, and an LCD multimedia projector can play significant roles in the related instruction classroom. A room available for individual student-coordinator conferences is an important feature of the cooperative FCS occupations education classroom. Here there is privacy for discussing personal problems or problems encountered at the training station. Glass partitions or large glass windows in the conference room allow the activities going on in the classroom to be in full view of the teacher-coordinator who is participating in an individual conference.

If there are other cooperative education plans, such as health occupations and industrial education in the school, conceivably this specially designed classroom could be shared with these teacher-coordinators. Also, this type of classroom is ideal for adult instruction. Additional chairs could be added if a different configuration of the tables were arranged.

CURRICULUM

The curriculum for cooperative family and consumer sciences occupations plans consists of world-of-work information and technical skills information. World-of-work information that is taught in the related classroom includes (1) determining the availability of jobs; (2) understanding how to apply for a job; (3) interviewing; and (4) understanding work responsibilities, work standards, ethics, policies and regulations on the job, and interpersonal relations. The technical skills information that is taught includes that which is needed by student-learners to obtain initial entry-level and advanced-level employment in the FCS occupations.

SOURCES OF INSTRUCTIONAL MATERIALS

The consultant or supervisor for family and consumer sciences education in the state can suggest sources of instructional materials. Teacher-coordinators can also contact the following companies and organizations, among others, and identify their specific needs for instructional materials for related classroom instruction in FCS occupations. These sources can provide titles, descriptions, and price lists of textbooks, manuals, and audio-visual aids. Each year the January-February issue of *Techniques,* the journal of the Association for Career and Technical Education, contains a directory of suppliers.

Curriculum and Instructional Materials Center
Oklahoma Department of Career and
   Technology Education
1500 West Seventh Avenue
Stillwater, OK 74074-4364
Phone: (800) 654-4502
Fax: (405) 743-5154
Web site: www.okcareertech.org/cimc

Delmar / Thomson Learning
P.O. Box 15015
Albany, NY 12212-5015
Phone: (800) 477-3692
Fax: (518) 464-7000
Web site: www.delmar.com

EIMC / Educational Resources
P.O. Box 7218
University of Texas
Austin, TX 78713-7218
Phone: (512) 471-7716
Fax: (512) 471-7853

EMC/Paradigm Publishing
Customer Care Center
875 Montreal Way
St. Paul, MN 55102
Email: educate@emcp.com
Web site: www.emcp.com

Goodheart-Willcox Publisher
18604 West Creek Drive
Tinley Park, IL 60477-6243
Phone: (800) 323-0440 or (708) 687-5000
Fax: (888) 409-3900
Email: custserv@goodheartwillcox.com
Web site: www.goodheartwillcox.com

International Furnishings and Design Association
191 Clarksville Road
Princeton Junction, NJ 08550
Phone: (609) 799-3423
Fax: (609) 799-7032
Email: info@ifda.com
Web site: http://ifda.com

Multistate Academic and Vocational
   Curriculum Consortium
1500 West Seventh Avenue
Stillwater, OK 74074-4364
Phone: (800) 654-4502
Fax: (405) 743-5154
Web site: www.mavcc.org

Prentice Hall School
Upper Saddle River, NJ 07458
Phone: (201) 592-2425 or (866) 326-4259
Web site: www.phschool.com/career_technical
   /index.html

South-Western Publishing Company
5101 Madison Road
Cincinnati, OH 45227
Phone: (800) 543-0487
Fax: (513) 527-6940
Email: swinformation@thomsonlearning.com
Web site: www.swep.com

The alert teacher-coordinator should also be able to obtain some materials free, such as those donated by local employers and associations. If school policy permits, the teacher-coordinator could ask student-learners to purchase some manuals of their trade, just as they would buy uniforms and shoes. This is a wise investment of some of the money they earn.

STUDENT ORGANIZATION

The FCCLA student organization is the third dimension of the cooperative occupational education plan. It is an integral part of the total instructional program, including the related classroom and the training station. Chapter 14 lists suggestions on how students may benefit from participation in a FCCLA student chapter.

The student organization is a good place to stress parliamentary procedures, net appearance and appropriate dress, social etiquette, and successful completion of individual and group projects.

SECURING DESIRED LEARNING ON THE JOB

Because the needs of each student-learner for instruction in his/her chosen occupation vary, the teacher-coordinator must determine the specific experiences needed. The technique that accomplishes this is the training plan that is developed for each student-learner, showing what is to be learned on the job. This plan has to be worked out by the teacher-coordinator in cooperation with the on-the-job supervisor or employer. This plan will be in the form of a task breakdown of the phases of the occupation that the student-learner is expected to learn.

Training may be given in occupations with which the coordinator may not be familiar and about which assistance will have to be sought. In all instances, the advisory committee can help. This committee should include employers, supervisors, and employees in family and consumer sciences wage-earning occupations. The employer should be encouraged to have a hand in developing the training plan. A detailed discussion of advisory committees is presented in Chapter 6. The training plan may be worked out in any one of a number of forms, the most effective being those designed for particular student-learners in their individual training stations. An example is shown in Figure 20–1.

As in other cooperative education plans, the step-by-step training plan (schedule of processes) should make the FCS wage-earning occupations training plan sharply distinctive from general work experience plans.

The plan indicates the specific career-technical objectives to be emphasized as well as the place where they are to be learned—school or job—and from whom they are to be learned—related instruction teacher or on-the-job supervisor. Frequently, the plan is attached to the training agreement (see Chapter 11) to make up a complete training plan. The details of the breakdown and the comprehensiveness or completeness of the training plan will depend upon the opportunities offered at the training station and the ability of the student-learner to benefit from the training.

Training stations and training plans can be developed for a wide range of student-learners, from slow achievers to fast ones. A competency-based SSLO (see Chapter 11) that is suitable to the learning abilities of the individual student-learner should be selected each evaluation period.

MEASURING STUDENT-LEARNERS' GROWTH

In the final analysis, measurement of the student-learners' growth and development at the end of the school year will be determined by the extent to which the objectives have been reached. The objectives of an effectively developed training plan will include:

1. Increasing the skills of the individual student-learners to perform successfully at the entry-level position.

2. Helping student-learners to develop an understanding of and an appreciation for the fundamentals of the information and knowledge related to the occupation.

CASE-IN-POINT

Recognize the Social Type with Whom You Work

Jane Thompson, a student-learner at the food facility in the veterans' hospital, came to her teacher-coordinator with a problem. Jane said that she was having difficulty working under her supervisor, Mr. Wright. She felt that Mr. Wright expected too much from her. She didn't mind working hard, but there seemed to be no way to keep up. Jane's teacher-coordinator promised to visit the training station to size up the situation. In the meantime, he reviewed four social types with Jane to help her in adjusting to Mr. Wright's method of operation.

The coordinator reminded Jane that supervisors and managers tend to be one of four basic social types—*amiable, analytical, aggressive,* or *expressive.** Of course, any individual might be a combination of one or more of these types. From visitations to the training station and a discussion with Jane, the teacher-coordinator determined that Mr. Wright was definitely an aggressive type. The two discussed the characteristics of the aggressive type, who tended to be independent, candid, decisive, pragmatic, determined, efficient, and objective. The aggressive types normally like employees who are available and who exhibit cooperativeness, loyalty, supportiveness, patience, and respect. Both Jane and her coordinator decided that when Mr. Wright realized she was productive and doing her best to be cooperative, he would be more understanding. So both laid out a plan to tackle the situation over time.

Knowing that other trainees might be having similar problems, they decided to study and discuss the various social types in class for the benefit of all the students.

---

*John R. Darling and Raymond E. Taylor, "Upward Management: Getting in Step with the Boss," *Business,* College of Business Administration, Georgia State University, University Plaza, Atlanta, Vol. 36, No. 2, April–May–June 1986, p. 3.

STEP-BY-STEP TRAINING PLAN

The _____(Name of Business Firm)_____ will permit _____(Name of Student)_____ from (Name of School)_____ High School to enter its establishment as an employee under the supervision of (Name)_____ for the purpose of gaining knowledge and experience in the occupational area of __19.09 Clothing Management Production, and Services,_____ so that the student may prepare for a career as a(n) Alteration Tailor_____.

Approximate Time	Learning Activities	Training in Class	Experience on the Job	Evaluation	Individual Study Assignment
2 weeks	**Orientation by teacher-coordinator**				
	1. Conducts personal evaluation.	X			
	2. Stresses importance of good school and job attendance.	X			
	3. Discusses necessity of positive attitude.	X			
	4. Discusses desirable work habits.	X			
	5. Explores job opportunities.	X			
Continuous	**Theory**				
	1. Selects, uses, and maintains tools and equipment.	X	X		
	2. Arranges and stores supplies.	X	X		
	3. Reads and interprets written instructions.	X	X		
	4. Uses and maintains sewing machines.	X	X		
	5. Fits, marks, or pins alterations.	X	X		
	6. Makes alterations.	X	X		
	7. Presses and cares for garments.	X	X		
Continuous	**Using and maintaining sewing machine**				
	1. Recognizes types of machines and jobs they perform.	X			
	2. Learns parts and adjustments of machines.	X	X		
	3. Threads and operates machine.	X	X		
	4. Cleans and oils machine.	X	X		
12 weeks	**Fitting and measuring**				
	1. Pins or marks to shorten garment.	X	X		
	2. Pins or marks to lengthen garment.	X	X		
	3. Pins or marks to remove fullness.	X	X		
	4. Pins or marks to adjust for tightness.	X	X		
	5. Pins or marks to adjust sleeve length and then shapes sleeve.	X	X		
	6. Pins or marks to adjust shoulder seams.	X	X		

FIGURE 20–1. (Courtesy, Vocational Education Program Area, School of Education, Indiana University)

STEP-BY-STEP TRAINING PLAN (Continued)

Approximate Time	Learning Activities	Training in Class	Experience on the Job	Evaluation	Individual Study Assignment
12 weeks	**Making alterations**				
	1. Pins or marks.	X	X		
	2. Stitch as pinned or marked.	X	X		
	3. Removes old seams.	X	X		
	4. Presses.	X	X		
4 weeks	**Stitching seams, darts, and hems**				
	1. Pins or marks.	X	X		
	2. Stitches and cuts when necessary to eliminate drawing.	X	X		
	3. Presses when alterations are completed.	X	X		
1 week	**Sewing in zipper application**				
	1. Selects correct type and color.	X	X		
	2. Puts in zipper.	X	X		
	3. Presses enclosure.	X	X		
1 week	**Hemming slacks**				
	1. Measures from inseam.	X	X		
	2. Cuts off excess.	X	X		
	3. Hems with blind stitch machine.	X	X		
	4. Steam presses.	X	X		
1 week	**Cuffing Slacks**				
	1. Measures from inseam.	X	X		
	2. Adds measurement required for turning cuff.	X	X		
	3. Cuts off excess.	X	X		
	4. Hems with blind stitch machine.	X	X		
	5. Turns cuff and then tacks on each side through top of seam.	X	X		
	6. Steam presses cuff.	X	X		

FIGURE 20–1 (Continued)

STEP-BY-STEP TRAINING PLAN (Continued)

Approxi-mate Time	Learning Activities	Training in Class	Experience on the Job	Evalua-tion	Individual Study Assignment
2 weeks	**Adjusting waistline on slacks**				
	1. Measures waistline for adjustment.	X	X		
	2. Opens top of belt, leaving old seam in place until adjustment is made (removes belt carriers as necessary).	X	X		
	3. Tapers seam allowance to blend into original seam as fitting requires.	X	X		
	4. Removes old seam.	X	X		
	5. Opens tapered seam and presses flat.	X	X		
	6. Turns belt over seam, turning raw edges of seam under at an angle.	X	X		
	7. Stitches belt down by using a tack stitch.	X	X		
	8. Presses.	X	X		
1 week	**Attaching pocket and flaps**				
	1. Selects type according to style of garment.	X	X		
	2. Pins or tacks placement of flap or pocket.	X	X		
	3. Stitches and turns pocket.	X	X		
	4. Steam presses.	X	X		
1 week	**Making buttonholes**				
	1. Discusses types.	X	X		
	a. Hand-made				
	b. Machine-made				
	(1) Keyhole				
	(2) Straight				
	(3) Bound				
	2. Selects type for garment.	X	X		
	3. Spaces and marks.	X	X		
	4. Stitches buttonhole.	X	X		
	5. Cuts opening.	X	X		
	6. Presses.	X	X		
1 week	**Adding buttons**				
	1. Selects size and type.	X	X		
	2. Spaces buttons.	X	X		
	3. Sews on buttons.	X	X		
	a. Hand sews with and without shank				
	b. Machine sews with and without shank				

FIGURE 20–1 (Continued)

3. Acquainting the student workers with sources of additional information and skills development to enable each of them to progress in his/her chosen occupation.

4. Laying groundwork of aesthetic and moral values that will support what the students have learned previously under other teachers to enable them to assume whatever leadership role in business or industry they are capable of pursuing.

Student measurement takes the form of evaluated projects that have become progressively more difficult and revealing, such as performance tests during class time and on the job; written tests on classwork and on individual study in the classroom and on the job; ratings of job growth by the job sponsor; and personal growth revealed by participation in student activities, including the student organization.

Positive growth in several or all of the areas mentioned can be measured by criteria such as the following:

1. The student-learners have attained an acceptable or better level of technical background and skills adequate to perform in their chosen trades, or specialties.

2. The student-learners can use resource manuals and training manuals in their chosen fields.

3. The student-learners have attained an appreciation of the concept of work to the extent that they have the potential to succeed in their chosen fields or in a related family of occupations.

4. The student-learners can communicate acceptably with their supervisors and co-workers

5. The employer–employee appreciation activity, or any other public display or function that is student-centered but coordinator-advised, reflects the skills acquired and the technical knowledge gained by the student-learners during the cooperative occupational education experience.

6. The student-learners can produce the quality of work desired in quantity that meets the standards of the job.

Figure 20–2 illustrates an employer rating sheet for use each grade period. It is to be filled out by the teacher-coordinator in consultation with the on-the-job supervisor and sometimes in the presence of the student. The most effective technique is for the teacher-coordinator to make an appointment at the training station to hand-carry the evaluation form rather than to mail it or send it by the student. Figure 20–3 shows an alternative type of form.

Questions and Activities

1. Prepare a 20-minute talk that you could give before a civic club or professional group explaining the cooperative plan for family and consumer sciences occupations. Prepare charts or other visual aids to supplement your talk. (Add the notes of this talk to your files.)

2. What are the minimum qualifications for becoming certified as a coordinator of the cooperative plan for family and consumer sciences occupations? (Refer to your state's certification plan.)

3. What topics for related instruction would you add to those suggested in this chapter? If you were the teacher-coordinator, which areas would you emphasize the most?

4. If you were preparing a long-range (three-to five-year) plan for purchasing equipment and supplies for a cooperative family and consumer sciences classroom, what essentials would you request the first year? What would be some desirable items you would put in your plan for purchase later on?

5. Analyze the learning outcomes that could be expected in a specific prospective training station in your local community. Determine the possibility of a student-learner's progress toward his/her career objective in this training station by the end of the school year. (Refer to Chapter 11 for suggestions on how to proceed.)

6. How do the activities of the FCCLA student organization provide educational experiences for student-learners?

7. Study current references on classroom equipment and floor plans. Draw a room layout to scale (½ inch equals 1 foot) and prepare an equipment list.

8. State the learning outcomes of a training plan in behavioral terms. (Refer to Chapter 11 for suggestions.)

For references pertinent to the subject matter of this chapter, see Reading Resources.

EMPLOYER RATING REPORT OF STUDENT PROGRESS

Student's Name _____ Date _____

Days Absent _____ Days Late _____

DATE TO BE RETURNED _____

Please state specifically your impression of the following skills and traits.

	Superior	Above Average	Average	Below Average	Unsatis-factory
Skill performance					
Using and maintaining sewing machine					
Fitting and measuring					
Making alterations					
Stitching					
Sewing in a zipper					
Hemming					
Cuffing					
Adjusting waistline					
Attaching pocket and flaps					
Making buttonholes					
Adding buttons					
Production ability					
Volume of work					
Quality of work					
Steadiness					
Organization of work					
Initiative					
Fundamentals					
Grammar					
Spelling					
Punctuation					
Arithmetic					
Command of English					
Handwriting					

FIGURE 20–2

EMPLOYER RATING REPORT OF STUDENT PROGRESS (Continued)

	Superior	Above Average	Average	Below Average	Unsatis-factory
Business techinques: **Ability to:**					
Meet people					
Work harmoniously with others					
Use telephone					
Use references (dictionary, etc.)					
Follow instructions					
Personal traits					
Appearance					
Attitude					
Health					
Personal hygiene					
Speech					
Adaptability					
Tact					
Accuracy					
Punctuality					
Manners					
Self-confidence					

Other comments regarding personality and business qualifications in unusual strength or weakness:

_____

_____

_____

Name of Employer _____

Signature of Supervisor _____

FIGURE 20–2 (Continued)

STUDENT'S PERIODIC RATING AND PROGRESS REPORT

Name (last, first, middle initial)   Grade   Employment

Training Establishment   Tel. No.   Report for Month of
_____ _____

Times Absent from Work	Times Absent from School	Wage Rate Beg. _____ End _____ of Month	Job Instructor	Average Work Week Hrs. _____	Expected On-the-Job Schedules of Hours:
					Monday ... to ... Friday ... to ...
			Employer	Total Wage This Month $_____	Tuesday ... to ... Saturday ... to ... Wednesday ... to ... Sunday ... to ... Thursday ... to ...

Note: This report is to provide information that will guide the school and the cooperating business agency to provide the most effective type of training for the student. The trainer is to check each unit where improvement is needed. Also the rating of the student should be checked below.

ON-THE-JOB TRAINER'S RATING OF THE STUDENT

Complete the rating chart below. Also check (X) all units of instruction on the job in which the student must show improvement to attain satisfactory performance.

Out-standing	Above Av.	Av.	Below Av.	Fail-ing	Traits or Characteristics	Comments by Trainer:
					1. Personal appearance	
					2. Interest	
					3. Punctuality	
					4. Dependability	
					5. Initiative	
					6. Ability to cooperate with others	Signed Date
					7. Progress — quality of work	Follow-up:
					8. Progress — quantity of work	
					9. Accuracy	
					10. Ability to follow instructions	
					11. Suitability for this occupation	
					12. Courtesy	
						Signed Date

Mark "X" in the space that best describes each item

RELATED CLASSROOM INSTRUCTION

Related Training Topics Studied by Student:

FIGURE 20–3. (Courtesy, Dade County Public Schools, Dade County, Florida)

21

The Plan in Marketing Occupations

Marketing teacher-coordinators teach students the concept of social marketing by having students conduct an "Adopt-a-Family" project.

KEY CONCEPT: Marketing education is a career-technical program of instruction designed to prepare youth and adults for careers in marketing and its related occupations.

GOALS

After successfully completing the study of this chapter, answering the questions, and carrying out the activities, the student should be able to:

- Identify and list major occupational clusters within the instructional area of marketing education.

- Identify and list the goals of marketing education.

- Identify and explain the curriculum patterns for marketing education.

- Describe the curriculum framework for marketing education.

- Plan and develop a marketing education instructional facility.

- List the major sources of instructional materials for marketing education.

- Describe student-learner benefits gained from membership in DECA.

- Evaluate the marketing education plan.

- Discuss the levels of employment used in marketing education

The families of marketing operations occupations, as they are classified in the occupational taxonomy of the National Center for Education Statistics,[1] help teacher-coordinators to identify the career goals of students and to determine the needed competencies that form the basis of instruction. The major occupational families or cluster contained within this overall category of marketing operations include:

52.00   Business/Commerce, General

52.02   Business/Managerial Operations

52.07   Entrepreneurial and Small Business Operations

52.08   Finance and Financial Management Services

52.09   Hospitality Administration/Management

52.10   Human Resource Management and Services

52.11   International Business

52.14   Marketing

52.15   Real Estate

52.17   General Sales, Merchandising, and Related Marketing Operations

52.18   Specialized Sales, Merchandising, and Related Marketing Operations

52.19   Insurance

52.99   Business, Management, Marketing, and Related Support Services

In another federal publication, the instructional program of marketing education (ME) is clearly defined.

> [Marketing] Education includes various combinations of subject matter and learning experiences related to the performance of activities that direct the flow of goods and services, including their appropriate utilization, from the producer to the consumer or user. These activities include selling, and such sales-supporting functions as buying, transporting, storing, promoting, financing, marketing research and management.

> [Marketing] Education is comprised of programs of occupational instruction in the field of . . . marketing. These programs are designed to prepare individuals to enter, or progress or improve competencies in, [marketing] occupations. Emphasis is on the development of attitudes, skills, and understanding related to marketing merchandising and management. Instruction is offered at the secondary, post-secondary, and adult education levels and is structured to meet the gainful employment and entrepreneurship at specified occupational levels. [Marketing] occupations are found in such areas of economic activity as retail and wholesale trade, finance, insurance, real estate, services and service trades, manufacturing, transportation, utilities, and communications.[2]

The Goals of the Marketing Education (ME) Program

Using a program approach, ME educators should begin by defining the occupations to be served, as discussed previously. Next, they should consider the groups to be served and should accept the operational philosophy of the Carl D. Perkins Vocational and Technical Education Act of 1998, which outlined these groups.[3] Further, not only must their instructional program consist of preparatory, job-entry and career development, but it must also be responsive to the needs for upgrading, advancement, and retraining.[4] With these service views in mind, ME educators should strive to fulfill the three basic "people-oriented" goals. They are:

---

[1] Robert L. Morgan and E. Stephen Hunt, *A Classification of Instructional Programs—2000*, National Center for Education Statistics, Presidential Building, 400 Maryland Avenue, S.W., Washington, DC 20202, 2000, pp. 41–43.

[2] U.S. Department of Health, Education, and Welfare, Office of Education, *Standard Terminology for Curriculum and Instruction in Local and State School Systems*, Bull. OE–23052, Washington, D.C.; U.S. Government Printing Office, 1970. The word "marketing" has been substituted for the word "distribution," which was used in the original definition. The field was previously called "Distributive Education."

[3] Cf. p. 68.

[4] Cf. p. 68.

Small group projects are popular for students who are learning management skills.

1. To provide instruction in marketing.

2. To assist student-learners in improving their techniques.

3. To help student-learners develop an understanding of the economic and social responsibilities of those permitted to engage in marketing in a free, competitive society.

A total program of marketing education is a "people-oriented" program. It serves youth and adults—those employed in marketing and those who want to be; those who as employees, supervisors, and managers want to upgrade themselves; and those who seek advancement and promotion. To do this, a marketing education plan may include a secondary plan, a part-time adult plan, and a full-time mid-management curriculum beyond the high school. The program is designed in a sequence of courses to obtain career development. The marketing tech-prep approach integrates academic courses with marketing courses. Students gain communication, math, science, and marketing skills, which are essential in today's technical workplace. These skills are taught on an entry level in the introductory courses and advanced in the second year.

A total program provides for many needs of the competent marketing worker or manager; basic operating skills, including safety; communications skills; understanding of economics; understanding of principles of marketing and merchandising; human relations skills; personal–social skills; healthy occupational attitudes; and technical information about products and processes.

The basic goal of marketing education is to aid student-learners who have career goals in marketing. It is a career and technical education program for those whose abilities range from lowest rank-and-file to management. Its success is measured by the development of occupational competence in each individual to the limits of that individual's abilities.

The Curriculum in Marketing Education

In the early years of marketing education, the secondary school cooperative plan was the entire marketing education preparatory curriculum. This situation still prevails in many localities. However, the "total program" philosophy has started a trend toward a broadened approach to education for marketing at the secondary level. This was enhanced by provisions of the Vocational Education Act of 1963, as amended in 1976, which provided federal acceptance of the project method of preparatory instruction. Those individuals using a program approach cannot and should not think of marketing education as a course. The program approach implies a series or a sequence of course or instruction experiences, with specific areas of subject matter concerned with marketing clearly defined and in accord with the students' career goals.

While the total program approach to planning a program of education for marketing is paramount, ME educators must be wary that this end does not obscure the needs of some students. There are many situations, forces, and motivations at work in the school, home, and personal lives of students that dramatically affect their commitment to marketing as a career and their ability to profit from instruction. The total program approach must not be construed as an attempt to fit all students into the same mold, holding the same standards for all. ME educators must provide programs that are extremely flexible so that every student who can profit from instruction may have the opportunity to do so. "Employability" is the key end-result, and educators must tailor their programs to individuals having varying levels of ability, with full cognizance of the wide range of job demands in marketing.

TWO CURRICULUM APPROACHES

Two basic curriculum approaches that involve the cooperative marketing education plan exist at the secondary level. One approach is traditional; the other is contemporary.

The traditional approach.—In the traditional approach born out of 1936 federal legislation, the secondary marketing education curriculum and the cooperative plan are synonymous; that is, the cooperative plan is used during all phases of instruction. The curriculum is based on any one of these three plans.

1. *Plan A:* Two years of instruction of one period daily plus cooperative experience in the occupational laboratory both years.

2. *Plan B:* One year of instruction of two periods plus cooperative experience for that one year.

3. *Plan C:* Two years of instruction of one period daily in the classroom and a related class plus cooperative experience in the occupational laboratory the twelfth year.

In these plans, it is atypical to have any required prerequisite courses, although some Plan C programs require a prerequisite eleventh-year course.

The contemporary approach.—The contemporary approach to the marketing education curriculum utilizes either the cooperative plan or the project laboratory (plan) as the capstone of the curriculum. Therefore, the curriculum would consist of a preparatory (pre-occupational) course of at least one year, such as introduction to marketing and merchandising, retailing, and sales techniques, followed by one of the following:

1. Cooperative (employment) instruction of at least one period for one year in school plus at least 450 hours (30 weeks times 15 hours) of instruction on the job (the occupational laboratory).

2. Project-related instruction plus the project laboratory where job experiences are simulated in the classroom.

The contemporary approach to the secondary marketing education curriculum makes the following phases of the curriculum mandatory.

1. General education in the common areas of language arts and communications, mathematics and

science, social science, physical and health education, and practical arts for general education. For general education purposes, the marketing education student might well take courses such as general business and economics.

2. Business courses providing background in skills such as bookkeeping/accounting, keyboarding, word processing, business math, and business organization and management.

3. Introductory courses in marketing, including retailing, sales techniques, and merchandising.

4. Employment instruction through the cooperative experience, including the related class of at least one period daily. This class might well be called "Marketing Education."

The courses described are not prescriptions. Rather, they are intended as illustrations of an approach to marketing education that conceives of a sequence of courses for which the cooperative experience becomes the capstone vocational experience.

In some situations, the marketing education curriculum includes more specialization than is shown in the general marketing education curriculum outlined here. In area career-technical schools particularly, there are more specialized curricula, such as food distribution, soft goods (or fashion) merchandising, petroleum merchandising, travel and tourism, sports marketing, and advertising and sales promotion. These specialized curricula are advanced courses based on core preparation in the marketing occupations.

SEQUENCE OF COURSES

Perhaps one of the most complete offerings in marketing education is in the commonwealth of Virginia, where students are introduced to marketing occupations early. For example, in Richmond, students explore various occupations, including market-

ing occupations, in the sixth, seventh, and eighth grades.

Then, students may continue their marketing education in high school by taking the following sequence.[5]

Tenth Grade	Preparatory Marketing Education	1 credit
Eleventh Grade	Cooperative Marketing Education	2 credits
	or	
	Retailing and Sales Techniques	1 credit
Twelfth Grade	Cooperative Marketing Education	2 credits

A popular program in many states is one shown as Plan C in this chapter, where students enroll in retailing, marketing, and/or sales techniques in the eleventh grade as prerequisite to cooperative marketing education in the twelfth grade.

Eleventh Grade	Marketing Practices and Principles	
	or	
	Retailing and Sales Techniques	1 credit
Twelfth Grade	Related Classroom Instruction in Marketing Education	1 credit
	On-the-Job Training in Marketing Education	1 credit

Some states have moved to competency-based models. For example, Indiana developed and used such a model divided into three levels. One level for each year of a three-year model is used, or the levels can be refined for a two-year program.

The competencies common to a variety of marketing occupations form the bases for the first two years. Common competencies for marketing management positions supply the foundation for the third year. Modified learning objectives, suggested learning activities, and appropriate materials are matched to the competencies.

---

[5]By 1980, several other states had implemented three-year secondary marketing education programs.

Tech-Prep (2 + 2) and Marketing Education

Marketing educators, state supervisors, teacher educators, guidance counselors, local CTE administrators, and others have explored a variety of ways to implement the requirements of tech-prep legislation in marketing education programs. The Carl D. Perkins Vocational and Technical Education Act of 1998 continues to support tech-prep programs. Marketing education will continue to strive to make these programs viable and beneficial to students in the 21st century.

In the Milwaukee City Schools, five teachers from South Division High School developed an integrated studies curriculum entitled "Technology Studies," which articulates two vocational and three academic courses of instruction. English, math, science, and marketing are integrated with manufacturing technology. The overriding curriculum goal is to provide options for graduates. These options include continuing academic education (two or four years), continuing career and technical education, full-time employment, continuing education full or part time, apprenticeship, or government/military service.

The first year of tech studies is designed for the tenth grade. The eleventh and twelfth grade curriculum is designed to be progressively developed during the preceding teaching years. The utilization of existing computer hardware and innovative software is an extremely integral component of each subject area. In addition, the tech studies allows participation in a shadowing program for eleventh and twelfth graders and an internship (patterned after traditional co-op) program either during summer months or during the school year.[6]

In Michigan Title III, tech-prep education "must occur with other state educational reforms which include school improvement plans, accreditation reviews, annual educational reports and a core curriculum, and related efforts such as technology education, employability skills development, and applied academics. The premise underlying development of a Tech Prep program is that these components are integrated and linked together in a systematic manner. The outcome is a Tech Prep graduate who has acquired advanced technical skills leading to an associate degree or a two-year certificate."[7]

Occupational tech-prep consortiums of educators, business, and labor were formed "to create a system of technical education beginning in the middle or secondary school and culminating with an associates degree or two-year certificate that prepares youth and adults for entry into technical careers. The system must identify minimum learning competencies for each program within an occupational cluster. Each cluster must have a sequence of learning that includes the pre-requisite competencies and the specialty, supportive, and general education course competencies for each certificate of associate degree."[8]

The Michigan Tech-Prep Model calls for a four-step process to formulate a tech-prep curriculum. These steps are: (1) conduct a program review, (2) restructure curriculum, (3) establish articulation agreements between secondary and post-secondary institutions, and (4) implement a tech-prep program by using local action plans. As a result, marketing education programs are "shifting their focus on preparing students for jobs requiring entry-level skills, primarily in retailing, to providing the skills and concepts needed for upward mobility. Programs are being expanded to serve employment in entrepreneurship, professional sales, with a focus on marketing careers."[9]

The restructured tech-prep curriculum is clustered (offering a wider range of instruction), com-

[6]Kent LaVelle, "Marketing Education for the Tech Prep/Integrated Studies Educational Initiative," paper presented at Tech Prep for Marketing Careers Conference, Denver, February 1, 1993.

[7]Thomas Benton and Carl A. Woloszyk, "Restructuring Marketing Education Program Through Action Plans," paper presented at Tech Prep for Marketing Careers Conference, Denver, January 31, 1993, p. 3.

[8]*Ibid.*, p. 3.

[9]*Ibid.*, p. 7.

MICHIGAN MARKETING EDUCATION CLUSTER PROGRAM OVERVIEW

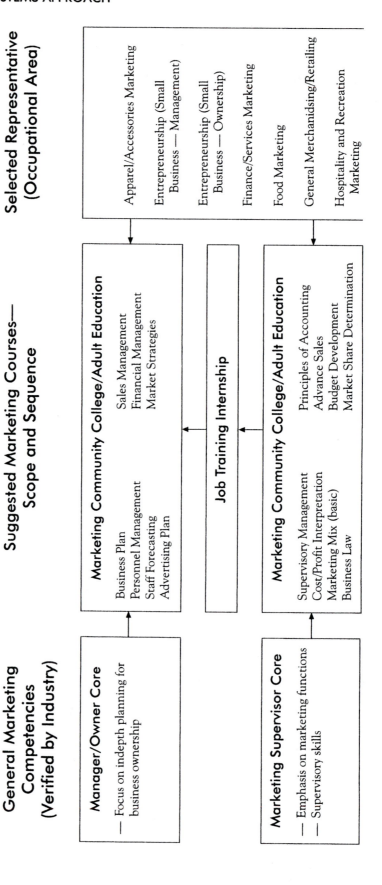

General Marketing Competencies (Verified by Industry)

Manager/Owner Core
— Focus on indepth planning for business ownership

Marketing Supervisor Core
— Emphasis on marketing functions
— Supervisory skills

Suggested Marketing Courses— Scope and Sequence

Marketing Community College/Adult Education

Business Plan
Personnel Management
Staff Forecasting
Advertising Plan
Sales Management
Financial Management
Market Strategies

Job Training Internship

Marketing Community College/Adult Education

Supervisory Management
Cost/Profit Interpretation
Marketing Mix (basic)
Business Law
Principles of Accounting
Advance Sales
Budget Development
Market Share Determination

Selected Representative (Occupational Area)

Apparel/Accessories Marketing

Entrepreneurship (Small Business — Management)

Entrepreneurship (Small Business — Ownership)

Finance/Services Marketing

Food Marketing

General Merchandising/Retailing

Hospitality and Recreation Marketing

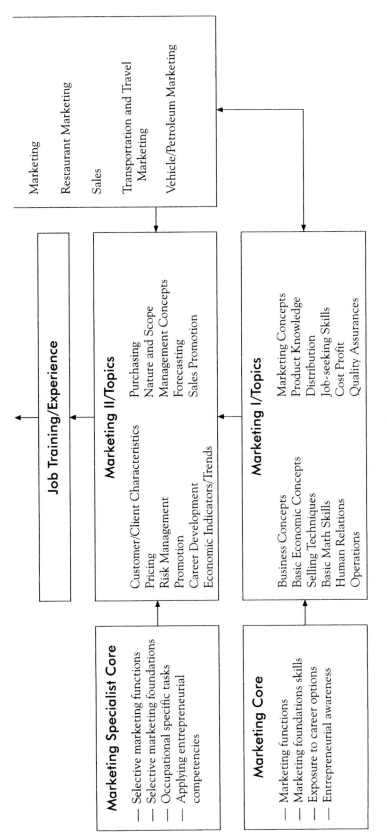

FIGURE 21-1

petency-based (utilizing MarkED competencies and National Curriculum Framework), and carefully articulated with post-secondary schools. Furthermore, the program requires a change in delivery options. Courses may now run 2½ hours, be taught at different locations, and have more variety in subjects and teaching methods (learning manager approach utilized).

Basing Instructional Content on Competency Needs

Since some major conferences in 1963, the marketing education curriculum has been philosophically based on the premise of developing competencies that relate to the task analysis of each student's career objectives. The Nelson paper of 1964 was a landmark and has been followed by substantial task analysis research, such as that of Crawford, Carmichael, Ertel, and Samson.[10]

Additional research, including validation of the Crawford study, has been conducted by the *Marketing Education Resource Center (MarkED)*. The competencies identified by Crawford and MarkED were the original basis for the first national competency-based curriculum framework developed in 1986 by the national Marketing Education Curriculum Committee. Figure 21–4 illustrates the competency-based system utilized in marketing education.

By using a *competency-based curriculum,* the student-learner benefits in several ways. These include:

1. The student-learner derives competencies from occupational roles in marketing—learning is relevant to marketing employment and advancement.

[10]For example, see Lucy Crawford, A Competency Pattern Approach to Curriculum Construction in Distributive Teacher Education, Vol. I–IV, ED–032383–032386, Blacksburg, Virginia: Virginia Polytechnic Institute and State University, 1967, and Kenneth Ertel, Clusters of Tasks Performed by Merchandising Employees Working in Three Standard Industrial Classifications of Retail Establishments, ED–023911.

2. Before entering into the employment area, the student-learner knows what is involved in an occupation category.

3. The student is able to identify expectations—doesn't have to ask "Why am I doing this?"

4. The student-learner receives individualized and personal instruction within an occupational category.

5. The student-learner improves his/her ability to advance within a marketing occupation and to succeed within the business situation.

6. The combination of individualized instruction and a competency-based curriculum should bring about a more highly motivated learner as well as a better prepared worker.

Providing Effective Related Instruction

As indicated earlier, the term "related instruction" refers to the in-school instruction that is correlated with the period of occupational experience on the job. The nature of related instruction will differ, depending on whether or not the cooperative phase of instruction is preceded by required preparatory (pre-employment) courses in marketing. If no preparatory courses are required, then the related instruction phase must "begin from scratch"; if the basic principles have been covered in a preparatory class, the related instruction phase can concentrate on more advanced content. However, the objectives of instruction are similar in both cases, even though the distribution of topics will differ. Instruction must be concerned with the learners' need to acquire occupational skills, information, and job intelligence basic to all marketing occupations, as well as with their need for specialized skills and information requisite to their career objectives in marketing.

TECH PREP (2 + 2): A MICHIGAN

Subject		High School		
	Freshman	Sophomore	Junior	Senior
Math	Applied Math I	Applied Math II	Algebra II	
English	English I, II, and III and Applied Communication (4 units)			
Science	Applied Biology/ Chemistry	Principles of Tech I	Elective	
Humanities	Geography, History, and Government (4 units)			
Other	Health/PE			Economics
Other				Psychology
Technical Core	Intro to Business	Computer Basics	Accounting	
Technical Core		Technical Systems		
Technical Speciality				Entrepre-neurship
Technical Speciality			Marketing Education	Marketing Education

FIGURE 21–2

MARKETING EDUCATION MODEL

Post-Secondary Associate Degree*			
Freshman	**Sophomore**	**Junior**	**Senior**
Business Math			
English I/ Writing	English II	Speech	
		Economics	
Intro to Business	Business Communication	Psychology/ Consumer Behavior	Marketing Research
	Principles of Accounting	Accounting Applications	Small Business Management
Sales	Principles of Accounting	Fashion Merchandising	Principles of Management
	Retailing		Advertising

*Not a transfer model

FIGURE 21–2 (Continued)

TECH-PREP 2 + 2 PROGRAM
Recommended Course Sequence for Careers in Marketing/Management

Advertising	Hospitality/Tourism	Fashion Merchandising
Buying	Public Relations	Risk Management
Banking & Finance	Sales Promotion	Sales/Retailing
Entrepreneurship	Market Research	Product Development

Grade	Required Courses	Credits	Elective Courses	Credits
9	Communication Skills Health Physical Education Government Algebra *or* Geometry Science	1.0 0.5 0.5 1.0 1.0 1.0 5.0	Keyboarding 5000*	1.0
10	Communication Skills, English E102 Communication Skills, Speech S102 Science Applied Math II *or* Geometry *or* Advanced Algebra Global Requirement Physical Education	0.5 0.5 1.0 1.0 1.0 0.5 4.5	Introduction to Marketing 5500 WordPerfect 5020 Other Choices	1.0 0.5 1.5
11	Communication Skills, (See Choices) U.S. History Physical Education	1.0 1.0 0.5 2.5	Marketing 5510* Marketing Work Experience 9660 Accounting I 5310* Lotus 1-2-3 5030 Other Choices	1.0 1.0 1.0 0.5 3.5
12	Communication Skills, (See Choices) U.S. History Physical Education	1.0 1.0 0.5 2.5	Marketing 5510* Marketing Work Experience 9660 Accounting I 5310* Lotus 1-2-3 5030 Other Choices	1.0 1.0 1.0 0.5 3.5

"Other Choices" means you can take other electives in physical education, comm. arts, foreign language, social studies, science, health, family and consumer education, marketing, technology education, music, art, math, or business education.

* This class is offered as dual credit (college credit) through Fox Valley Technical College. One Fine Arts credit must be included somewhere in grades 9–12. Graduation requirements equal a total of 22 credits minimum.

November 1992

FIGURE 21–3. (Courtesy, Appleton, Wisconsin, Area School District)

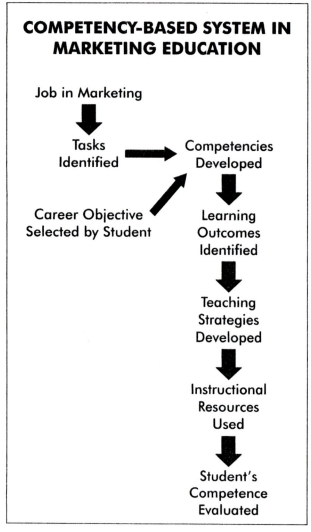

FIGURE 21–4. (From DECA transparency, Reston, Virginia)

CHARACTERISTICS OF RELATED INSTRUCTION

The classroom phase of related instruction in cooperative marketing education is ideally suited to effective teaching in several respects.

First, *student-learners* are a selected group, with distinctive motivation. Each student-learner has expressed a career interest in the field of marketing, with the desire for a job either immediately upon graduation from high school or after further study in college. Each is of work age and has his/her parents'

consent to work part time at a training station while he/she is attending school. With the aid of guidance counselors and with access to various school records, the teacher-coordinators select the students. With the same care and preparation, the teacher-coordinators select merchandising and service firms as training stations and assign student-learners to part-time jobs with pay.

Ideally, these students have had general business in grade 9, keyboarding and/or record keeping in grade 10, and marketing, sales techniques, and retailing in grade 11. The marketing, sales techniques, and retailing courses may be considered to be preparatory training under the Vocational Education Act of 1963 if the state plan so stipulates.

Second, *teacher-coordinators* are the ones who have specialized in coordinating the learning experiences in the classroom with those encountered on the part-time jobs at the training stations. Teacher-coordinators have an unusual opportunity for achieving meaningful learning outcomes. In addition, they arrange with each supervisor on the job, or job sponsor, to carry out a carefully analyzed training program, and periodically they visit both the student-learners and the training station sponsors on the job.

Teacher-coordinators have met certain specific educational and experience requirements in order to be certified as coordinators and to qualify their high schools for reimbursement from CTE funds.

Third, this *classroom portion of related instruction* devotes a given amount of the daily class period to instructing all the students, in a more or less formal manner, on the subject matter of marketing; in addition, related instruction devotes a given amount of time to instructing each student individually on the specific subject matter needed on his/her job and for advancing his/her career objective.

THE BASIC AREAS OF RELATED INSTRUCTION

The marketing education curriculum is divided into two primary parts: Foundations and Functions. The **Foundations** are fundamental to an under-

Examples of Career Applications

Advertising	Hospitality Marketing	Real Estate
Customer Service	Importing/Exporting	Restaurant Management
e-Commerce	International Marketing	Retail Management
Entrepreneur	Marketing Research	Sales Management
Fashion Merchandising	Product Management	Service Marketing
Financial Services	Professional Sales	Sports Marketing
Food Marketing	Public Relations	Travel/Tourism Marketing

FIGURE 21–5. The curriculum framework for marketing education. (Reprinted with permission. Copyright © 2000 by Marketing Education Resource Center [MarkED], Columbus, Ohio)

standing of marketing and can be viewed as prerequisites or co-requisites for marketing. The content of the marketing foundations must be mastered for marketing-specific content to have relevance to student learning. There are four broad-based foundational areas: Business, Management, and Entrepreneurship; Communication and Interpersonal Skills; Economics; and Professional Development. The **Functions** define the discipline of marketing as applied in business operations. They address marketing from the perspective of how it is practiced. Each function is viewed from its relationship to the marketing of a good, service, or idea. The seven functions are Distribution, Financing, Marketing-Information Management, Pricing, Product/Service Management, Promotion, and Selling.

Standards associated with the foundations and functions are identified below. Since students can demonstrate their understanding in a variety of ways, the foundational and functional standards are supported by a series of performance indicators: learner expectations for each standard. Standards are fully supported by industry-validated performance indicators updated regularly by the Marketing Education Resource Center (MarkED). These performance indicators provide the basis for marketing curriculum models in most states and for virtually all competitive events sponsored by the National DECA student association. Examples of performance indicators follow each of the foundational and functional standards. The complete list of performance indicators for the foundational and functional standards can be found in the *National Curriculum Planning Guide*. Information related to marketing education standards and indicators has been reprinted with permission of the Marketing Education Resource Center (MarkED).[11]

---

[11]Copyright © 2000 by Marketing Education Resource Center (MarkED), Columbus, OH. www.mark-ed.org.

Foundations of Marketing

1. **Business, Management, and Entrepreneurship**

   Understands fundamental business, management, and entrepreneurial concepts that affect business decision making

   a. Identify ways that technology impacts business

   b. Explain the nature of marketing strategies

   c. Demonstrate leadership characteristics

   d. Monitor variables associated with business risk

   e. Demonstrate procedures for controlling a business's fiscal activities

   f. Identify potential business ventures based on community, market, and opportunity analysis

   g. Formulate a business plan

2. **Communication and Interpersonal Skills**

   Understands concepts, strategies, and systems needed to interact effectively with others

   a. Communicate clearly and concisely in writing

   b. Use appropriate technology to facilitate marketing communications

   c. Make decisions

   d. Treat others fairly at work

   e. Demonstrate interpersonal skills in team working relationships

   f. Apply interpersonal skills to develop good customer relationships

3. **Economics**

   Understands the economic principles and concepts fundamental to marketing

   a. Explain the concept of economic resources

b. Interpret the impact of supply and demand on price

c. Identify factors affecting a business's profit

d. Determine factors affecting business risk

e. Explain the concept of productivity

f. Evaluate the influences on a nation's ability to trade

4. **Professional Development**

Understands concepts and strategies needed for career exploration, development, and growth

a. Analyze employer expectations in the business environment

b. Identify employment opportunities in marketing and business

c. Utilize resources that can contribute to professional development

Marketing Functions

1. **Distribution**

Understands the concepts and processes needed to move, store, locate, and/or transfer ownership of goods and services

a. Explain the relationship between customer service and distribution

b. Select distribution channels and channel members

2. **Financing**

Understands the financial concepts used in making business decisions

a. Describe the role of financing in marketing and business endeavors

b. Calculate exchange rates

c. Use budgets to meet the financial needs of a business

3. **Marketing-Information Management**

Understands the concepts, systems, and tools needed to gather, access, synthesize, evaluate, and disseminate information for use in making business decisions

a. Determine the need for marketing information

b. Analyze the environments in which businesses operate

c. Demonstrate procedures for gathering marketing information using technology

4. **Pricing**

Understands concepts and strategies utilized in determining and adjusting prices to maximize return and meet customers' perceptions of value

a. Determine pricing objectives, policies, and strategies

b. Use technology to assist in setting prices

5. **Product/Service Management**

Understands the concepts and processes needed to obtain, develop, maintain, and improve a product or service mix in response to market opportunities

a. Plan a product/service mix

b. Analyze product-liability risks

c. Select materials/products/services to purchase

d. Describe factors used by marketers to position products/businesses

6. **Promotion**

Understands the concepts and strategies needed to communicate information about products, services, images, and/or ideas to achieve a desired outcome

a. Explain the communication process used in promotion

b. Write promotional messages that appeal to target markets

c. Utilize publicity

d. Develop a promotional plan

7. **Selling**

Understands the concepts and actions needed to determine client needs and wants and respond through planned, personalized communication that influences purchase decisions and enhances future business opportunities

a. Develop an understanding of customers/clients

b. Utilize selling techniques to aid customers/clients in making buying decisions

c. Determine/minimize risks in selling to a customer

d. Utilize strategies to build and maintain clientele

Employment Levels in Marketing Education

Employment levels in marketing education vary in complexity, ranging from simple to complex, in terms of assigned responsibilities and extent of skills and knowledges required. The marketing activities that take place at each level of employment can still be classified within the broad categories of marketing functions. Pre-baccalaureate marketing education is directed at instruction in the following five **levels of employment**. The marketing duties and the scope of competencies for each curriculum are also specified.[12]

1. **Entry-level** marketing occupations involve standard or routine activities with limited need for

decision-making skills. These jobs may be obtained by individuals with no previous marketing education, experience, or commitment to a marketing career. Competencies required to secure and hold such jobs are good personal appearance, general business behavior, and basic skills such as math and communications.

2. **Career-sustaining** marketing occupations involve performing more complex duties, using routine decision-making skills, and having limited control of the working environment. Individuals in career-sustaining positions should have a basic understanding of the marketing concept and its foundations and have indicated an initial interest in marketing as a potential career.

3. **Marketing specialist** occupations involve frequent use of decision-making and leadership skills. Jobs at this level require mastery of marketing skills across functions or extensive technical knowledge skills in one function. Individuals in marketing specialist occupations need to have a thorough understanding of the marketing concept and its functions.

4. **Marketing supervisor** involves a high level of competence in decision-making and leadership. Individuals in this role are responsible for planning, coordinating, and supervising people. They may also be involved in the planning, coordinating, and supervising of marketing-related activities.

5. **Manager-entrepreneur** implies competence in a variety of tasks related to owning one's own business or managing a department within an organization. Individuals at this level are responsible for the success or failure of a unit within an organization or of a small business enterprise. A comprehensive understanding of marketing and management competencies is required to function in this role.

There are higher levels of employment in marketing that involve the preparation of individuals for top-level positions, such as executives of large companies. These positions usually require baccalaureate degrees or extensive experience and thorough com-

[12]Marketing Education Curriculum Committee, Marketing Education Resource Center, Columbus, Ohio, 1987.

prehension of the theories of marketing and their application to corporate activity.

Instructional Facilities and Equipment

The cooperative marketing education plan requires an instructional laboratory conducive to career-technical instruction in marketing in the same way that laboratories are necessary for instruction in chemistry, music, and physical education.

THE ME CLASSROOM

While it is true that marketing education trains individuals for all areas of marketing, it is equally true that in the secondary cooperative plan, the majority of the student-learners will be placed in retail stores. Therefore, the marketing education laboratory classroom should simulate merchandising conditions, providing space and equipment necessary for practice activities in areas such as visual merchandising, sales techniques, store systems, and stockkeeping. In addition, the classroom should be equipped to facilitate both large- and small-group instruction through discussion, role-playing, demonstration, individual project study, and other methods. To do this, the laboratory needs tables and chairs plus files and library shelves. An adjoining but separate office is desirable so that the coordinator can administer the plan even though the classroom is in use.

The question sometimes raised by administrators is "Isn't this classroom expensive when it is used only one or two hours a day?" Actually, the room can be used during other hours by the marketing education preparatory classes, such as those in marketing, general retailing, and sales techniques. In addition, if display cases and files are locked, the room is ideal for classes in general business, business organization, and accounting. It also makes an excellent adult education classroom.

The marketing education classroom is ideally located on the first floor, where hallway display cases can be seen by visitors. A location close to, or part of, the business education suite is desirable to foster and to maintain a close relationship between office personnel and marketing education faculty members.

Equipping the classroom.—The classroom should be furnished so as to provide the atmosphere of the marketing business as much as possible through the use of appropriate display equipment, a store counter, and other store equipment and supplies. In most states, a staff member from the state board of CTE (state supervisor or consultant in marketing education) or a teacher-educator can be called on to assist in recommending equipment to fit the needs of a specific community or school system. The local marketing education advisory committee should also be asked for advice. The teacher-coordinator might make a long-range (three- to five-year) plan for purchasing the following equipment and supplies.

Furniture, Built-in

Bulletin board, writing board, display case, storage lockers, bookcases, shelves, full-length mirror, and wash basin.

Furniture, Portable

Teacher's desk and chair, bookcases, telephone (individual or the departmental phone), individual tables with chairs to match (tablet arm chairs are not suitable), two waste baskets, card-size filing cabinet, letter-size filing cabinet, counter displays, shelves, magazine rack, spotlights for window display, extension cords, peg board and equipment, and clock.

Machines and Tools

Computerized cash register, calculator, computers and printers, business software, sign maker, paper cutter, staplers (hand and gun), paper punch, hammer, T-square, saw, tin snips, screwdriver, and display tools, including brushes, and pens.

Audio–Visual Aids

3' × 4' flannelboard, digital camera, audiotape recorder, TV, VCR, DVD player, overhead projector, and LCD multimedia projector.

Securing Desired Learning Experiences on the Job

The step-by-step training plan should make the *cooperative* marketing education *plan* sharply distinctive from a *work experience plan,* with respect to occupational education and career development. This step-by-step plan makes a special attempt to relate the classroom instruction specifically to on-the-job training so that the student can gain all the potential learning advantages therefrom. It serves as the basis for guiding the student-learner through worthwhile educational experiences on the job.

The step-by-step training plan indicates the specific objective to be emphasized and whether it is to be stressed in the classroom or at the training station or both. It is derived jointly by the teacher-coordinator and the training station sponsor from a realistic analysis of the tasks, duties, and responsibilities of the student-learner on the part-time job. The development of these specific training plans, together with follow-up activities on them, becomes one of the most important activities of the teacher-coordinator during coordination time.

Successful practitioners in marketing education differ in their approach to the preparation of a training plan. In some school districts, the related class instruction teacher may not be the same person who coordinates the job training experiences. In some cases it would be difficult to make the best use of training plans or to facilitate the operation of the cooperative method as theoretically conceived. Even in states where teacher-coordinators teach the classroom phase and coordinate the job training, the methods of preparing the training plan differ. For example, the Marketing Education Resource Center

recommends the use of training plan and employer evaluation, as illustrated in Figure 21–6. Such a training plan assesses what occurred during a student-learner's experiences at a training station. It may serve as an evaluation of the effectiveness of the training station, or the teacher-coordinator may use it to prepare a training plan for some future student-learner in the particular training situation. In addition, training plans may be expressed in behavioral terms or as competency statements.

EVALUATING THE STUDENT-LEARNERS' PROGRESS

Evaluation of the student-learners in regard to the classroom instruction is based on performance on (1) written subject matter assignments, (2) work in assigned manuals, (3) oral presentations, and (4) individual and group projects. The evaluation of the student-learners' on-the-job learning and performance progress is based on (1) observations of the teacher-coordinator, (2) a check of the student-learners' weekly job reports, (3) the teacher's discussion with the training station sponsors regarding the training profile or the step-by-step training plan, and (4) employee rating sheets filled out by the training station sponsors.

GUIDING THE DECA CHAPTER

The general principles of utilizing student organization activities as learning experiences are covered thoroughly in Chapter 14. However, there are some chapter practices that are specific to the marketing education program and to the competencies of marketing and management.[13]

Chapter meetings are often held during the related class period, a satisfactory time because the student organization program is part of the instruc-

---

[13]See William A. Stull and Charles Beall, "Integrating Competency-based DECA Competitive Events into the Marketing and DE Curriculum," *The Balance Sheet,* Cincinnati, Ohio: South-Western Publishing Co., January, 1980, p. 154.

TRAINING PLAN
and employee evaluation

for: _____

School: _____

CAREER-SUSTAINING LEVEL

General Merchandise Retailing

RATING SCALE

Ratings are assigned by the instructor in conjunction with the training sponsor (where appropriate).

1. Can perform task satisfactorily without supervision.
2. Can perform task with supervision—needs additional work.
3. Cannot perform task satisfactorily.

(✓) Check for on-the-job concentration during current evaluation period.

Final Teacher Approval	✔	TASK OR SKILL (Related MarkED Competencies)	On-the-Job Evaluations			
			1	2	3	4
		NOTE: Tasks listed below have been selected for on-the-job evaluation; complete list is available in Student Competency Record.				
		Human Resource Foundations: Mathematics	*	*	*	*
		Solve addition problems (MA:097)				
		Solve subtraction problems (MA:098)				
		Solve multiplication problems (MA:099)				
		Solve division problems (MA:100)				
		Solve mathematical problems involving fractions (MA:101)				
		Solve mathematical problems involving percentages (MA:377)				
		Read charts and graphs (MA:113)				
		Make change (MA:386)				
		Operate register/P.O.S. terminal (MA:095)				
		Calculate tax, discounts, and misc. charges for purchases (MA:089)				
		Complete sales checks (MA: 090)				
		Complete charge sale transactions (MA:092)				
		Accept customer/client checks (MA:091)				
		Verify identification (MA:050)				
		Process returns, refunds, and exchanges (MA:094)				
		Process layaway/COD sales (MA:093)				

FIGURE 21–6. (Source: MarkED; copyright © 1986 by MarkED, Inc., Columbus, Ohio)

		RATING SCALE
Ratings are assigned by the instructor in conjunction with the training sponsor (where appropriate).		1. Can perform task satisfactorily without supervision. 2. Can perform task with supervision—needs additional work. 3. Cannot perform task satisfactorily.

(✔) Check for on-the-job concentration during current evaluation period.

Final Teacher Approval	✔	TASK OR SKILL (Related MarkED Competencies)	On-the-Job Evaluations			
			1	2	3	4
		NOTE: Tasks listed below have been selected for on-the-job evaluation; complete list is available in Student Competency Record.				
		Human Resource Foundations: Communication Skills	*	*	*	*
		Use proper grammar and vocabulary (CO:004)				
		Address people properly (CO:005)				
		Use telephone in businesslike manner (CO:114)				
		Listen to and follow directions (CO:119)				
		Human Resource Foundations: Interpersonal Skills	*	*	*	*
		Handle customer inquiries (HR: 004)				
		Direct customer/client to other locations (HR:031)				
		Handle difficult customers (HR:045)				
		Interpret business policies to customers/clients (HR:030)				
		Handle customer/client complaints (HR:046)				
		Human Resource Foundations: Management	*	*	*	*
		Orient new employees (MN:078)				
		Operational Concepts	*	*	*	*
		Handle company equipment properly (OP:696)				
		Operate calculator (OP:133)				
		Complete housekeeping responsibilities (OP:077)				

FIGURE 21–6 (Continued)

RATING SCALE	
Ratings are assigned by the instructor in conjunction with the training sponsor (where appropriate).	1. Can perform task satisfactorily without supervision. 2. Can perform task with supervision—needs additional work. 3. Cannot perform task satisfactorily.

(✓) Check for on-the-job concentration during current evaluation period.

Final Teacher Approval	✔	TASK OR SKILL (Related MarkED Competencies)	On-the-Job Evaluations			
			1	2	3	4
		NOTE: Tasks listed below have been selected for on-the-job evaluation; complete list is available in Student Competency Record.				
		Distribution: Physical Distribution	*	*	*	*
		Determine processing priorities (DS:006)				
		Check incoming stock (DS:007)				
		Reconcile shipping/receiving discrepancies (DS:008)				
		Process returned/damaged product (DS:009)				
		Process returns to vendors (DS:010)				
		Price mark stock (stamps, tags, tickets, etc.) (DS:011)				
		Maintain stock levels (DS:023)				
		Complete stock counts (DS:025)				
		Financing: Credit	*	*	*	*
		Facilitate credit applications (FI:003)				
		Promotion: Visual Merchandising/Display	*	*	*	*
		Maintain displays (PR:052)				
		Care for live plants (PR:053)				
		Dismantle displays (PR:054)				
		Decorate business/department (PR:068)				
		Risk Management: Safety Considerations	*	*	*	*
		Use safety precautions (RM:010)				

FIGURE 21–6 (Continued)

	RATING SCALE
Ratings are assigned by the instructor in conjunction with the training sponsor (where appropriate).	1. Can perform task satisfactorily without supervision. 2. Can perform task with supervision—needs additional work. 3. Cannot perform task satisfactorily.

(✓) Check for on-the-job concentration during current evaluation period.

Final Teacher Approval	✔	TASK OR SKILL (Related MarkED Competencies)	On-the-Job Evaluations			
			1	2	3	4
		NOTE: Tasks listed below have been selected for on-the-job evaluation; complete list is available in Student Competency Record.				
		Selling: Process and Techniques	*	*	*	*
		Open the sales presentation (SE:869)				
		Question/probe for information (SE:024)				
		Suggest products substitutions (SE:871)				
		Demonstrate product (SE:893)				
		Use feature/benefit selling (SE:873)				
		Selling: Product Knowledge	*	*	*	*
		Obtain product information from sources on/with the item (SE:011)				
		Use company promotional material for selling information (SE:012)				
		Obtain product information from appropriate individuals (SE:010)				
		Selling: Support Activities	*	*	*	*
		Arrange alterations (SE:825)				
		Arrange delivery of purchases (SE:023)				
		Wrap/package products (SE:887)				
		Process special orders (SE:009)				
		Sell gift certificates (SE:016)				
		Process telephone orders (SE:835)				

FIGURE 21–6 (Continued)

GENERAL EVALUATION

	RATING SCALE
Ratings are assigned by the instructor in conjunction with the training sponsor (where appropriate).	1. Excellent, above average. 2. Acceptable, but improvement possible. 3. Not acceptable, needs significant improvement.

TASK OR SKILL (Related MarkED Competencies)	Grading Period Ending			
	1	2	3	4
Self-Understanding: Personality Traits	*	*	*	*
Maintain appropriate personal appearance (HR:263)				
Maintain positive attitude (HR:274)				
Demonstrate interest and enthusiasm (HR:303)				
Demonstrate responsible behavior (HR:022)				
Demonstrate honesty and integrity (HR:312)				
Demonstrate orderly and systematic behavior (HR:267)				
Demonstrate initiative (HR:317)				
Demonstrate self-control (HR:294)				
Self-Understanding: Self-Development	*	*	*	*
Use feedback for personal growth (HR:295)				
Adjust to change (HR:301)				
Set personal goals (HR:014)				
Interpersonal Skills	*	*	*	*
Foster positive working relationships (HR:269)				
Show empathy for others (HR:028)				
Use appropriate assertiveness (HR:021)				
General Work Habits	*	*	*	*
Promptness and attendance				
Dependability				
Overall quality of work				

Validating Signatures

1. _____
   (student employee)   (date)

   _____
   (employer)   (date)

   _____
   (teacher-coordinator)   (date)

2. _____
   (student employee)   (date)

   _____
   (employer)   (date)

   _____
   (teacher-coordinator)   (date)

3. _____
   (student employee)   (date)

   _____
   (employer)   (date)

   _____
   (teacher-coordinator)   (date)

4. _____
   (student employee)   (date)

   _____
   (employer)   (date)

   _____
   (teacher-coordinator)   (date)

FIGURE 21–6 (Continued)

tional program. However, some teacher-coordinators hold meetings in the evening, while some utilize the "zero" period—that is, the half-hour prior to the time school actually begins in the morning. Local student chapter activities should involve community business resources to further career objectives in marketing. For example, a fashion buyer, an insurance agent, a purchasing agent, or an industrial sales representative may be invited to present a short talk on careers in his/her particular field; a visual merchandising person may demonstrate good design; a chamber of commerce representative may discuss current economic problems, such as downtown renewal and the local tax base.[14]

Financing student organization activities provides many opportunities for putting into practice the principles of marketing and merchandising. Some specific money-making activities used by marketing education student organizations are: (1) operating the concession stands at athletic events, with profits split with the athletic committee; (2) operating the school store; (3) engaging in sales promotions, such as selling candy and selling desk calendars (with advertisements); (4) holding dances; and (5) sponsoring skating parties.

National DECA can provide local chapters with a list of approved sales and marketing sales projects. These reputable companies offer an acceptable margin of profit as well as items that have maximum acceptance by the public.

Another highlight of the school year is preparing for competency-based competitive events (both series and written), individual participating events, and chapter events. A competency-based event is a "learning activity designed to evaluate a student's development of the essential competencies needed for entry or advancement in marketing occupations in which the student competes with other students on predetermined standards."[15] In their activity, students compete at either the master employee level

(entry level) or the supervisory (mid-management or supervisory) level. Each activity consists of a series of three to five individual events (based on instructional areas) within a single occupational category.

National DECA
Official Competitive Events List, 2001–2002

High School Division

AL: Associate Level
ML: Management Level

Individual Series Events:

1. Advertising Campaign Event
2. Apparel and Accessories Marketing Series, AL
3. Apparel and Accessories Marketing Series, ML
4. Business Services Marketing Series
5. Finance and Credit Services Series
6. Financial Services Marketing
7. Food Marketing Series, AL
8. Food Marketing Series, ML
9. Full Service Restaurant Management Series
10. General Marketing Series, AL
11. General Marketing Series, ML
12. Hospitality Services Marketing Series
13. Marketing Management Series
14. Quick Serve Restaurant Management Series
15. Retail Merchandising Series, AL
16. Retail Merchandising Series, ML
17. Sports and Entertainment Marketing Management
18. Technical Marketing Event
19. Travel and Tourism Marketing Management
20. Vehicles and Petroleum Marketing Series

Individual Written Events:

1. Business Personal and Financial Services Marketing Written Event
2. Entrepreneurship Written Event
3. Food Marketing Written Event
4. General Marketing Written Event
5. Hospitality and Recreation Marketing Written Event
6. Specialty Store Retailing Written Event

---

[14]Other ideas are presented in Chapter 14.

[15]National DECA, Reston, Virginia, 1995.

Individual Participating Events:

1. e-Commerce Business Pilot Event
2. Entrepreneurship Participating Event (Organizing a Business)
3. Individual Free Enterprise Event
4. Fashion Merchandising Promotion Plan Event
5. International Business Plan Event
6. Virtual Business Challenge

Chapter Projects:

1. Chapter Free Enterprise Project
2. Civic Consciousness Project
3. Creative Marketing Research Project
4. DECA Quiz Bowl
5. Free Enterprise Event
6. Learn and Earn Project
7. Public Relations Project
8. 7UP Challenge

Delta Epsilon Chi Division

Business Simulations:

1. Apparel and Accessories Marketing
2. Financial Services Marketing/Management
3. Food Marketing/Management
4. Hospitality Marketing/Management
5. Marketing Management
6. Restaurant and Food Service Marketing/Management
7. Retail Merchandising
8. Travel and Tourism Marketing/Management

Case Studies:

1. Business-to-Business Marketing
2. Human Resource Management Decision-Making
3. International Marketing
4. Marketing Decision-Making/Merchandising
5. Sales Manager Meeting
6. Sports and Entertainment Marketing

Prepared Business Presentations:

1. Advertising Campaign
2. Design
3. Entrepreneurship

4. Sales Promotion Plan
5. Sales Representative

Special Activities:

1. Delta Epsilon Chi Quiz Bowl
2. Culinary Management Institute
3. e-Commerce Management Institute

SOURCES OF INSTRUCTIONAL MATERIALS

Today's marketing education teachers can choose from an abundance of instructional materials. Textbooks, workbooks, games and simulations, computer programs, quizzes and tests, and audio-visuals are produced by several major publishers and a variety of state and national curriculum materials centers. Teacher-coordinators can obtain information on available materials by contacting the following sources.

DECA Related Materials
1908 Association Drive
Reston, VA 20191
Phone: (703) 860-5000
Fax: (703) 860-4013
Web site: www.deca.org

Glencoe/McGraw-Hill
Customer Service Department
P.O. Box 543
Blacklick, OH 43004-0544
Phone: (800) 334-7344
Fax: (614) 755-5682
Web site: www.glencoe.com

Marketing Education Resource Center (MarkEd)
1375 King Avenue
P.O. Box 12279
Columbus, OH 43212
Phone: (800) 448-0398
Fax: (614) 486-1819
Email: service@mark-ed.com
Web site: www.mark-ed.com

Multistate Academic and Vocational
Curriculum Consortium
1500 West Seventh Avenue
Stillwater, OK 74074-4364
Phone: (800) 654-4502
Fax: (405) 743-5154
Web site: www.mavcc.org

South-Western Publishing company
5101 Madison Road
Cincinnati, OH 45227
Phone: (800) 543-0487
Fax: (513) 527-6940
Email: swinformation@thomsonlearning.com
Web site: www.swep.com

In addition, materials can be obtained from businesses and industries, as well as current trade and business publications, such as *Advertising Age*, *Business Week*, and *Chain Store*. Thus, it is easy to see why DECA is an important part of the student's competency development. It is the responsibility of the marketing education teacher-coordinator and/or DECA advisor to insure that the student-learners are being taught in the classroom and on the job those competencies that will make them employable. DECA can be a real co-curricular activity when it is properly integrated into classroom and on-the-job instruction. Remember the adage "Teach your subject and you will not have to teach the contest."

EVALUATING THE ME PLAN

Evaluation of the ME plan involves both short-range evaluation of each year's progress and long-range evaluation covering a span of years, such as a five-year period. (General methods of evaluating the cooperative plan are discussed in Chapter 15.)

Continuous evaluation.—Teacher-coordinators should continually be evaluating themselves as they teach and coordinate; noting in what areas they need improvement. In addition, at the end of each year, teacher-coordinators should make a written report to their administration in which they summa-

---

CASE-IN-POINT

Jenny McAfee graduated with a B.S. in Marketing Education from the state college near her home. Soon after her graduation, Jenny married and stayed home to be a homemaker and then a mother. Once her two boys were in school, Jenny sought employment and was hired as a marketing education coordinator in a county high school.

Having been away from the field for eight years, Jenny felt rather "out of it." She made an appointment with her college marketing education teacher-educator, who reviewed with her recent changes in the field and suggested sources where she could look for help in developing her curriculum.

Using the state marketing education curriculum guide, her former teacher-educator recommended that she write to the Marketing Education Resource Center (MarkED) for a catalog of available materials. Since her state was a member, she was entitled to a discount on her purchases.

From the Marketing Education Resource Center, Jenny learned she could specialize the general marketing curricula of her state. For a small cost, the Marketing Education Resource Center would send her a computerized listing of competencies needed by her students according to their individual career choices. Jenny also found this list and accompanying learning activity packets (LAPS) helpful in preparing her students for DECA competitive events. In addition, she could order questions prepared for previous DECA competitive events.

Jenny is now a firm believer in the Marketing Education Resource Center and the Texas Instructional Materials Center, both of which supply excellent instructional materials to supplement the necessary textbooks needed in the marketing education classroom.

---

rize the year's progress. They should also make a brief annual follow-up to determine the employment status of the previous year's graduates.

The cooperative marketing education plan is continually evaluated by the advisory committee, which meets regularly. The effectiveness of this committee

MARKETING EDUCATION PROGRAM REVIEW

Date _____ Consultant _____ Co. No. _____

Corp. _____ Corp No. _____ School_____

School No. _____ Teacher _____ Action Recommended:

Probation _____ Approved _____ Disapproved _____

Reasons _____

_____

_____

_____

Teaching Schedule

Period	Course Title	Soph., Jr., Sr.	Intro., Prep, or Related	Lab, Co-op, or Store	No. DECA Members
1	_____	____	____	_____	_____
2	_____	____	____	_____	_____
3	_____	____	____	_____	_____
4	_____	____	____	_____	_____
5	_____	____	____	_____	_____
6	_____	____	____	_____	_____
7	_____	____	____	_____	_____

Program Structure: 1 yr. _____, 2 yrs. _____, 3 yrs. _____, Lab _____, Co-op _____, Comb. _____

FIGURE 21–7. (Courtesy, Superintendent of Public Instruction, Indianapolis, Indiana)

MARKETING EDUCATION PROGRAM REVIEW (Continued)

ME Textbooks Used:

Sophomore _____

Junior _____

Senior _____

Instructional Resource Materials Used:

Sophomore _____

Junior _____

Senior _____

	Yes	No	Comments
1. Students			
1R. Credit — 1 credit/semester/hour of classroom instruction, 2 hours for on-the-job training.			
2R. Follow-up procedure organization on a 1- and a 5-year basis.			
3G. Placement provided by LEA.			
4R. Classes open to all, regardless of sex.			
5R. Students who are disadvantaged and handicapped included in regular program.			
6G. Student organization (DECA) functioning (soph., jr., sr.).			
7G. Co-op education — weekly training station report by student; initialed by coordinator and filed.			

FIGURE 21–7 (Continued)

MARKETING EDUCATION PROGRAM REVIEW (Continued)

	Yes	No	Comments
2. **Program**			
8R. Enrollment — student–teacher ratio.			
9R. Safety taught — classroom/training station.			
*10R. Written curriculum available.			
*11G. Curriculum written conceptually or objectives stated in terms observable/ measurable and on file in the LEA.			
*12R. Program related to occupations as listed in the *Dictionary of Occupation Titles* and in the *Occupational Outlook Handbook*.			
*13R. Career objectives on file in the LEA.			
*14R. Training agreement for each student on file with the coordinator — signed.			
*15R. Training plan developed by coordinator and employer and on file with the coordinator.			
16R. Employment complies with all state and federal laws.			
17R. Related class — 1 credit/semester in addition for on-the-job training.			
18R. Students given time during school schedule for on-the-job training.			
19R. Students employed not less that 15 hours/week.			
DEG. Textbooks and instructional materials up-to-date and readily accessible to students. _____ LAPS, _____ MAPS, _____ OMPM, _____ Texas manuals, _____ others.			
*DEG. Instructional methodologies used: _____ competency-based, _____ individualized, _____ lecture/textbook, _____ large-small.			
*DEG. Methods of evaluation: _____ on-the-job, _____ laboratory, _____ classroom, _____ store, _____ projects, _____ DECA, _____ LAPS, _____ other.			

FIGURE 21–7 (Continued)

MARKETING EDUCATION PROGRAM REVIEW (Continued)

	Yes	No	Comments
3. Staff			
20R. Vocational certification appropriate to program being taught.			
21G. Extended contract with listing of activities: (no. of weeks) _____ lab program, _____ co-op program.			
22R. Coordination activities — 1/2 hr./week/student during student release time.			
23G. Teacher-coordinator teaches the related class and performs coordination functions.			
*24G. Coordinator visits training station regularly.			
*25G. Coordinator files weekly coordination activities.			
DEG. Professional membership in IVA, AVA, IMEA, ISTA/NEA/AFT, other.			
4. Advisory committee			
26R. Organized and functioning.			
27G. Minutes on file with administrator.			
28G. Response on file.			
5. Facilities and equipment			
29R. Minimum space and facility for size:			
_____ Classroom/laboratory/store (1,700) _____ Storage (150) _____ Office/counseling (150) _____ Small-group area (150) _____ File cabinets _____ Desk _____ Merchandising units _____ Shelving _____ Other			

FIGURE 21–7 (Continued)

MARKETING EDUCATION PROGRAM REVIEW (Continued)

	Yes	No	Comments
6. **Administration/Financial**			
30G. Identified budget for _____ travel allowance for supervision, _____ state-called meetings, _____ consumable supplies and teaching aids.			

List three (or more) aspects of the program that are the most satisfactory or commendable.

1. _____

2. _____

3. _____

List three (or more) aspects of the program that need the most improvement.

1. _____

2. _____

3. _____

_____

*Copies are to be available for the consultant to review.

FIGURE 21-7 (Continued)

as an evaluative body depends on the leadership strength of the teacher-coordinator and on the cooperation gained from the committee.

Teacher-coordinators may informally study the work of graduates on the job by visiting business establishments in the community. Periodically, they may profit from making a formal survey.

In this type of plan, the regular school class discussions afford unusual opportunities for student participation in the evaluation of the plan. Teacher-coordinators should always be in command of the discussions, with the students giving their main attention to their respective learning tasks.

Long-range evaluation.—Teacher-coordinators should make an intensive and formal evaluation of their programs at least once every five to seven years. This evaluation would include outside assistance from the state department staff and teacher-educators. Figure 21–7 is an example of an evaluation form utilized by state consultants in Indiana. In addition, the review should involve other school personnel, such as administrators, and the advisory committee representing the business community.[16]

The following questions will assist local authorities in determining the quality and status of the marketing education plan.

Advisory Committee

Does the committee meet regularly?

Does a good *esprit de corps* exist?

Are there frank discussions and helpful suggestions?

Does the teacher-coordinator command the respect of businesspersons as the expert in education?

Follow-Up on Graduates

Do the graduates continue to work for the same firms, or do they change jobs soon?

Do employed graduates continue in some phase of marketing?

Are student-learners helped or hindered in job selection by the school business program?

Is the scope of the student-learners' education in marketing unnecessarily narrowed by their specified on-the-job training?

Is there sufficient emphasis upon related school work?

How many students of college calibre continue their studies in marketing programs in colleges?

How valuable is their background information and work experience in their academic work in college?

Does work experience aid them in defraying college expenses?

Teacher-Coordinator Qualifications

What changes, if any, should be made in the qualifications of the teacher-coordinator?

Does the coordinator have sufficient background and knowledge to handle adult education responsibilities as well as the secondary program?

Is the teacher-coordinator active in local civic and professional business groups?

Questions and Activities

1. Prepare a 10-minute talk that you could give before a civic club or professional group, explaining the secondary or post-secondary marketing education plan. Prepare charts or other visual aids for your talk. (Add the notes of this talk to your files.)

2. What are the minimum qualifications for becoming certified as a marketing education coordinator in a reimbursable program in your state?

3. How adequate is the list of specific topics for related instruction as suggested in this chapter? If you were

---

[16]Many states now have developed their own evaluative criteria (see Chapter 15).

the teacher-coordinator, which areas would you emphasize the most? Why?

4. If you were preparing a long-range (three-to five-year) plan for purchasing equipment and supplies for a marketing education classroom, what essentials would you request the first year? What would be some desirable items that you would put in your plan for purchase later on? (Add the notes of these ideas to your files.)

5. Analyze the learning outcomes that could be expected in a specific prospective training station in your local community. Determine the possibility of a student-learner's progress toward his/her career objective in this training station by the end of the year. (Refer to Chapter 11 for suggestions on how to proceed.)

6. How do DECA activities provide educational experiences for student-learners?

7. From examples you have studied, prepare a form that you could use in a local marketing education program for each of the following: a rating sheet on a student-learner at the training station and a weekly report form on which the student-learner could report his/her training station activities. (Add a copy of each of these to your files.)

8. Review published individual training manuals and report on these and other ways of providing individual instruction.

9. Prepare a brief evaluation of a local marketing education plan.

10. Explain what is meant by the "total program" approach in marketing education.

11. Prepare arguments pro and con on the topic "Specialized ME Cooperative Plans in an Area such as Food of Fashion Merchandising Have No Place in the High School."

12. Discuss the merits of the step-by-step training plan (or schedule of processes) in organizing a training station and the merits of the profile approach to evaluating a training station experience.

13. How might you, as a teacher-coordinator, participate in a local marketing education adult training and development program?

14. Study current references on classroom equipment and floor plans. Draw a room layout to scale (1/4 inch equals 1 foot) and prepare an equipment list.

15. State the learning outcomes of a training plan in behavioral terms. (Refer to Chapter 11 for suggestions.)

For references pertinent to the subject matter of this chapter, see Reading Resources.

22

The Plan in Trade and Industrial Occupations

A training station where goods and services are technically related to a trade; the student-learner placed here should have industrial education background and interests.

KEY CONCEPT: Trade and Industrial occupations cut across many business, trade, and cooperative plan lines. After making a careful competency analysis of job requirements, considering the career objective of each student, and carefully developing a realistic step-by-step training plan for each student, the teacher-coordinator should be able to determine the proper plan of instruction and the proper occupational classification.

GOALS

After successfully completing the study of this chapter, answering the questions, and carrying out the activities, the student should be able to:

- Identify and list the major occupational clusters within the instructional area of trade and industrial occupations education.

- Identify and list appropriate training station and career objectives in trade and industrial occupations.

- Develop an appropriate curriculum for a local plan in cooperative occupational education for trade and industrial occupations education.

- List the major sources of instructional materials for trade and industrial occupations.

- Devise appropriate training plans for students at various training stations.

- Evaluate other student-learners in the classroom and on the job.

Trade, Industrial, and Technical Occupations

Trade and industrial education is a plan of instruction designed to prepare students for entry-level jobs in the various industrial–technical occupations and skilled trades. Apprenticeship programs are not considered as cooperative plans in the sense that the term is used in this text, although cooperative education in some states and local areas is accepted as part of the apprenticeship period by the local apprenticeship council.

This occupational preparation aids student-learners in acquiring general skills and knowledges applicable to all industrial occupations and also in acquiring specialized skills and knowledges applicable to a particular trade and to the training stations where they are placed. In addition, this background prepares student-learners for advancement on the job.

Some occupations in areas for which secondary students might train include: automotive repair, automotive technology, baking, bricklaying, carpentry, cabinet making, cleaning and pressing, drafting, electrical appliance servicing, electric motor repair, electrical wiring, farm machinery servicing, furniture upholstery, machine trades, meat cutting, painting and decorating, graphics, plumbing, printing, radio, television, and VCR servicing, residential wiring, sheet-metal work, and welding.

Cooperative industrial education is one of the original forms of cooperative education in U.S. secondary schools. The provisions of the trade and industrial sections of the Smith–Hughes Act definitely established the character of the cooperative industrial education plan. The wide scope of the plan offers great flexibility. However, in some states and localities, unionization of shops and/or state and safety labor laws have made some occupations, in reality, closed to the placement of student-learners.

In general, the patterns in cooperative industrial education follow the three plans described in the preceding chapter for marketing education. The major difference is that student-learners are more likely to be farmed out to various industrial classes for at least one period of related instruction because of their needs for specialized content appropriate to an individual trade.

TYPICAL COOPERATIVE INDUSTRIAL EDUCATION PATTERNS

There are two basic types of cooperative industrial education plans. One is the *single-trade plan*, in which all student-learners prepare for a given trade. For example, some large high schools or technical schools have cooperative drafting plans. However, most schools operate a second type of cooperative industrial education plan, one in which the student-learners represent a variety of trade and industrial occupations—draftspersons, automotive technicians, building trades workers, etc. In some states, these plans are called *diversified occupations plans* because of the many different trades represented. Much confusion could be eliminated if the term "diversified occupations plans" always referred to plans involving a variety of all occupations rather than a variety of only industrial occupations.

The trade and industrial occupations plan is concerned with preparing individuals for gainful employment as semi-skilled or skilled workers, technicians, or sub-professionals in recognized industrial–technical occupations and in new or emerging occupations. The instructional programs may prepare persons for initial employment, for upgrading within an occupation, or for retraining when they have been displaced through technological advancement.

Related training for apprentices or journeypersons already engaged in industrial–technical occupations is normally a function of the trade and industrial–technical program of career and technical education. Also included may be instructional programs for those service and semi-professional occupations that are primarily trade and industrial in nature. Trade and industrial occupations are classified as either custom or service. Machinists, tool-and-die makers, cabinet workers, carpenters, and printers work in custom occupations. Automotive technicians, computer technicians, cosmetologists, and appliance repairers work in service occupations.

The cooperative plan of instruction has enabled industrial educators to offer career and technical education in many occupations for which school laboratory instruction cannot be provided because of the high cost or the complexity of equipment necessary or because only a small number of students are interested in a given trade.

THE SCOPE OF THE INSTRUCTIONAL PROGRAM

The trade and industrial CIP Codes 46 to 49 include the following clusters of occupations.[1]

10.03	Graphic Communications
12.04	Cosmetology and Related Personal Grooming Services

46 Construction Trades

46.00	Construction Trades, General
46.01	Mason/Masonry
46.02	Carpenters
46.03	Electrical and Power Transmission Installers
46.04	Building/Construction Finishing, Management, and Inspection
46.05	Plumbing and Related Water Supply Services
46.99	Building/Construction Trades, Other

47 Mechanics and Repairer General

47.00	Mechanics and Repairers, General

[1]Robert L. Morgan and E. Stephen Hunt, *A Classification of Instructional Programs—2000*, National Center for Education Statistics, Presidential Building, 400 Maryland Avenue, S.W., Washington, DC 20202, 2000, pp. 8, 9, 31–33.

47.01	Electrical/Electronics Maintenance and Repair Technology
47.02	Heating, Air Conditioning, Ventilation, and Refrigeration Maintenance
47.03	Heavy/Industrial Equipment Maintenance
47.04	Precision Systems Maintenance and Repair Technologies
47.06	Vehicle Maintenance and Repair Technologies
47.99	Mechanic and Repair Technology, Other

48 Precision Production Trades

48.00	Precision Production Trades, General
48.03	Leatherworks and Upholsterers
48.05	Precision Metal Workers
48.07	Woodworking
48.08	Boilermaking/Boilermaker
48.99	Precision Production Trades, Other

49 Transportation and Materials-Moving Workers

49.01	Air Transportation Services
49.02	Ground Transportation Services
49.03	Marine Transportation Services
49.99	Transportation and Materials Moving Services, Other

THE FUNCTIONS APPROACH TO TRADE, INDUSTRIAL, AND TECHNICAL OCCUPATIONS

A useful way for educators to determine the occupations for which the cooperative industrial education plan (either pattern) is suitable is the "functions approach." This line of reasoning and analysis is useful also in determining the content of related instruction in school and the needed experiences on the job. As discussed previously, the functions approach to industrial education is based on the existence of common elements in each of two major branches of industry—**goods-producing**, or custom and **goods-servicing**.[2] These functions are described as follows:

1. Goods-producing activities of industrial products

    a. Research
    b. Product development
    c. Planning for production
    d. Manufacturing
        (1) Custom, or unit
        (2) Mass production, or continuous

2. Goods-servicing activities of industrial products

    a. Diagnosis
    b. Correction
        (1) Adjustment
        (2) Replacement
        (3) Repair
    c. Testing

The following list of descriptive titles of occupations that may appear in local cooperative industrial training plans illustrates that while some of the occupations are goods-producing, most of them are goods-servicing because of the nature of the tasks performed and because of the need for more service occupations and fewer custom occupations.

Air conditioning and refrigeration mechanic
Airplane mechanic
Appliance mechanic
Auto body and fender mechanic
Auto mechanic
Beautician
Brick and stone mason
Building construction mechanic
Cabinet maker
Carpenter
Cement mason
Commercial artist

Cosmetologist
Custodian
Diesel mechanic
Drafter
Dry cleaning and pressing specialist
Electrician
    Communications
    Maintenance
    Wiring
Floor layer
Foundry worker
Hair stylist
Instrument Repairer
Jeweler and watchmaker
Maintenance and service worker
Needle trades worker
Operating engineer
Optical and lens specialist
Ornamental iron worker
Painter and paper hanger
Patternmaker
Photodeveloper and engraver
Plumber and pipefitter
Printer
Repairer, general
Repairer, radio, TV, and VCR
Upholsterer
Welding specialist, arc
Welding specialist, combination

Technicians

Automotive technician
Chemical technician
Electronics technician
Highway engineering aide
Inspector, production
Laboratory assistant

THE RELATIONSHIP BETWEEN APPRENTICESHIP PROGRAMS AND COOPERATIVE INDUSTRIAL EDUCATION PLANS

The apprenticeship approach to industrial training dates back to the early guilds of the Middle Ages. Still much used as a means of providing skilled-trades training, it involves considerable on-the-job experi-

[2]William M. Bateson and Jacob Stern, "The Functions of Industry as the Basis for Industrial Education Programs," *Journal of Industrial Teacher Education*, Vol. 1, Fall 1963, p. 7.

ence, coupled with a limited amount of classroom theory instruction.

Are apprenticeship programs and cooperative industrial education programs the same, and, if not, is there a relationship between them? Apprenticeship programs are typically for youth who have left or graduated from the full-time school. They involve full-time employment, plus some related instruction in school for a few hours a week. They are contracts—apprenticeship agreements between employers, unions, and local apprenticeship councils. These contracts are for definite periods of time, usually not less than three years and, in some cases, as many as six or seven years.

The apprenticeship system in the skilled trades is one with an old and honorable history. In a sense, apprenticeship is the original "cooperative plan," since it involves (1) cooperative agreements between the employers and the school and (2) instruction both on the job and in school. On the other hand, there are some real differences between apprenticeship and cooperative education. Cooperative industrial education is generally of much shorter duration, usually no more than two years, and it prepares the students for entry-level jobs. The student-learners in a cooperative plan are students who work part time and go to school part time, while apprentices work full time. The related instruction in school in the cooperative plan involves more hours than for the apprenticeship program. The students continue their general education, which is not part of the apprenticeship program. Apprenticeship is intended to prepare them to qualify as journeypersons upon completion of training.

The question of whether there is a relationship between apprenticeship programs and cooperative plans is one that cannot be answered categorically yes or no. In some states and in some local school districts, apprenticeship councils have approved the policy of giving student-learners credit for their cooperative experience toward their apprenticeship period. In some cases, year-by-year equivalency is allowed; in other cases, the cooperative years are counted only as

half-years. In some areas of the country, cooperative student-learners are approved by the apprenticeship councils when they first become cooperative students; in other cases, the students are not accepted as apprentices until they finish their cooperative plans. In still other states, there is no relationship whatsoever between a cooperative plan and an apprenticeship program.

DIFFERENTIATING BETWEEN OCCUPATIONS THAT HAVE INDUSTRIAL COMPETENCIES

When classifying a given training station or a given training situation, the local coordinator (or coordinators in areas having more than one cooperative plan) is often faced with the problem of determining where the line falls between trade and industrial, marketing, business, health, and family and consumer sciences occupations. After making a careful competency analysis of the job requirements, considering the *career objective* of the student, and carefully developing a realistic step-by-step training plan, the coordinator will determine the proper classification. Usually, determining what related instruction content will contribute most to the majority of the learning experiences of a given student-learner will not be difficult. For example, if students, for the most part, need instruction in retailing, merchandising, and management, they are best classified as marketing education students, for the content of general related and specific instruction can best be taught by a marketing education teacher, who has the professional and technical background in these areas. If students need instruction in office skills, knowledges, and understandings, including the operation of data and word processing software and hardware, then most certainly the related instruction and the coordination can best be handled by the business education coordinator. If students need instruction in industrial skills, knowledges, and understandings, such as would be found in technical laboratories, building trades, repair trades, and the technical service and repair of data processing and word processing equipment, then the related instruction and coordination can best be handled by the industrial coordinator, who has this professional and technical background. In many cases, students may need skills in all major content areas.

CURRICULUM

Cooperative industrial education plans at the secondary level commonly follow the half-day school and half-day work principle found in cooperative plans in other CTE areas. However, the nature of the trades or the occupations involved in the plans gives rise to some variations in how the plans are organized.

In the single-trade pattern, such as drafting, the related class is usually one period, and the student-learners are likely to be enrolled in at least one other related industrial subject. Also, the learners may be required to have completed prerequisite courses, such as shop math and general science, and beginning industrial education courses, such as Drafting I and II. Thus, this type of plan is similar to the cooperative office plan when it is the capstone of a curriculum. Such plans may be for one year or two years; more commonly, they are for one year, except in career-technical high schools or area career-technical schools.

In the diversified industrial occupations plan, the two-year pattern with one hour daily of directly related instruction each year is the most common. In this plan, many of the students will not have completed any industrial courses. Such plans, then, are similar to many cooperative marketing education plans of the past in which the cooperative plans were the real beginning of the students' CTE instruction. Some cooperative industrial occupations plans operate for one year only with two periods of related instruction to meet the needs of students who have made late career choices just before their senior year.

It is not possible to prescribe an exact curriculum pattern that will meet all the experiences needed in cooperative industrial education. However, a general picture can be presented. In the large high school, where there is an adequate number of students in each of three categories (Option 1, Option 2, and Option 3), three possible plans of pre-employment education for industrial occupations might be considered. (See Table 22–1).

The alternative sequences shown in Table 22–1 indicate the possibility of handling the very slow achievers in a program different from the usual way the average achievers are handled in the usual cooperative industrial training plan represented by the middle sequence. It is conceivable that a few of the slow achievers would be ready to enter the regular cooperative plan by the beginning of the eleventh year after remedial work on fundamental industrial processes and development of basic skills—reading, writing, and computation—as applied to industrial trades.

RELATED INSTRUCTION

Although this text has emphasized that many administrative elements of the cooperative plan of instruction are common to all programs, regardless of the occupations represented, the instructional function is quite unique to each field. In the cooperative industrial education plan, this difference is due to (1) the nature of the competencies that are required in the various occupations, (2) the types of competencies that can be developed in school for various occupations, and (3) the variety of different types of occupations that are represented.

The nature of related instruction in cooperative industrial education.— In cooperative industrial education, the job is the place where many technical (manipulative) skills are to be learned. This is because of the highly specialized equipment that is required and the difficulty of simulating an industrial job in the classroom. The actual on-the-job experience provides a monetary incentive and the realization that a student's efforts are vital in the produc-

Table 22–1
Grade Level and Industrial Occupations
Pre-employment Education Plans

Grade Level	Option 1	Option 2	Option 3
12	Continues on supervised work experience and may transfer to the regular cooperative plan. The 1990 legislation continues impetus for plans for individuals in special populations	Cooperative industrial training[1] composed of one period of related instruction (two if a senior-year program only), 450 hours on the job (15 hours' minimum per week for 30 weeks)	Cooperative industrial training or cooperative electronics[2]
11	"How to work"—6 months' to 2 years' supervised work experience with variable credit	Cooperative industrial training composed of one period of related instruction and 450 hours on the job	Mathematics Science Cooperative industrial training
10	Industrial arts: Survey of occupations Remedial work	Selects two supplementary courses in industrial arts, *e.g.,* electricity or electronics.	Selects two supplementary courses in industrial arts, *e.g.,* electricity or electronics.
9	Grooming and personal development Remedial arithmetic Remedial writing Remedial reading	In a general shop: Woods, 6 weeks Electricity, 6 weeks Drawing, 6 weeks Graphic arts, 6 weeks Metals, 6 weeks Power mechanics, 6 weeks	In a general shop: Woods, 6 weeks Electricity, 6 weeks Drawing, 6 weeks Graphic arts, 6 weeks Metals, 6 weeks Power mechanics, 6 weeks

[1]Large schools might provide cooperative industrial education through homogeneous groups according to craft, *e.g.,* cooperative drafting, cooperative sheet metal, cooperative electronics.

[2]Typically, these electronics students would be taking science and math courses to enable them to begin work at a higher level, or while they are working, to enter a post-secondary technical curriculum.

tion or service that the school environment doesn't provide. The classroom instruction in school must teach the technical information related to the trade, as well as the general information required of all workers.[3] In cooperative industrial education, this is the teacher-coordinator's major classroom function. Therefore, the related class must have many reference materials on technical information. The teacher-coordinator organizes the technical material, and the student-learners meet daily for one period for the purpose of assimilating the information.

It is at this point that the teacher-coordinator assumes the task of guiding and directing the individualized study of each student-learner. Individualized instruction is a must. Another aspect of instruction in cooperative industrial education is the diversity of trades that are represented. If all students represent one trade such as drafting, then the related instruction can proceed on a group basis.[4] If the students

[3]Contrast this with the business plan in which core knowledges and manipulative skills are taught in school and applied on the job, or with the marketing area in which manipulative skills are a small part of the occupation but in which human relations skills and marketing concepts are very important.

[4]In this case, the single-trade industrial plan is similar to the business or marketing plan, in which the various business or marketing occupations all have a common body of content as their core.

represent a variety of trades, then the related instruction in technical information obviously must be given on an individual study basis. On the other hand, each student-learner can be enrolled in a class that provides supplementary career-technical instruction for the student-learner's trade area. In that class, such as machine trades, building construction, or electronics, the student-learner would be enrolled with non-cooperative trainees, and the class would probably be a laboratory where manipulative skills are taught.[5]

Thus, the general principles of instruction of cooperative industrial education are that (1) in-school instruction stresses technical information rather than skills and (2) instruction becomes primarily individual rather than group when diversity of occupation exists.

OBJECTIVES OF RELATED INSTRUCTION

There are three areas of instructional education. The first area of objectives describes the information, concepts, and attitudes that are basic to all industrial occupations and that the student-learners use to advance successfully toward their career objectives. These instructional objectives are taught primarily through group instruction. This is known as general related instruction (the topics are given later in this chapter).

The second area of objectives emphasizes the student-learners' building of occupational skills and knowledges applicable to the particular initial jobs in which they are placed. Each student's individual learning outcomes are stated in the training plan outline, which indicates the work processes in which the

student-learner will be trained.[6] These learning outcomes are stressed in the informal individual conferences of student-learner and teacher-coordinator and/or the training station sponsor.

The third area of objectives deals with the improvement of the students' general education as it relates to the job. The teacher-coordinator should certainly contribute to the improvement of each student's ability to read, write, compute, and communicate.

INSTRUCTIONAL METHODS, MATERIALS, AND FACILITIES

The instructional methods in cooperative industrial education plans involve two types of instruction necessary for developing career-technical competence. These are *general related instruction* and *specific related instruction*.

General related instruction.[7]—The term "general related instruction" is applied to instruction in topics considered of common importance to workers in industrial occupations.[8] In some cooperative industrial education plans, one period per day from the two-hour block is devoted to such instruction. Group instruction methods, with emphasis upon readings, lectures, and discussions, are appropriate for general related instruction, which is primarily conceptual and attitudinal in nature.

While the topics in the general related class may vary from program to program, the following topics are some common ones.

---

[5]In this case, the plan would be similar to some cooperative business plans, in which student-learners would be enrolled one period in a supplemental skills course, such as advanced shorthand and transcription.

[6]In single-trade programs, some of these objectives are accomplished by in-school instruction through group study, since all student-learners have common needs because they are in the same trade.

[7]Some trade and industrial educators call this "guidance related instruction."

[8]Many state departments have developed guides and course outlines for general related instruction. Contact the state department or a teacher-educator in your state for some suggestions.

Units of Study

School and Industrial Relations

- The Industrial Cooperative Training Plan
- What Makes a Good Trainee?
- School Rules and Regulations
- Responsibilities of the Student-Learner
- Grooming and Dressing for the Job
- Choosing a Means of Earning a Living
- Getting a Job and Growing in It
- Job Versus Training Station
- Employer–Employee Relations
- Paying for Services Through Taxes
- Private Insurance Plans
- Social Security
- Legal Regulations for Young Workers
- Labor Unions
- Wage and Hour Records
- Filing Your Income Tax Report
- Safety
- Career Opportunities in Industry

Recurring Topics

Personal Development

- Understanding Ourselves
- Keeping Out of Trouble
- Character and Personality
- Successful Job Relations
- Practice of Manners
- Effective Study Habits
- Budgeting Your Income
- Saving and Investing
- Credit and Money Management
- Understanding Contracts
- Consumer Legal Problems
- Using Negotiable Instruments
- Planning Your Career

Human Relations

- Understanding Others
- Courtesy to Others
- Cooperation

Student Organization Activities

- Aims and Purposes
- *Robert's Rules of Order*
- Making a Speech
- Learning Opportunities
- Recreational Opportunities

Specific related instruction.—The term "specific related instruction" is used to denote instruction given to students in the related class in the technical information requisite to their trades. At least half of the time available for related instruction should be devoted to these needs. Whereas general topics of value to all class members may be handled through group instruction, the technical requirements of the numerous industrial occupations must be handled through individual instruction. The technical requirements of the various trades and occupations differ. The rate and extent of learning vary considerably among student-learners. To be effective, technical instruction in cooperative industrial education must be individualized, and student-learners must accept the responsibility for applying themselves, with the guidance of their teachers.

Individual instruction or supervised study.—The most frequently used vehicles of individual study in the cooperative industrial education plan are the job manual, the student notebook in which the student plans and studies for occupational growth, a technical manual from a manufacturer, a technical manual or guide prepared for a specific occupation (sometimes prepared by committees of supervisors and teacher-coordinators), an apprenticeship manual, a trade magazine, and one or more textbooks.

Job-study guides are available for a considerable number of occupations. Coordinators should contact their state departments or their teacher-education institutions for a list of guides and their sources. The guides contain not only study questions but also suggested references and occasionally test questions. Some coordinators have been experimenting with the use of correspondence courses. However, their use is so limited that no firm recommendation can be made with regard to their effectiveness.

SPECIAL LABORATORIES AND CLASSROOMS

The classroom for related instruction for the cooperative industrial education plan is usually a regular classroom if a single-trade cooperative plan is involved. Whenever a diversified-trades plan is involved, a classroom suited to both group instruction and individual instruction is desirable.

Movable tables might well be arranged in a "U" shape, with chairs outside the "U." Writing board space, bulletin board space, reference book shelves, magazine racks, and filing cabinets are needed. Technological advancements are significantly impacting all career and technical education, including industrial education. Computers, a digital camera, an audiotape recorder, a TV, a video camcorder, a VCR, a DVD player, an overhead projector, and an LCD multimedia projector can play significant roles in the related instruction classroom.

A conference room available for individual student-coordinator conferences is an important feature of the cooperative education classroom. It will provide privacy for the discussion of personal problems or problems encountered at training stations. Glass partitions or large glass windows in the conference room will allow the activities going on in the classroom to be in full view of the teacher-coordinator while he or she is participating in an individual conference.

SOURCES OF INSTRUCTIONAL MATERIALS

Teacher-coordinators can contact the following companies and organizations, among others, and identify their specific needs for instructional materials for related classroom instruction in cooperative industrial education. These sources can provide titles, descriptions, and price lists of textbooks, manuals, and audio-visual aids. Each year the January-February issue of *Techniques*, the journal of the Association for Career and Technical Education, contains a directory of suppliers.

American Technical Publishers, Inc.
1155 West 175th Street
Homewood, IL 60430-4600
Phone: (800) 323-3471
Fax: (708) 957-1137
Email: service@americantech.net
Web site: www.americantech.org

Delmar / Thomson Learning
P.O. Box 15015
Albany, NY 12212-5015
Phone: (800) 477-3692
Fax: (518) 464-7000
Web site: www.delmar.com

Glencoe/McGraw-Hill
Customer Service Department
P.O. Box 543
Blacklick, OH 43004-0544
Phone: (800) 334-7344
Fax: (614) 755-5682
Web site: www.glencoe.com

Goodheart-Willcox Publisher
18604 West Creek Drive
Tinley Park, IL 60477-6243
Phone: (800) 323-0440 or (708) 687-5000
Fax: (888) 409-3900
Email: custserv@goodheartwillcox.com
Web site: www.goodheartwillcox.com

Multistate Academic and Vocational
  Curriculum Consortium
1500 West Seventh Avenue
Stillwater, OK 74074-4364
Phone: (800) 654-4502
Fax: (405) 743-5154
Web site: www.mavcc.org

Prentice Hall School
Upper Saddle River, NJ 07458
Phone: (201) 592-2425 or (866) 326-4259
Web site: www.phschool.com/career_technical
/index.html

V-Tecs (Vocational-Technical Consortium
of States)
1866 Southern Lane
Decatur, GA 30033-4097
Phone: (404) 679-4501, Ext. 543
Fax: (404) 679-4556
Web site: www.v-tecs.org

Teacher-coordinators should always be looking for materials that they can get at no expense, such as those donated by local employers and associations. If school policy permits, it may be wise to ask student-learners to purchase some manuals of their trade, just as they would buy tools and safety shoes. This is a prudent investment of some of the money they earn.

STUDENT ORGANIZATION

The industrial education student organization (SkillsUSA–VICA) is the third dimension of the cooperative education plan. It is an integral part of the total instructional program, including the related classroom and the training station. (Refer to Chapter 14 for suggestions on how students can benefit from participation in a local industrial education student organization chapter.)

The student organization is a good place to stress *Robert's Rules of Order*, neat appearance and appropriate dress, social etiquette, and successful completion of individual and group projects. If the number of student-learners warrants it, a special chapter for cooperative industrial education students might be appropriate.

SECURING DESIRED LEARNING ON THE JOB

The needs of each industrial student-learner for instruction in his/her chosen trade requires that the teacher-coordinator determine what the specific

CASE-IN-POINT
SkillsUSA–VICA "Turns On" Bud Wilcox

Bud Wilcox, who is a cooperative industrial training student-learner, is taking his on-the-job training at All-Tran Transmission Service. All-Tran repairs or rebuilds all makes of transmissions. Bud is doing well at his training station and is interested in the related instruction in the classroom. He shows potential for working his way into management and supervisory responsibilities after graduation. However, he is hesitant about expressing himself verbally—a skill he will need to develop if he wants to manage or supervise. Bud's teacher-coordinator decides to search for ways to give Bud experience in expressing himself. He helps Bud to plan a short five-minute presentation about his training station to share with the other classmates in the related class. Bud's presentation goes well and he enjoys it.

Through encouragement from the teacher-coordinator, the SkillsUSA–VICA president appoints Bud to appear with him before the advisory committee to help explain the cooperative plan and its accomplishments. The president talks about the industrial education student organization; Bud explains his training station experience; and the teacher-coordinator tells about classroom instruction. Bud and the president wear their special jackets—those bearing the SkillsUSA–VICA emblem. All goes well. Both students exhibit pride in their CTSO, and the advisory committee members are impressed.

The local TV station provides free time for SkillsUSA–VICA representatives to explain the cooperative plan, and Bud participates. He is now an enthusiastic promoter of the organization and the cooperative plan and does not hesitate to talk about them to others. SkillsUSA–VICA served as a good vehicle to give Bud verbal experiences.

learnings need to be for each student. The technique that accomplishes this is the development of a *training plan*.

For each individual student-learner, there must be a training plan showing what is to be learned on the job. This plan has to be worked out by the teacher-coordinator in cooperation with the employer. This plan will be in the form of a job analysis of the phases of the occupation that the student-learner is expecting to learn.

Training will be given in many occupations with which the coordinator cannot be familiar and about which assistance will have to be sought. In most instances, a craft or advisory committee can help. This committee may consist of the employer or supervisor and an employee in the establishment, or it may consist of people engaged in the occupation outside the educational agency. The employer should be encouraged to have a hand in developing the training plan. This training plan may be worked out in any one of a number of forms, but the three-column variety will probably be found the most usable and convenient. The first, or left-hand, column contains a listing by units of the manipulative skills to be learned on the job, while the related information that correlates with these job units appears in the second column. In the third column are references to specific books, magazines, etc., which indicate exactly where the desired related information can be found.

When the training plan is completed, it becomes a part of the training agreement. Training plans are discussed under the following head, and examples are shown in Figures 22–1 and 22–2.

When a schedule of processes or other job analysis has been worked out for each student-learner, the teacher-coordinator still faces the problems of learning to guide each student's individual study, making necessary instructional materials available, and developing a classroom suitable for individual study.[9]

Using the step-by-step training plan.—As in other cooperative education plans, the step-by-step training plan (schedule of processes) should make the cooperative industrial training plan sharply distinctive from general work experience plans.

---

[9]Training stations and training plans can be developed for a wide range of student-learners, from slow achievers to fast achievers. A competency-based SSLO, described in Chapter 11, can be selected each evaluation period, suitable to the learning abilities of the individual student-learner.

This plan indicates the specific career-technical objectives to be emphasized as well as the place where they are to be learned—school or job—and from whom they are to be learned—related instruction teacher or job sponsor. Figure 22–1 illustrates the areas of experience and training section of a step-by-step training plan for a training station in automotive technology. The plan is attached to the training agreement to make up a complete training plan. A teacher-coordinator may have the student-learner record these areas of experience and training on an analysis and progress chart. Figure 22–2 illustrates a part of an analysis and progress chart for an auto mechanic trainee. This chart provides a record of the instruction and the performance. Figure 22–3 outlines a training plan for a drafter. This plan is shown in some detail. The detail of the breakdown and the comprehensiveness or completeness of the training plan will depend upon the opportunities offered at the training station and the ability of the student-learner to benefit from the training.

MEASURING STUDENT-LEARNERS' GROWTH

In the final analysis, measurement of the student-learners' growth and development at the end of the school year will determine the extent to which career objectives have been reached. The objectives of the teacher-coordinator of a cooperative industrial training plan will have included the following:

1. Increasing the skills of the individual student-learners to perform successfully in the entry position.

2. Helping student-learners to develop an understanding for and an appreciation of the fundamentals of technology and the basic management principles that will enable them to grow and to be successful in a family of related occupations.

3. Acquainting the young workers with sources of additional information and skills development to enable them to progress in their chosen occupations.

4. Laying groundwork of aesthetic and moral values that will support what the students have

LEARNING OUTCOMES* FOR A STUDENT-LEARNER WITH CAREER OBJECTIVE OF AUTOMOBILE MECHANIC

Expected Learning Outcomes	Training in Class	Experience on the Job
1. Orientation to automobile mechanics		
a. To learn about the nature of the automobile repair business.	X	X
b. To gain an overview of automobile design.	X	X
c. To gain an appreciation of automobile repair hand tools.	X	X
d. To learn shop hazards and safe practices.	X	X
2. Competencies related to the engine parts		
a. To learn the physical principles involved.	X	
b. To learn to clean and repair the motor.		X
3. Competencies related to the fuel system.		
a. To learn the principles involved in fuel system operation.	X	X
b. To learn to clean and repair the fuel system.		X
4. Engine lubrication		
a. To learn the principles of lubrication.	X	
b. To learn the procedures for necessary complete lubrication.		X
5. Cooling system		
a. To learn the principles used in the cooling system.	X	
b. To learn the procedures necessary for complete lubrication.		X
6. Electrical system		
a. To learn the principles involved in the operation of the electrical system.	X	
b. To learn to maintain and repair the system.		X
7. The clutch		
a. To learn how the clutch operates.	X	X
b. To learn to maintain and repair the clutch.		X
8. The transmission		
a. To learn how manual and automatic transmissions operate.	X	X
b. To learn to maintain and repair transmissions.		X
9. Propeller shafts and universal joints		
a. To learn to check, remove, repair, and replace universals.		X
b. To learn to check, remove, repair, and install the slip joint.		X

FIGURE 22–1

LEARNING OUTCOMES* (Continued)

Expected Learning Outcomes	Training in Class	Experience on the Job
10. Rear axle and differential		
a. To learn to remove, adjust, and replace axle shaft and bearings.		X
b. To learn to remove, repair, and replace the differential assembly.		X
11. Front axles and steering gear		
a. To learn to make preliminary wheel alignment check.	X	X
b. To learn to adjust steering and balance wheels.		X
12. Springs and shock absorbers		
a. To learn to check, remove, repair, and replace springs.	X	X
b. To learn to check, adjust, remove, and replace shock absorbers.		X
13. Brakes		
a. To learn to make minor brake adjustments.	X	X
b. To learn to check and repair brakes.		X
c. To learn to test, maintain, and repair hydraulic systems.	X	X
14. Tires		
a. To learn to remove, repair, and remount tires.	X	X
15. General lubrication		
a. To learn the principles involved in the lubrication system.	X	X
b. To learn how to perform a complete lubrication job.		X
16. Automobile repair business management		
a. To learn the fundamental problems of organizing the business.	X	X
b. To learn the fundamental problems of staffing.	X	X
c. To learn the fundamental problems of controlling the budget, financing, and maintaining proper inventory.	X	X

---

*The learning outcomes could be stated in behavioral terms. For example, number 3 could be stated as follows:

    a. To be able to apply the principles involved in the fuel system operation.

    b. To be able to clean and repair the fuel system.

FIGURE 22–1 (Continued)

STERLING TOWNSHIP HIGH SCHOOL
ANALYSIS AND PROGRESS CHART

John Doe   Automobile Mechanic

Student-Learner   Training Station Position

Job Training	Observed	Instruction	Performed	Grade	Related Information	Assignment No.	Date Started	Date Completed	Grade
- - - - -					- - - - -				
Engine lubrication									
98. Check and fill crankcase					98. Composition and properties of engine lubricating oils				
99. Clean oil pan					99. Necessity of crankcase ventilation				
100. Check and adjust splash lubricating system					100. Function of engine lubricating system				
101. Check, repair, and adjust pressure-feed lubricating system					101. Types of lubricating systems				
102. Check and repair oil cooler					102. Types and functions of oil coolers				
103. Check and repair damaged oil line					103. Types and functions of oil filters				
104. Check and adjust or replace oil-pressure gauge					104. Theory of oil-pressure gauge operation				
105. Correct excessive oil consumption					105. Testing instruments used in lubrication system service				
106. Diagnose lubrication trouble					106. Introduction to engine lubrication trouble diagnosis				
- - - - -					- - - - -				

FIGURE 22–2

STEP-BY-STEP TRAINING PLAN
DRAFTER

A. **Title of Job:** Drafter (a two-year program)

B. **Job Description:** Prepares clear, complete, and accurate working plans and detailed drawings from rough or detailed sketches or notes for engineering or manufacturing purposes according to specified dimensions; makes final sketch of the proposed drawing, checking dimension of parts, materials to be used, the relation of one part to another, and the relation of the various parts to the whole structure; inks in all lines and letters on pencil drawings; makes charts for representation of statistical data; makes designs from sketches.

C. **Career Objective: Drafter**

D. **Areas of Training and Experience:**

    1. Instruments and Equipment
    2. Applied Geometry
    3. Freehand Lettering and Sketching
    4. Orthographic Projection
    5. Dimensions
    6. Sectional and Auxiliary Views
    7. Working Drawings
    8. Screw Threads, Gears, Welding
    9. Pictorial Drawings
  10. Piping Drawings and Schematics
  11. Development Drawing
  12. Revolution and Rotation
  13. Architectural Drawing
  14. Structural Drafting
  15. Topographic Map Drawing
  16. Reproduction of Drawings

E. **Detail of Areas of Training and Experience:**

Planned Learning Outcomes	Training in Class	Experience on the Job
1. Using instruments and equipment		
a. Learns types and uses of drafting instruments.	X	
b. Selects instruments and equipment.	X	X
c. Uses tracing paper and cloth.	X	
d. Cleans and adjusts drafting equipment.		X
e. Sharpens drafting pencils.		X
f. Sharpens compass leads.	X	
g. Places and fastens drawing paper.		X
h. Draws horizontal lines.	X	
i. Draws vertical lines.	X	X
j. Scribes circles.	X	
k. Understands inclined lines and the use of triangle.	X	X
l. Erases lines.	X	X
m. Draws lines to scale.	X	X
n. Draws irregular curves.	X	X
o. Uses inking equipment.	X	X

FIGURE 22–3

STEP-BY-STEP TRAINING PLAN (Continued)

Planned Learning Outcomes	Training in Class	Experience on the Job
2. Learning applied geometry		
a. Bisects straight lines.	X	X
b. Bisects angles.		X
c. Bisects arcs.		X
d. Divides a straight line into a given number of equal parts.	X	X
e. Constructs regular pentagons.	X	X
f. Constructs hexagons.	X	X
g. Constructs polygons.	X	X
h. Draws ellipses.	X	X
i. Locates the center of a circle through three given points.		X
j. Draws lines tangent to given circles.		X
3. Learning freehand lettering and sketching		
a. Letters freehand.	X	X
b. Sketches freehand.	X	X
c. Learns isometric sketching.	X	
4. Learning orthographic projection		
a. Pencils a drawing.	X	X
b. Prepares three-view orthographic drawings.	X	X
c. Learns orthographic reading.	X	
d. Practices orthographic projection.	X	
e. Lays out title blocks.	X	X
5. Learning dimensions		
a. Makes arrowheads.	X	X
b. Places dimensions.		X
c. Learns material symbols (section).	X	
d. Learns shop terms.	X	
e. Learns manufacturing precision.	X	
f. Learns precision and tolerance.	X	
g. Learns American Standard Association Classification of Fits.	X	
h. Checks drawings.	X	X
6. Learning sectional and auxiliary views		
a. Makes full-section drawings.	X	X
b. Makes half-section views.		X
c. Represents broken section views.		X
d. Constructs detailed or removed section views.		X
e. Illustrates phantom section views.		X
f. Constructs auxiliary views.	X	X

FIGURE 22–3 (Continued)

STEP-BY-STEP TRAINING PLAN (Continued)

Planned Learning Outcomes	Training in Class	Experience on the Job
7. Learning to make working drawings		
a. Learns types and uses of working drawings.		X
b. Makes detailed drawings.	X	
c. Makes assembly drawings.	X	X
d. Makes layout drawings.	X	X
e. Prepares bill of materials list.	X	X
8. Learning to draw screw threads, gears, welding symbols		
a. Draws screw threads.	X	X
b. Executes welding drawings.	X	X
c. Makes welding drawings symbols.	X	X
d. Represents and specifies gears.	X	X
9. Learning to make pictorial drawings		
a. Develops isometric projection drawing.		X
b. Sketches pictorial drawings.		X
c. Illustrates drawings.	X	X
d. Shades drawings.	X	X
e. Learns special shading methods.	X	
10. Learning to make piping drawings and schematics		
a. Develops pipe threads.	X	X
b. Specifies pipe and fittings.	X	X
c. Makes pipe drawings.	X	X
d. Dimensions pipe drawings.		X
e. Makes pipe schematic drawings.	X	X
f. Makes electrical schematic drawings.		X
11. Learning development drawing		
a. Lays out development drawings.	X	X
b. Classifies lines and surfaces.	X	X
c. Makes intersection drawings.	X	X
d. Practices development drawing.	X	
12. Learning revolution and rotation		
a. Makes revolution and drawings.	X	X
b. Practices development drawing.	X	
13. Learning architectural drawing		
a. Learns architectural terms.	X	
b. Learns types of house architecture.	X	
c. Makes preliminary studies.	X	X

FIGURE 22–3 (Continued)

STEP-BY-STEP TRAINING PLAN (Continued)

Planned Learning Outcomes	Training in Class	Experience on the Job
13. Learning architectural drawing (Continued)		
d. Prepares plot plans.	X	X
e. Makes presentation drawings.	X	X
f. Uses architectural symbols and conventions.	X	X
g. Prepares floor plans.	X	X
h. Dimensions architectural drawings.	X	X
i. Letters architectural drawings.	X	X
j. Makes wall sections.	X	X
k. Lays out elevations.	X	X
l. Makes detail drawings.	X	X
m. Makes perspective drawings.	X	X
n. Draws specifications.		X
14. Learning structural drafting		
a. Learns structural terms.	X	
b. Specifies structural steel.	X	X
c. Dimensions structural drawings.	X	X
d. Makes structural working drawings.		X
e. Makes structural detail drawings.	X	X
f. Learns structural notations and detailing information.	X	
g. Makes timber-structure drawings.	X	X
h. Makes masonry-structure drawings.	X	X
15. Learning topographic map drawings		
a. Develops topographic drawings.	X	X
b. Uses topographic symbols.		X
c. Lays out plats.	X	X
d. Learns classification of maps and topographic drawings.		X
16. Learning reproduction of drawings		
a. Operates and maintains reproduction equipment.		X
b. Makes Ozalid prints.	X	X
c. Makes blueprints.		X

Specific Reference

1. Latest guide for related instruction for ICT. (Some states have manuals.)

Career Objective References

1. Allen, H.M., and Fred Brasfield. *Computer Aided Drafting,* Denton, Texas: RonJon Publishing, 1991.
2. Feirer, John L. *Carpentry and Building Construction.* New York: Macmillan Publishing Co., Inc., 1991.
3. French, Thomas E., *et al. Mechanical Drawing,* Tenth Edition. New York: McGraw-Hill Book Company, 1985.

FIGURE 22–3 (Continued)

STUDENT'S PERIODIC RATING AND PROGRESS REPORT

Name (last, first, middle initial)　　　　　Grade　　　　　Employment

Training Establishment　　　　　Tel. No.　　　　　Report for Month of

_____ __

Times Absent from Work	Times Absent from School	Wage Rate Beg. _____ End _____ of Month	Job Instructor	Average Work Week Hrs. _____	Expected On-the-Job Schedules of Hours:
			Employer	Total Wage This Month $_____	Monday ... to ... Friday ... to ... Tuesday ... to ... Saturday ... to ... Wednesday ... to ... Sunday ... to ... Thursday ... to ...

Note: This report is to provide information that will guide the school and the cooperating business agency to provide the most effective type of training for the student. The trainer is to check each unit where improvement is needed. Also the rating of the student should be checked below.

ON-THE-JOB TRAINER'S RATING OF THE STUDENT

Complete the rating chart below. Also check (X) all units of instruction on the job in which the student must show improvement to attain satisfactory performance.

					Mark "X" in the space that best describes each item	Comments by Trainer:
Out-standing	Above Av.	Av.	Below Av.	Fail-ing	**Traits or Characteristics**	
					1. Personal appearance	
					2. Interest	
					3. Punctuality	
					4. Dependability	
					5. Initiative	
					6. Ability to cooperate with others	Signed　　　Date
					7. Progress — quality of work	Follow-up:
					8. Progress — quantity of work	
					9. Accuracy	
					10. Ability to follow instructions	
					11. Suitability for this occupation	
					12. Courtesy	
						Signed　　　Date

RELATED CLASSROOM INSTRUCTION

Related Training Topics Studied by Student:

_____

_____

_____

FIGURE 22–4. (Courtesy, Dade County Public Schools, Dade County, Florida)

learned previously under other secondary teachers to enable them to assume whatever leadership role in industry they are capable of pursuing.

Student measurement takes the form of evaluated projects that have become progressively more difficult and revealing, such as performance tests during class time and on the job; ratings of job growth by the job sponsor; and growth revealed by participation in student activities, including the student organization.

Positive growth in several or all of the areas mentioned can be measured by criteria such as the following:

1. The student-learners have attained an acceptable or better level of technical background and have developed adequate skills to perform in their chosen trade.

2. The student-learners can use resource manuals and training manuals in their chosen trade.

3. The student-learners have attained an appreciation of the concept of work to the extent that they have the potential to succeed in their chosen trades or in a related family of occupations.

4. The student-learners can communicate effectively with their supervisors and co-workers.

5. The employer–employee appreciation event, or any other public display or function that is student-centered but coordinator-advised, reflects the skills acquired and the technical knowledge gained by the student-learners during the cooperative educational experience.

6. The student-learners can produce the quality of work desired in the quantity that meets the standards of the trade.

TRAINING STATION PERFORMANCE BY STUDENTS

The most valid evaluation forms for on-the-job performance are those developed by the teacher-coordinator for individual student-learners, covering the experiences and learning outcomes applicable for a particular grading period. This completed form, preferably filled out by the teacher-coordinator face-to-face with the supervisor and perhaps in the presence of the student-learner, can be the basis for discussion with the student-learner and the job supervisor. Any letter grade given as the result of the rating and the discussion should be determined by the teacher-coordinator.

Questions and Activities

1. Prepare a 20-minute talk that you could give before a civic club, trade group, or union explaining the cooperative industrial education plan. Prepare charts or other visual aids to supplement your talk. (Add the notes of this talk to your files.)

2. What are the minimum qualifications for becoming certified as a cooperative industrial education teacher-coordinator in a reimbursable program in your state?

3. How adequate is the list of topics for related instruction suggested in this chapter? If you were the teacher-coordinator, which areas would you emphasize the most? Why? Use a specific-type program as a case for your analysis.

4. How does the instruction in a cooperative industrial education plan differ from that of another occupational field? Illustrate how differences occur because of factors such as degree of on-the-job skills development, prior skill readiness, and demand for cognitive learnings in these occupations.

5. If you were preparing a long-range (three-to five-year) plan for purchasing equipment and supplies for a cooperative industrial training classroom, what essentials would you request the first year? What would be some desirable items you would put in your plan for purchase later on? (Add the notes of these ideas to your files.)

6. Analyze the learning outcomes that could be expected in a specific prospective training station in your local community. Determine the possibility of a student-learner's progress toward his/her career objective in this training station by the end of the school year. (Refer to Chapter 11 for suggestions on how to proceed.)

7. Discuss the merits of the step-by-step training plan (or schedule of processes) in organizing a training station and the merits of the profile approach to evaluating a training station experience.

8. How do cooperative industrial education student organization activities provide educational experiences for student-learners? How do they differ from other occupational areas?

9. From examples you have studied, prepare a form that you could use in a local cooperative industrial training plan for each of the following: a rating sheet on the student-learner at the training station and a weekly report form on which the student-learner could report his/her training station activities. (Add a copy of each of these to your files.)

10. Using census data or state employment service sources, make a series of charts showing industrial employment trends in one county or standard metropolitan area of your state.

11. Study current references on classroom equipment and floor plans. Draw a room layout to scale (1/4 inch equals 1 foot) and prepare an equipment list.

12. What are the differences between a single-trade plan and a diversified-trades plan? Suggest how these differences affect student selection, training station placement, related instruction, prerequisite courses, and coordination activities.

13. State the learning outcomes of a training plan in behavioral terms. (Refer to Chapter 11 for suggestions.)

For references pertinent to the subject matter of this chapter, see Reading Resources.

Glossary of Key Terms

Glossary of Key Terms

Academic sanction—A penalty involving school grades, credit, or graduation.

Accrediting associations—Associations that certify that secondary and post-secondary schools meet established standards.

Administrative laws—Legislative enactments, such as congressional acts, state statutes, and city ordinances.

Adult education—The process of educating and training adults, conducted by school corporations, community private industry councils, and post-secondary institutions.

Advisory committee—A group serving strictly in an advisory capacity, with the educational policy remaining under the control of the superintendent of schools. It usually consists of 7 to 12 persons—teachers, businesspersons, labor leaders, parents, and students. If a steering committee is used, some members may be asked to serve on it.

Age certificate—A document that is issued for minors 17 to 21 verifying their age.

Alternating cooperative experiences—A series of full-time experiences in industry that alternate with a period of full-time study.

Alumni—Former graduates of cooperative occupational education plans.

Annual contract—A legal agreement for reimbursement for vocational education, which is filed by the district administrator at the beginning of the school year.

Annual report—An account of actual expenditures of funds, in which the local superintendent requests reimbursement at the end of the fiscal year, based on the annual contract.

Apprentice—A person who learns a trade by working under the guidance of a skilled master.

Apprenticeship—A training program authorized by the National Apprenticeship Act of August 16, 1937, and administered by the Bureau of Apprenticeship.

Area Redevelopment Act of 1962—An emergency measure born out of a recession, which authorized $4.5 million annually to be used for vocational education until 1965. It recognized the critical need for training that arises from unemployment and underemployment in economically distressed areas.

Area career-technical school/center—A high school, a department of a high school, a technical institute or career-technical school, a department or a division of a junior–community college or a university used exclusively or principally to provide career and technical education to students who are entering the labor market.

Association for Career and Technical Education—An organization composed of career-technical educators. Members receive *Techniques*, the career and technical education journal; insurance; registration with its placement office; annual conventions; and professional relationship services with other associations and the government. All career-technical educators are eligible to join.

Assured jobs program—A program that provides work on a voluntary basis for those who can meet an income test—such as that required for JTPA assistance—and who have received unemployment compensation for more than 15 weeks.

At-risk populations—Certain segments of society, all who have disabilities and/or disadvantages, such as members of minority groups, women, persons who are economically and/or academically disadvantaged, and those who are physically and/or mentally disabled.

Bilingual education—Formal learning and training for individuals with limited English proficiency to prepare them for occupational entry and to provide them with instruction in the English language so that they will be equipped to pursue such occupations in an English language environment.

Business Professionals of America (BPA)—The career-technical organization for those secondary and post-secondary students enrolled in career-technical business education programs.

Calendar of events—An outline of activities that serves as a guide for school officials who have the responsibility for getting the cooperative plan underway. Also, it may be a guide for the teacher-coordinator after the initiation of the cooperative plan.

Career and technical education (CTE)—The process of educating and training individuals for occupational competency.

Career development competency—A type of occupational competency in which an individual realizes the need for experience and further study in order to accomplish an identified career goal.

Career Education Act of 1978—Federal legislation that established the comprehensive career development concept, which viewed the individual as progressing through various planned experiences, a series of dimensions that total a complete cycle. These dimensions begin with career awareness at an early age, add employability skills, and end with educational awareness.

Career objective—A career-technical goal toward which a cooperative student prepares and through which the student expects to obtain full-time employment, with opportunities for monetary compensation and for advancement.

Career-technical student organization (CTSO)—A cooperative education co-curricular student group, whose activities are directly related to instructional goals. In competitive events/activities, students must demonstrate either occupational knowledge or general knowledge related to specific careers.

Carl D. Perkins Vocational and Applied Technology Education Act of 1990—Federal legislation that amended the 1984 Perkins Act. Its purpose was to make the United States more competitive in the world economy by developing more fully the academic and occupational shifts of our population.

Carl D. Perkins Vocational and Technical Education Act of 1998—Federal legislation that amended the 1990 Perkins Act. Its major purposes are to further develop the academic, vocational, and technical skills of career and technical education students through high standards and established accountability related to specified core indicators.

Carl D. Perkins Vocational Education Act of 1984—Federal legislation that amended the Vocational Education Act of 1963 and replaced the amendments of 1968 and 1976. It changed the emphasis of federal funding in vocational education from primarily expansion to program improvement and at-risk populations.

Changing occupation—An existing occupation that has experienced a change in duties, skills, or tasks, significant enough to require training beyond a short demonstration, but not significant enough to reclassify it into another existing occupation or to create a new occupation.

Chapter work/activities—A series of well-balanced, varied student organization experiences.

Cluster concept—The development of instructional programs designed to provide skills, knowledges, and understandings common to several related occupations.

Combination plans—Cooperative plans that involve more than one broad occupational grouping.

Community survey—A detailed study designed to secure information from the student body and the business community, thus aiding in the decision-making as to whether a cooperative plan should be initiated.

Competency—A skill, an attitude, and/or a knowledge needed by an individual to master an occupation.

Competency-based curriculum—A curriculum built around the skills, aptitudes, and/or knowledges needed by an individual to master an occupation.

Comprehensive Employment and Training Act (CETA) of 1973—Federal legislation intended to continue the goals of the previous MDTA legislation, plus expand the previous services MDTA made available. It recognized the high unemployment level at that time, the increasing number of welfare recipients, the increasing number of economically disadvantaged rural and urban communities with hard core unemployed, the significantly large number of youth unable to find part-time or full-time employment, and the increasing number of people in minority groups who were unskilled and unemployed.

Comprehensive Employment and Training Act Amendments of 1978—Federal legislation that authorized the extension of the activities and the programs of the 1973 law. It also included provisions to improve the delivery, accountability, and coordination of the various employment and training programs and services.

Consulting—Working with businesspersons in a community to help solve business problems.

Continuing education—The process of educating and training adults through short-term activities, such as conferences, clinics, and workshops, to help keep adults up-to-date in their employment skills.

Cooperative Education Association—An organization composed primarily of post-secondary and college or university cooperative education personnel. Members receive a quarterly journal and other services. There is an annual convention.

Cooperative occupational education plans—Organized and supervised work experience plans that correlate school and job learnings in tandem with defined career objectives. They incorporate related classroom instruction, supervised on-the-job training, and a student organization, under the direction of a certified teacher-coordinator.

Cooperative plan of instruction—A strategy/structure/system of instruction that achieves predetermined goals and competencies in a given occupational field, which includes on-the-job training.

Coordination—The relationship of all those activities outside the classroom that are performed by the teacher-coordinator in organizing, administering, operating, and improving the cooperative plan.

Correlated instruction—The direct relationship between the study in school and the training activities on the job, both of which are based upon the student-learner's career objective.

Correlation—The relationship between professionally trained individuals in the school laboratory and occupationally proficient individuals in the work laboratory who share the teaching–learning responsibility.

Counselor-coordinator—An educator who handles a work exploration program.

Cost-benefit studies—Studies that get at the economics of input-output by asking: Was the vocational education program worth its cost to the individual? Was it worth its cost to the taxpayer and to society?

Critical factor technique—A technique for assessing factors thought to be vital by a local district. The technique is based on the theory that a few factors are crucial and that if any of these are missing or badly handled, the plan is in trouble, no matter how high the quality of a host of minor factors.

Debriefing meetings—After the students have completed their cooperative experiences, get-togethers in which the students and their coordinators discuss the students' reactions to and impressions of their experiences and review their employers' evaluations.

DECA—An Association of Marketing Students—The national career-technical organization for secondary and post-secondary students who are enrolled in marketing education programs.

Demographic trends—Trends that deal with characteristics of human populations that are concerned with size and density, distribution, and vital statistics, such as age, sex, income.

Directed occupational experience—Training and/or activities that include two complementary components: an actual occupational activity and a seminar designed to supplement that experience.

Disabled—Referring to those students, ages 3 to 21, who are disabled mentally, educationally, and/or physically. They may be in public elementary and secondary schools or they may have been placed in private schools by public agencies.

Disadvantaged—Characterizing individuals who are economically and/or academically disadvantaged to the extent that they cannot actively participate in vocational programs.

Due process—The course of legal proceedings established to protect each individual's rights and liberties. *Substantive due process* refers to the legislation (rule) itself: Is the rule's purpose within the power of the lawmaker to pursue and is it rationally related to accomplishing its purpose? *Procedural due process* refers to the implementation of the rule: Has the rule been implemented fairly?

Education for All Handicapped Children Act—Federal legislation that applies to students, ages 3 to 21, who are disabled and in public elementary and secondary schools and to those who have been placed in private schools by public agencies.

Employer–employee appreciation event—A meal, a program, or an activity, which is an expression of appreciation by the student-workers to their employers.

Employment certificate—A work permit that is required by many states to be on file by employers of minors 16 and under.

Engaged-time analysis—Analysis of the time students spend during the instructional period, both in the related class and on the job.

Equal educational opportunity—Concept interpreted by most courts to mean that school districts must take affirmative action to overcome the deficiencies of certain groups of students.

Experiential learning—The acquisition of a wide range of learning experiences. From involvement in such learning experiences, the teacher-coordinator may learn about potential training stations, may acquire additional knowledge about occupational fields, and may gain expanded local support.

Exploratory work experience plans—Plans that are designed to help young people learn firsthand about an occupation by either observing it or engaging in productive work.

Extended contract—A legal agreement for teacher-coordinators that continues beyond the regular school year, sometimes covering the entire summer.

Family, Career, and Community Leaders of America (FCCLA)—The national career-technical organization for junior and senior high school students enrolled in family and consumer sciences occupations education. The organization's goal is to help youth assume active roles in society as wage earners, community leaders, and family members.

Farm-out system—An arrangement whereby related instruction is assigned to a teacher in an occupational course not taught by the teacher-coordinator.

Federal Revenue Act—An act intended to encourage employers to provide training and jobs to targeted populations who possess little if any job skills.

Feedback—Information gathered through contact with the job supervisor to the student-learner relative to training or events on the job.

Future Business Leaders of America (FBLA)—A national career-technical organization for students enrolled in secondary business courses. Students do not have to be in a career-technical program to belong.

General employability competence—The possession of the fundamental personal traits necessary for employment. These include reliability, well-groomed personal appearance, pleasant personality, cooperation, and positive attitude toward work.

General work experience plans—Plans that are primarily intended to help students explore occupations and to give them experience at working. These may be either remunerative or non-remunerative.

George–Barden Act of 1946—Federal legislation that authorized an appropriation of $28.5 million annually for the further development of vocational education. It is also known as the Vocational Education Act of 1946. It replaced the George–Deen Act of 1936.

George–Deen Act of 1936—Federal legislation that authorized an annual allotment of $12 million for agriculture, home economics, and trade and industrial education. Marketing occupations were recognized for the first time, and $1.2 million was authorized for them annually.

Goals 2000: Educate America Act—Federal legislation creating a National Standards Board to certify voluntary performance standards for each industry or career field.

Health Occupations Students of America (HOSA)—The national career-technical organization for secondary and post-secondary students who are enrolled in health occupations education.

Higher education institutions—Two-year post-secondary institutions and four-year colleges and universities.

in loco parentis—Educational concept in which the courts view teachers as standing in place of the parents of the students under their supervision and as such they have not only the right but also the duty to act on the behalf of their students.

Internal evaluation—Evaluation of the program by the teacher-coordinator, school administrators, employers, and students.

Internship—A period of occupational experience by a person who has completed all the academic requirements for admission to a profession prior to certification (licensing) as a recognized practitioner. It is also called a *practicum*.

Job adjustment competency—Type of occupational competency whereby an employee must successfully develop the interpersonal relationships needed in any new employment situation.

Job development—The process of creating jobs (training opportunities, part-time employment, or full-time employment) that did not exist or whose existence was not known to the school.

Job intelligence competency—Type of occupational competency that involves an individual's acquiring the skills, knowledges, and attitudes necessary to perform the duties needed in a given occupation.

Job Training Partnership Act (JTPA) of 1982—Federal legislation that replaced CETA and enlarged the role of state governments and private industry in federal job training programs, imposed performance standards, limited support services, and created a new program of retraining for displaced workers.

Job Training Partnership Act (JTPA) of 1992—An act designed to establish programs to prepare youth and adults facing serious barriers to employment for participation in the labor force by providing job training and other services that will result in increased employment and earnings, increased educational and occupational skills, and decreased welfare dependency, thereby improving the quality of the workforce and enhancing the productivity and competitiveness of the nation.

Journeyperson—A skilled worker who has completed an apprenticeship program.

Learning agreement—A written contract between the student and the coordinator or faculty advisor detailing the goals or the objectives that the student proposed to accomplish while on the cooperative assignment.

Learning manager—The cooperative teacher whose effectiveness in the cooperative classroom depends upon that teacher's depth of understanding of the learning process and knowledge of how career-technical learning is implemented by proper procedures and techniques in the classroom and on the job.

Learning-time analysis—Same as "engaged-time analysis."

Levels of employment—The five steps of marketing employment—entry-level, career-sustaining, marketing supervisor, and manager-entrepreneur to which instruction in prebaccalaureate marketing education is directed.

Mainstreaming or inclusion—The placing of students who are disabled in the "least restrictive environment" and then developing an individualized education program (IEP) for every student who has a disability.

Mandatory evaluation—Evaluation required by the state and/or federal government that provides funds.

Manpower Development and Training Act (MDTA) of 1962—Federal legislation that authorized funds for training and retraining of unemployed and underemployed adults.

Marketing Education Resource Center (MarkED)—A national center of resource materials and research activities located at The Ohio State University, which provides competency-based training materials, seminars, and promotional aids.

Method of instruction—A procedure used by the instructor in the classroom, such as lecture, case problem, or project.

Monthly enrollment report—A monthly account of enrollment information, which aids the state staff in determining the amount of training being accomplished.

Morrill Act of 1862—Federal legislation that appropriated public lands for the establishment of a college of vocational education in each state. It is also known as the Land-Grant College Act.

National Defense Education Act of 1946—Federal legislation that authorized an appropriation of $15 million annually for four years to support programs limited to the training of highly skilled technicians in occupations recognized as necessary for the national defense. The provisions became Title III of the George–Barden Act of 1946.

National FFA Organization (FFA)—The national career-technical organization of secondary students in agriculture programs.

Needs analysis—The evaluation of training needed by an individual or an organization.

Net increase in labor force—Growth in the labor force over a specific period of time.

New occupation—An occupation that has come into existence in a particular geographic area in the last 10 years, has employment levels large enough to be measured, and requires tasks, skills, and duties not included in any currently existing occupation.

News release—A news article structured around answering the five main "w" questions—Who? What? When? Where? Why?

1978 Adult Education Amendments—Amendments that specified particular segments of the population that Congress believed had benefited the least from past adult education programs because of inadequate methods used by states to inform them of available services.

Non-additive credit—Unit of work earned at the workplace rather than on campus.

Occupational competency—Type of competency whereby an individual must have enough specific education and training to be employable, but with a broad enough background to be able to move horizontally as well as vertically in a cluster of related occupations for advancement or promotion.

Open-interviewing—The process of allowing students to interview with more than one organization for employment during their internship programs.

Optional scheduling technique—The method of preparing and following an elective schedule of classes; the cooperative plan related courses are omitted in favor of other courses.

Orientation experiences—Those activities designed to give students the opportunity to sample different professions.

Parallel cooperative—Periods of experience, often full time for a term/semester, or halftime in some schools located in metropolitan areas where school and work can be conveniently combined. Usually occurs in the latter half of the curriculum when the student has completed the basic or core courses in his/her major field.

Plan of instruction—A strategy that uses appropriate teaching methods, such as the lecture and the field trip, for the learning outcomes sought.

Preparatory courses—Those studies in a plan of instruction that develop a student's occupational awareness and those skills, concepts, and attitudes needed before that student can be placed as a cooperative student-learner.

Private Industry Council (PIC)—Group in each SDA that is responsible for establishing goals for accomplishing performance standards, including elements such as placement data of clients in unsubsidized employment, job retention figures, and reductions in welfare payments. The majority of the members are private sector representatives. Other members are from educational institutions, organized labor, and economic development agencies.

Process-oriented evaluation—Evaluation of the procedures used in career and technical education programs to teach occupational skills and knowledge.

Product-oriented evaluation—Evaluation of the outcomes of career and technical education programs.

Professionalism—The use of high standards by teacher-coordinators in dealing with their peers in the teaching profession, with administrators, with student-learners whom they serve, and with adults in business and industry.

Program of instruction—A series of courses and/or experiences having educational objectives.

Program review—Assessment of the planning and operational processes, the success of program graduates, and the achievement of students.

Project laboratory (plan)—An outline of marketing activities taught through simulations provided in a school laboratory setting.

Public relations (PR)—Those activities, such as developing brochures, assisting in school activities, making speeches, publicizing the cooperative plan, and appearing on radio and television programs, designed to create favorable public opinion.

Recurring activities—Those coordination activities repeated during the school year in order to have an effective cooperative occupational education plan.

Reimbursement rate—The percentage of the cooperative plan budget that is returned to the local school system from state and federal funds.

Related class—In-school instruction that is relevant to those activities learned and performed by students on the job.

School store—A facility located in the school building or in a shopping center that is operated by marketing education students under the supervision of their teacher-coordinator. This facility provides marketing activities that are ordinarily carried out in marketing businesses.

School to Work Opportunities Act of 1994—Legislation designed to establish a national framework for creating state-wide school-to-work opportunity systems in conjunction with other comprehensive educational reform structures.

Sectional cooperative plan—A design whereby directly related instruction is given separately to students in both occupational areas, but in which one teacher instructs both classes and coordinates both groups.

Seminars—Education or training sessions of one or two meetings, totaling two to five hours.

Service delivery area—A unit of local government with a population of 200,000 or more or a consortium of contiguous local governments with an aggregate population of 200,000 or more that serves a substantial part of the labor market area. In addition, any rural concentrated employment program that serves as a prime sponsor under CETA may request to be a service delivery area.

Sex equity—The elimination of sex bias and sex stereotyping.

SkillsUSA–VICA—A non-profit national career-technical organization for secondary and post-secondary students enrolled in trade and industrial occupations programs.

Smith–Hughes Act of 1917—Federal legislation that provided an annual grant of approximately $7.2 million in perpetuity to the states for the promotion of vocational education in agricultural, trade and industrial, and home economics education.

Special delivery areas (SDA's)—Units of local government set up by each state governor. Each unit is based on a population of 200,000 or more, a consortium of contiguous local governments with an aggregate population of 200,000 or more that serves a substantial part of the labor market area, or a rural concentrated employment program that served as a prime sponsor under CETA.

Specialized plans—Cooperative plans in the various fields of cooperative education—agriculture, business, health, family and consumer sciences, marketing, and trade and industry.

Special needs populations—Individuals who are (1) physically and/or mentally disabled, economically and/or academically disadvantaged, and/or limited–English-speaking, (2) members of minority groups, and/or (3) gifted and talented.

Specific employability competence—The possession of definite job skills, attitudes, and/or knowledges necessary for employment.

Sponsor development—The process of enhancing the training abilities of job supervisors or job sponsors through meetings and learning activities.

SSLO (statement of student-trainee learning objectives)—A statement that combines the behavioral learning objectives concept employed by educators and the management by objectives tool used by business and industry.

State councils—Advisory groups, each of which advises its state's board of vocational education on policies that would strengthen that state's career and technical education program. Their policy objectives are (1) to achieve considerable public involvement in shaping CTE policy, (2) to provide independent assessments of career-technical programs and activities, and (3) to establish mechanisms that will improve CTE policies and the policy-making process.

State plan—A written plan submitted to the U.S. Secretary of Education for a two-year period that shows how that state proposes to use funds provided through a particular career and technical education act.

Steering committee—Usually a temporary advisory group that decides whether a cooperative education plan should be started in its school system. It is usually composed of representatives from education, business, labor, the news media, parents, and the community at large.

Student-learner—A person who has enrolled in the cooperative plan to prepare for an occupation or an area of occupations. This person is a *student* in the secondary or post-secondary school and a *learner* in an occupation in the supervised business laboratory experiences.

Supplementary instruction—Instruction designed to develop skills, concepts, and attitudes that are useful to student-learners but are not directly needed for employability.

Tax credit—Legislation that provides employers who participate in cooperative plans with eligibility for being reimbursed for excess costs involved in supplying training stations for student-learners.

Teacher-coordinator—A teacher, counselor, administrator, and public relations person.

Technology Student Association (TSA)—National student organization that is oriented to introducing students to high-skill technical occupations.

Tech-prep—An educational program that provides planning and demonstration grants to consortia of local educational institutions for the development and operation of four-year programs that are designed to provide a tech-prep educational program leading to a two-year associate degree or certificate.

Third-party evaluation—Evaluation of the program by a contracted outside entity.

Time-on-task analysis—Same as "engaged-time analysis."

Time-use analysis—Same as "engaged-time analysis."

Training agreement—A written document clarifying the specific responsibilities of the student-learner, the training sponsor, the parents, the teacher-coordinator, and the school.

Training plan—A step-by-step procedure listing what the student-learner is to study in school and on the job. It brings the student face to face with determining what career objective to pursue and with deciding what competencies to develop. Through the training plan, the employer becomes more definitely aware of the student learner's occupational goal and will usually try to lead that student toward achieving that objective by providing adequate work activities and on-the-job instruction.

Training station—A cooperating business, industry, or government agency, which has been selected according to criteria that measure its ability to provide opportunities for supervised educational experiences that will prepare the student for his/her intended career objective.

Training station sponsor—An experienced employee, supervisor, or manager who is directly responsible for the occupational learning experiences of the student-learner on the job, as indicated by the step-by-step training plan.

Value-added assessment—Assessment of the degree of improvement in terms of knowledge, skills, and attitudes. Basic to this evaluation scheme is the notion that "value-added" causes change.

Vocational Education Act of 1963—Federal legislation that for the first time mandated that vocational education meet the needs of individual students, not just the employment needs of industry. Its major purposes were to maintain, extend, and improve existing programs of vocational education and to provide part-time employment for young people who needed the earnings from such employment to continue their schooling on a full-time basis.

Vocational Education Act of 1976—Federal legislation that mandated vocational education to reflect the needs of the labor market more accurately and, in some cases, to become a partner with business, industry, and labor in attempting to change the workplace to be more responsive to the needs of its workers to become proactive rather than reactive.

Vocational Education Amendments of 1968—Federal legislation that authorized federal grants to states to assist them in maintaining, extending, and improving existing programs of vocational education and to provide part-time employment for young people who needed the earnings from such employment to continue their vocational training. Research, experimental demonstration programs, and exemplary programs and projects were encouraged.

Vocational Rehabilitation Act—Federal legislation that was designed to protect persons who are disabled and who are attending preschool, elementary, secondary, or adult education programs, as well as those individuals with disabilities who are in post-secondary education programs.

Weekly training report—A statement prepared by trainees or student-learners that indicates the areas in which they are gaining on-the-job instruction.

Work exploration program—A series of experiences that provide students with temporary exposure to the world of work, as students are placed in exploratory situations to observe and assist in occupational areas in which they have indicated an interest or have shown an aptitude.

Workforce Investment Act of 1994—Legislation written to consolidate, coordinate, and improve employment, training, literacy, and vocational rehabilitation programs in the United States.

Work observation plans—Non-remunerative, short-term work experiences that are intended to give students a feel for the work environment and a superficial understanding of the activities performed by various employees at their work stations.

Work-study plans—Work experience plans that are associated with secondary or post-secondary institutions, without correlation of school and job learnings in tandem with defined career objectives.

Reading Resources

Reading Resources

1. THE SCOPE OF INSTRUCTIONAL PROGRAMS USING THE WORK ENVIRONMENT

Bailey, T., & Merritt, D. (1993). *The school-to-work transition and youth apprenticeship: Lessons from the U.S. Experience.* New York: Manpower Demonstration Research Corporation.

Bragg, D. (1997). *Grubb's case for compromise: Can "education through occupations" be more?* Journal of Vocational Education Research, 22(2), 115–122.

Brown, S. (1993). Apprenticeship program pays students to stay in school. *Vocational Education Journal, 68(4)*, 34–35.

Campbell, C. P. (1991). Enhanced flexibility through nonformal vocational training. *Journal of Epsilon Pi Tau, 17(1)*, 45–52.

Coleman, M. (1993). The thrill of victory. *Vocational Education Journal, 68(4)*, 30.

Denby, S. (1991). What students want: American business and labor need to change their view of apprenticeship. *Vocational Education Journal, 66(7)*, 24–25.

Filipozak, B. (1992). Apprenticeship from high school to high skills. *Training, 29(4)*, 23–29.

Grubb, W. N., & Kruskouskas, E. (1993). Building bridges. *Vocational Education Journal, 68(2)*, 24.

Hamilton, S. F. (1990). *Apprenticeship for adulthood: Preparing youth for the future.* New York: Free Press.

Hamilton, S. F., & Hamilton M. A. (1992). Bridging the school-to-work gap. *School Administrator, 49(3)*, 8–15.

Harmon, H. L. (2000). Linking school-to-work and rural development. *Forum for Applied Research and Public Policy, 15(1)*, 97–100.

Hoerner, J., & Wehrley, J. B. (1995). *Work-based learning.* New York: Glencoe/McGraw-Hill.

Hogg, C. L. (1999). Vocational education: Past, present, future. In A. J. Pautler, Jr. (Ed). *Workforce education: Issues for the new century* (pp. 3–19). Ann Arbor, MI: Prakken Publications.

Jennings, J. F. (Ed.). (1995). *National issues in education: Goals 2000 and school-to-work.* Bloomington, IN: Phi Delta Kappa Educational Foundation.

Kaufman, P., Bradby, D., & Teitelbaum, P. (2000). *High schools that work and whole school reform: Raising academic achievement of vocational completers through the reform of school practice* (Report No. NCRVE-MDS-1295). Berkeley, CA: National Center for Research in Vocational Education. (ERIC Document Reproduction Service No. ED438418)

Lynch, R. L. (2000). *New directions for high school career and technical education* (Information series No. 384). Columbus, OH: ERIC Clearinghouse on Adult, Career, and Vocational Education, Center on Education and Work. (ERIC Document Reproduction Service No. ED444037)

McKernan, J. R. (1994). *Making the grade: How a new youth apprenticeship system can change our schools and save American jobs.* Boston: Little, Brown.

Paris, K. A., & Mason, S. A. (1995). *Planning and implementing youth apprenticeship & work-based learning.* Madison, WI: Center on Education and Work.

Pucel, D. J. (1999, October). *The changing roles of vocational and academic education in future high schools.* Paper presented at the Central Educational Science Research Institute, Beijing, China. (ERIC Document Reproduction Service No. ED434242)

Reis, R. (1996). The right prescription. *Vocational Education Journal, 71(1),* 44.

Rosenbaum, J. E., et al. (1992). *Youth apprenticeship in America: Guidelines for building an effective system.* Washington, DC: William T. Grant Foundation Commission on Youth and America's Future.

Schreiber, A. P., & Walter R. P. (1995). Three states tackle school-to-work reform. *Vocational Education Journal, 70(3),* 41.

Scott, J. L., & Sarkees-Wircenski, Michelle. (2001). *Overview of career and technical education.* Homewood, IL: American Technical Publishers.

Shapiro, D. (1999). *School-to-work partnerships and employer participation: Evidence on persistence and attrition from the national employer survey* (Report No. NCPI-TR-2-10). Stanford, CA: National Center for Postsecondary Improvement. (ERIC Document Reproduction Service No. ED440570)

Steinberg, A., Cushman, K., & Riordan, R. (1999). *Schooling for the real world: The essential guide to rigorous and relevant learning.* San Francisco: Jossey-Bass.

Thompson, M. J., & Hoberson, S. (1990). Apprenticeship and cooperative education. *Journal of Cooperative Education, 26(3),* 7–14.

Von Erden, J. (1991). Linking past and present, students and jobs. *Journal of Vocational Education, 66(7),* 30–32, 69.

Worthington, R. M. (1992). National perspective on cooperative education. *Workplace Education, 10(2),* 2, 4–7.

Worthington, R. M. (1993). Marketing vocational education. *Vocational Education Journal, 68(1),* 28–32.

2. THE DEVELOPMENT OF HUMAN RESOURCES

Barling, J., & Kelloway, E. K. (Eds.). (1999). *Young workers: Varieties of experience.* Washington, DC: American Psychological Association.

Cantor, J. A. (1990). A study of two potentially compatible activities: Job training and economic development. *Journal of Vocational Education Research, 15(2),* 55–75.

College Placement Council. (1994). *Developing the global workforce.* Bethlehem, PA: Author.

Crysdale, S., King, A. J. C., & Mandell, N. (1999). *On their own? Making the transition from school to work in the information age.* Ithaca, NY: CUP Services.

Hannah, G. (1993). Shift or drift. *Vocational Education Journal, 68(4),* 21.

Harrison, B. C. (1992). Developing human capita. *Vocational Education Journal, 67(5),* 28.

Kincheloe, J. L. (1999). *How do we tell the workers? The socioeconomic foundations of work and vocational education.* Boulder, CO: Westview Press.

National Association of State Boards of Education. (1995). *Framework for the future: Creating a school-to-work system for learning, livelihood, and life.* Alexandria, VA: Author.

National Center on Education and the Economy. (1995). *Building a system to invest in people: States on the cutting edge.* Washington, DC: Author.

Sathre, R. (1987). Putting special needs provisions into focus. *Vocational Education Journal, 62(7),* 38.

Thurston, L. P. (1999, September). *Practical partnerships: A cooperative life skills program for at-risk rural youth.* Paper presented at the annual conference of the National Rural Education Association, Colorado Springs, CO. (ERIC Document Reproduction Service No. ED439896)

Vernezze, M., & Henkel, M. (1993). Gateway to careers. *Vocational Education Journal, 68(4),* 26.

3. PUBLIC POLICY GOALS AND INSTITUTIONAL ROLES

American Vocational Association. (1998). *The official guide to the Perkins Act of 1998.* Alexandria, VA: Author.

Busse, R. (1992). The new basics. *Vocational Education Journal, 67(5),* 24.

Committee for Economic Development. (1990). *An America that works: The life-cycle approach to a competitive work force.* New York: Author.

Dornsife, C. (1992). *Beyond articulation: The development of tech-prep programs.* The Ohio State University, National Center for Research in Vocational Education.

Dykman, A. (1995). On the block. *Vocational Education Journal, 70(6),* 26.

Flude, M., & Sieminski, S. (Eds.). (1999). *Education, training and the future of work II: Developments in vocational education and training.* New York: Routledge.

Haddad, Sergio, (Ed). (1997). *Adult education—the legislative and policy environment.* Norwell, MA: Kluwer Academic Publishers.

Hettinger, J. (1999). The new Perkins . . . finally. *Techniques, 74(1),* 40–42.

Hoachlander, G., & Klein, S. (1999). Answering to Perkins. *Techniques, 74(2),* 46–48.

Knoblauch, B., & McLane, K. (1999). *Overview of the Individuals with Disabilities Education Act Amendments of 1997 (P.L. 105-17): Update 1999.* ERIC EC Digest E576. (ERIC Document Reproduction Service No. ED433668)

McCarthy, K. (1994, September). *School-to-work: A guide for state policymakers.* Denver: National Conference of State Legislatures.

McFarland, W. P. (1992). Meeting of the minds. *Vocational Education Journal, 67(5),* 26.

O'Brien, N. (1996). AVA efforts improve federal outlook for Voc. Ed. *Techniques, 71(6),* 49.

Olson, L. (1992). Final regulations on vocational education law narrow scope of local evaluations. *Education Week, 12(2),* 33.

Pollard, R. R. (1991). A comparison of attitudes toward tech-prep programs. *Community College Review, 19(3),* 34–42.

The push for Perkins funding. (1995). *Vocational Education Journal, 70(3),* 18.

Richardson, W. (1998). Work-based learning for young people: National policy, 1994–1997. *Journal of Vocational Education and Training: The Vocational Aspect of Education, 50(2),* 225–246.

What the governors think. (1995). *Vocational Education Journal, 70(6),* 32.

4. COOPERATIVE EDUCATION MODELS

Appelbaum, E., & Batt, R. (1993). *High-performance work systems: American models of workplace transformation.* Washington, DC: Economic Policy Institute.

Bice, J. (1995). STW roles for new and related services. *Vocational Education Journal, 70(10),* 47.

Buller, P. F., & Stull, W. A. (1990). Strategy and performance in cooperative education programs. *Research in Higher Education, 31(3),* 257–270.

Cantor, J. A. (1991). The auto industry's new model: Car companies and community colleges collaborate to provide high-technological training. *Vocational Education Journal, 66(7),* 26–29.

Caton, J., & Buck, C. (1990). Interning, co-oping, and programming alternatives: Successful partnerships and funding. *National Business Education Yearbook,* 142–150.

Coble, C. R. (1999). Going to scale: North Carolina's education partnerships. *Metropolitan Universities: An International Forum, 10(2),* 63–71.

Edgert, P., & Polkinghorn, R., Jr. (1999). California's collaboration to improve education. *Metropolitan Universities: An International Forum, 10(2),* 41–48.

Evanciew, C., & Rojewski, J. W. (1999). Skill and knowledge acquisition in the workplace: A case study of mentor-apprenticeship relationships in youth apprenticeship programs. *Journal of Industrial Teacher Education, 36(2),* 24–53.

Kerka, S. (1999). *New directions for cooperative education.* ERIC Digest No. 209 (Report No. EO-CE-99-209). Washington, DC: Office of Educational Research and

Improvement. (ERIC Document Reproduction Service No. ED434245)

Kranendonk, B. (1992). *Programmed for partnership: IBM's support for business education partnerships includes a school-to-work system.* Washington, DC: U.S. Department of Education.

Minear, D. J. (1999). Educational excellence through partnerships in Florida. *Metropolitan Universities: An International Forum, 10(2),* 49–56.

Saltmarsh, J. (1992). John Dewey and the future of cooperative education. *Journal of Cooperative Education, 28(1),* 6–16.

Stem, D. (1991). *Combining school and work: Options in high schools and two-year colleges.* Washington, DC: U.S. Department of Education, Office of Vocational and Adult Education.

Straub, R. (1990). Growing pains of a cooperative education management information system." *Journal of Cooperative Education, 26(3),* 80–84.

☛ See also chapters on specialized program areas.

5. COORDINATORS AND THEIR ROLES

Barton, P. E. (1996). *Cooperative education in high school: Promise and neglect. A policy issue perspective.* Princeton, NJ: Educational Testing Service, Policy Information Center. (ERIC Document Reproduction Service No. ED400413)

Hartley, N. K., & Wentling, T. L. (Eds.). (1996). *Beyond tradition: Preparing the teachers for tomorrow's workforce.* Columbia, MO: University Council for Vocational Education.

Heath, B. (1982, Spring). Developing a public relations plan for your MDE program. *Marketing Educator's News, 2(3),* 11.

Martins, Reynoldo L., Jr. (1992). Wanted: Leadership qualities. *Vocational Education Journal, 67(5),* 30.

Misko, J. (1998). *School students in workplaces: What are the benefits?* Leabrook, South Australia: National Centre for Vocational Education Research. (ERIC Document Reproduction Service No. ED422526)

Misko, J. (2000). *Getting to grips with work experience.* South Australia: National Centre for Vocational Edu-

cation Research. (ERIC Document Reproduction Service No. ED442007)

Missouri University, Instructional Materials Lab. (1996). *Redesigning cooperative vocational education.* Columbia: Author. (ERIC Document Reproduction Service No. ED404534)

Ohio State Department of Education, Division of Vocational and Adult Education. (1997). *Occupational work experience: Manual of operation.* Columbus: Author. (ERIC Document Reproduction Service No. ED419105)

Wadsworth, R. B. (1981, Fall). Public relations ideas for cooperative education. *Journal of Cooperative Education, 17(1),* 88–93.

Wichowski, C., Erwin, N. B., & Walker, T. J. (1997). *Content standards for the professional development of workplace learning teacher coordinators and teacher leaders.* Philadelphia: Temple University, Center for Vocational Education Professional Personnel Development. (ERIC Document Reproduction Service No. ED415375)

6. INITIATING THE PLAN

Burkhart, J. (1995). *Workplace education advisory council.* Denver: Colorado State Department of Education, State Library and Adult Education Office. (ERIC Document Reproduction Service No. ED399430)

Goldberg, M. F. (1998). *How to design an advisory system for a secondary school.* Alexandria, VA: Association for Supervision and Curriculum Development. (ERIC Document Reproduction Service No. ED421669)

Hong Vo, C. D. (1996). Selling self-interest. *Vocational Education Journal, 71(2),* 22.

Mercer, J. W., & Dillon, B. M. (1997). *Member handbook and leadership guide for Minnesota technical program advisory committees.* St. Paul: Minnesota State Council on Vocational Technical Education. (ERIC Document Reproduction Service No. ED408467)

Nolot, T. (1995). Shopping for jobs: Mall internship program opens doors for HVAC student. *Vocational Education Journal, 70(1),* 28.

Tenenbaum, I. M., Jackson, C., & Couch, B. (2000). *Handbook for advisory groups in career and technology education.* Columbia: South Carolina State Department of Education. (ERIC Document Reproduction Service No. ED441971)

7. COORDINATOR RESPONSIBILITIES AT THE SECONDARY LEVEL

Burrow, J. (1986, January–February). Parental involvement: An overlooked resource. *Balance Sheet, 67(3),* 4–6.

Catri, D. B. (1996). State and federally supported materials centers can be useful resources for educators–teachers' helpers. *Vocational Education Journal, 71(3),* 28.

Wanat, J. A. (1983). Help your employers to become cooperating partners. *Workplace Education, 2(1),* 4–5.

☛ See also chapters on specialized program areas and Chapter 5.

8. COORDINATOR RESPONSIBILITIES FOR ADULT TRAINING AND DEVELOPMENT

Cooper, M. J., & Droddy, F. (1999). Staff training recognizing adult training needs: Issues in education. *Journal of Early Education and Family Review, 6(4),* 12–17.

Eurich, N. P. (1991). *The learning industry: Education for adult workers.* Princeton, NJ: Carnegie Foundation for the Advancement of Teaching.

Finch, C. R., & Crunkilton, J. R. (1999). *Curriculum development in vocational and technical education: Planning, content, and implementation* (5th ed.). Boston: Allyn & Bacon.

Liddell, S., & Ashley-Oehm, D. (1995, January). *Adult workers: Retraining the American workforce.* Paper presented at National Conference of State Legislatures, Denver, CO.

O'Connor, P. J., & Reece, B. L. (1982). A practical model for adult MDE needs assessment. *Marketing & Distributive Educator's Digest, 7(2),* 24.

Price, W. T., Jr. (1982). Instructing the adult learner in marketing and distributive education. *Marketing & Distributive Educator's Digest, 8(1),* 29.

Price, W. T., Jr. (1983). Marketing and distributive education teacher involvement in adult training and development. *Marketing & Distributive Educator's Digest, 8(2),* 22.

Salopek, J. L. (2000). The young and the rest of us. *Training and Development, 54(2),* 26–30.

Smith, C., Payne, E., & Thorton, G. (1999). *Standards and guidelines for work-based learning programs in Georgia.* Atlanta: Georgia Department of Education.

Wang, C. (2000). Establishing a productive climate for adult learning in the small instructional group. *Performance Improvement, 39(5),* 13–17.

9. COORDINATOR RESPONSIBILITIES AT THE POST-SECONDARY AND COLLEGIATE LEVELS

Breslin, R. D. (1980). An internship program in the liberal arts. *Improving College and University Teaching, 29(4),* 171–176.

Klaupins, G., & Warberg, W. (1992). Course prerequisites for personnel cooperative education students. *Journal of Cooperative Education, 28(1),* 48–55.

Lazarus, F. C. (1992). Learning in the academic workplace: Perspectives of cooperative education directors. *Journal of Cooperative Education, 28(1),* 67–76.

Lopez, V. (1996). *People, programs, and partnerships: ABCs of advisory committees technical programs in community colleges.* Parma Heights, OH: Cuyahoga Community College. (ERIC Document Reproduction Service No. ED392478)

Rheams, P. A., & Saint, F. (1991). Renovating cooperative education programs. *New Directions for Community Colleges, 19(3),* 47–54.

Snell, M. (1981). A comparison of employers of cooperative work experience education programs on the secondary and post-secondary education levels. *Journal of Cooperative Education, 17(2),* 20–25.

Stull, W. A., & DeAyora, M. R. (1984). The benefits of faculty involvement in cooperative education in institutions of higher education in the United States. *Journal of Cooperative Education, 20(3),* 18–27.

10. PLANNING AND CARRYING OUT EFFECTIVE IN-SCHOOL INSTRUCTION

Barton, P. E., & Kirsch, I. S. (1990). *Workplace competencies: The need to improve literacy and employment readiness.* Washington, DC: U.S. Government Printing Office.

Baumgardner, M. (1980). Teaching strategies in cooperative education. *Journal of Business Education, 56(1),* 17–18.

Council of Chief State School Officers. (1991). *Connecting school and employment.* Washington, DC: Author.

Fratus, J. C. (1985). A long-term effort leads to co-op placements. *Vocational Education Journal, 59(5),* 45.

Grubb, W. N. (1995). *Education through occupations in American high schools: Volume 1, Approaches to integrating academic and vocational education.* New York: Teachers College Press.

Grubb, W. N. (1995). *Education through occupations in American high schools: Volume 2, The challenges of implementing curriculum integration.* New York: Teachers College Press.

Johnson, A. W., & Summers. A. A. (1993). *What do we know about how schools affect the labor market performance of their students?* Philadelphia: National Center on the Educational Quality of the Workforce, University of Pennsylvania.

Kincheloe, J. L. (1995). *Toil and trouble: Good work, smart workers, and the integration of academic and vocational education.* New York: Peter Lang.

Madsen, D. L. (1987). Designing an instructional delivery system. *Business Education Forum, 42(1),* 19.

Searle, G. A., & Collins, I. (1985). Quality components of cooperative vocational education. *Balance Sheet, 66(4),* 31–33.

Secretary's Commission on Achieving Necessary Skills. (1992). *Learning a living: A blueprint for high performance.* A SCANS report for America 2000. Washington, DC: U.S. Government Printing Office.

Secretary's Commission on Achieving Necessary Skills. (1992). *Skills and tasks for jobs.* A SCANS report for America 2000. Washington, DC: U.S. Government Printing Office.

Smith, T. E. C., Polloway, E. A., Patton, J. R., & Dowdy, C. A. (1995). *Teaching children with special needs in inclusive settings.* Boston: Allyn & Bacon.

State Higher Education Executive Officers. (1992). *Building a quality workforce: An agenda for postsecondary education.* Denver: Author.

Van Loon, C. (1985). A magic kingdom of training. *Balance Sheet, 67(1),* 15–16.

11. DEVELOPING TRAINING STATIONS AS INSTRUCTIONAL LABORATORIES

Bailey, T. R. (Ed.). (1995). *Learning to work: Employer involvement in school-to-work transition programs.* Washington, DC: The Brookings Institution.

Burns, W. (1986). Careers: Vocational aptitude test can help. *Industrial Education, 75(9),* 14.

Buys, N., Kendall, E., & Ramsden, J. (1999). *Vocational education and training for people with disabilities: Review of research.* Leabrook, South Australia: National Centre for Vocational Education Research. (ERIC Document Reproduction Service No. ED431874)

General Accounting Office. (1990). *Training strategies: Preparing noncollege youth for employment in the U.S. and foreign countries.* Washington, DC: Author.

Husted, S. W. (1994). America's most admired retail training stations. *Perspectives on Marketing, 9(5),* 6.

Kantrowitz, B. (1986, June 9). The kids and jobs: Good or bad? *Newsweek,* 4.

McCage, R. D. (1986). Using the V-tecs data bases. *Vocational Education Journal, 61(6),* 39.

National Alliance of Business. (1993). *Real jobs for real people: An employer's guide to youth apprenticeship.* New York: Author.

Rainforth, B. (1996). *Related services supporting inclusion: Congruence of best practices in special education and school reform.* Issue Brief. Alexandria, VA: Consortium on Inclusive Schooling Practices. (ERIC Document Reproduction Service No. ED404804)

Steinberg, L., Fegley, S., & Dornbush, S. M. (1993). Negative impact of part-time work on adolescent adjustment: Evidence from a longitudinal study. *Developmental Psychology, 29(2),* 1993, 171–180.

Stitt, T., & Stitt, B. (1983). Legality of the training agreement: Implications for the vocational coordinator. *Journal of Cooperative Education, 19(2),* 83–89.

Virginia Vocational Curriculum and Resource Center. (1993). *Vocational cooperative education guide for teacher-coordinators.* Richmond: Author. (ERIC Document Reproduction Service No. ED381632)

12. CORRELATING INSTRUCTION BETWEEN SCHOOL AND JOB LABORATORIES

General Accounting Office. (1991). *Transition from school to work: Linking education and worksite training.* Washington, DC: Author.

Kruger, L. B., & Reynolds, J. N. (1986). Planning for an effective cooperative education student handbook. *Business Education Forum, 40(4),* 63–67.

Missouri University, Instructional Materials Laboratory. (1996). *Redesigning cooperative vocational education.* Columbia: Author. (ERIC Document Reproduction Service No. ED404534)

Nellermore, D. A. (1992). Preparing students for employment—or, what managers really want. *Business Education Forum, 46(1),* 11–13.

Ordover, E. L., & Annexstein, L. T. (1999). *Ensuring access, equity, and quality for students with disabilities in school-to-work systems: A guide to federal law and policies.* Washington, DC: Center for Law and Education.

Weinstein, D., & Wilson, J. W. (1983). An employer description of a model employer cooperative education program. *Journal of Cooperative Education, 20(1),* 60–82.

☛ See also chapters on specialized program areas and Chapter 11.

13. THE MATURING OF THE COOPERATIVE PLAN

Barton, P. E. (1993). *Improving the transition from school to work in the United States.* Washington, DC: American Youth Policy Forum.

Batt, R., & Osterman, P. (1993). *A national policy for workplace training: Lessons from state and local experiments.* Washington, DC: Economic Policy Institute.

Center for Workforce Preparation. (1994). *New century workers: Effective school-to-work transition programs.* Washington, DC: U.S. Chamber of Congress.

General Accounting Office. (1993). *Transition from school to work: States are developing new strategies to prepare students for jobs.* Washington, DC: Author.

Georgia Department of Education. (2000). *The technology/ career education industry certification handbook.* Atlanta: Author.

Knapp, A. H. (1987). Test your advisory group savvy. *Vocational Education Journal, 62(1),* 23.

National Board for Professional Teaching Standards. (1997). *Vocational education: Standards for board certification.* Washington, DC: Author.

Martin, R. A. (1987). Preparing teachers for advisory committee work. *Vocational Education Journal, 62(1),* 19.

Mercer, J. W., & Latta, E. M. (1987). New roles for state councils. *Vocational Education Journal, 62(1),* 37–38.

Pautler, A. J., Jr. (Ed.). (1994). *High school to employment transition: Contemporary issues.* Ann Arbor: MI: Prakken Publications.

Thiers, N. (Ed.). (1995). *Successful strategies: Building a school-to-career system.* Alexandria, VA: American Vocational Association. (ERIC Document Reproduction Service No. ED404555)

Tucker, M. (1994, April). *A school-to-work transition system for the United States.* Washington, DC: National Center on Education and the Economy.

Woloszyk, C. A. (1996). *Vocational education's linkages with the business community of Michigan.* East Lansing: Michigan State University, Michigan Center for Career and Technical Education. (ERIC Document Reproduction Service No. ED404476)

Zentz, C. S., & Shry, C. L. (1984). Reaching out to the community. *VocEd, 59(6),* 21–22.

14. STUDENT ORGANIZATIONS AS AN INTEGRAL PART OF INSTRUCTION

Dykman, A. (1993). Not just "seed and feed." *Vocational Education Journal, 68(4),* 33.

Hall, T. (1993). VICA goes to college. *Vocational Education Journal, 68(4),* 28.

Hannah, G. (1993). Shift or drift. *Vocational Education Journal, 68(4),* 21–25.

Husted, S. W. (1987). Developing leadership camp for student officers. *Business Education Forum, 41(8),* 27.

Lankard, B. A. (1996). *Youth organizations: Myths and realities.* Columbus, OH: ERIC Clearinghouse on Adult,

Career, and Vocational Education. (ERIC Document Reproduction Service No. ED392895)

Moss, J. W. (1990). Changing the FFA. *The Agricultural Education Magazine, 63(5)*, 4–19.

National Coordinating Council for Career and Technical Student Organizations. (1999). *Career and technical student organizations: A reference guide.* Washington, DC: National Association of State Directors of Vocational and Technical Education Consortium.

Vaughn, P. R. (1999). *Handbook for advisors of vocational student organizations.* Winterville, GA: The American Association for Vocational Instructional Materials.

15. ACCOUNTABILITY THROUGH EVALUATION

Brustein, M. (1999). *Examination of the data requirements of the Workforce Investment Act and the Perkins Act of 1998: Report of the National Postsecondary Education Cooperative Working Group on Workforce Development.* Washington, DC: National Center for Education Statistics. (ERIC Document Reproduction Service No. ED447312)

Bunn, P. C., & Stewart, D. L. (1998). Perceptions of technical committee members regarding the adoption of skill standards in vocational education programs. *Journal of Vocational and Technical Education, 14(2)*, 7–17.

Chamberlain, V., Commings, M., & Swofford, H. (1985). A rating scale for coop students. *Vocational Education Journal, 60(1)*, 20–21.

Herman, A. M. (1998). *The Workforce Investment Act of 1998: A vision for youth.* Washington, DC: U.S. Department of Labor.

Hoachlander, E. G. (1991). Designing a plan to measure vocational education results: Here are tips for developing accountability systems to meet Perkins Act requirements. *Vocational Education Journal, 66(2)*, 20–21.

Hoachlander, E. G. (1995). What the numbers really mean: Perkins accountability requirements can lead to improved programs. *Vocational Education Journal, 70(3)*, 20.

King, G. C. (1986). The quest for excellence in cooperative education through program evaluation. *Workplace Education, 4(1)*, 4–5, 11, 22.

McCaslin, N. L., & Headley, W. S. (1993). *A national study of approved state systems of performance measures and standards for vocational education.* Columbus: The Ohio State University, Graduate School. (ERIC Document Reproduction Service No. ED360474)

Rabinowitz, S. N. (1995). Beyond testing: A vision for an ideal school-to-work assessment system. *Vocational Education Journal, 70(3)*, 27.

Resnick, L. B., & Wirt, J. G. (Eds.). (1996). *Linking school and work: Roles for standards and assessment.* San Francisco: Jossey-Bass.

Roberts-Gray, C. (1991). *Research and development performance standards for selected vocational education programs: Final report.* Austin: University of Texas, Extension Instruction and Materials Center.

Rothman, R. (1995). *Measuring up: Standards, assessment, and school reform.* San Francisco: Jossey-Bass.

Sarkees-Wircenski, M., & Wircenski, J. L. (1991–92). Performance standards and measures for vocational education programs: Meeting school reform through vocational programs. *Journal for Vocational Special Needs Education, 14(2–3)*, 28–32.

Seeman, H. (1983). A rationale and methodology for assessing and awarding credit for student's work experience in cooperative education. *Workplace Education, 2(2)*, 11–12.

Silberman, H. F. (1987). The national assessment of vocational education. *Vocational Education Journal, 62(7)*, 44.

VanCronkhite, J. B. (1990). Assessing future business/education linkage. *TechTrends, 5(4)*, 58–59.

Wentling, T. L., & Roegge, C. A. (1989). Development of a computer-aided evaluation system for vocational education programs. *Journal of Vocational Education Research, 14(1)*, 1–14.

16. LEGAL AND REGULATORY ASPECTS OF COOPERATIVE EDUCATION

Brustein, M. (1987). Maintenance: What the law says about it. *Vocational Education Journal, 62(7)*, 40.

Dumaine, B. (1993, April 5). Illegal child labor comes back. *Fortune, 129(6)*, 86–96.

General Accounting Office. (1991). *Child labor: Characteristics of working children.* Washington, DC: Author. (ERIC Document Reproduction Service No. ED334027)

Gentry, D. (1987). The governance controversy. *Vocational Education Journal, 62*(7), 47.

Kaufman, B. E., & Wills, J. L. (1999). *User's guide to the Workforce Investment Act of 1998. A companion to the law and regulations.* Alexandria, VA: Association for Career and Technical Education.

National Safe Workplace Institute. (1992). *Sacrificing America's youth: The problem of child labor and the response of government.* Chicago: Author. (ERIC Document Reproduction Service No. ED352448)

Radar, M. H., & Kurth, L. A. (1999). Federal workplace laws: Are business work experience programs in compliance? *Business Education Forum, 53*(4), 26–29.

Valentine, H., & Zikmund, D. G. (1998, December). *Legal issues involving student workers.* Paper presented at the American Vocational Association Convention, New Orleans, LA.

17. THE PLAN IN AGRICULTURAL OCCUPATIONS

Conrads, J. A. (1984). Business and vocational education . . . Marriage for the future. *The Agricultural Education Magazine, 57*(1), 9–12.

Cox, D. E. (1991). Supervised experience. *The Agricultural Education Magazine, 64*(6), 4, 6–16, 18–20, 22–23.

Gless, R. (1984). Developing a cooperative experience program. *The Agricultural Education Magazine, 56*(10), 22.

Herren, R. V., & Donahue, R. L. (2000). *Delmar's agriscience dictionary with searchable CD-ROM.* Albany, NY: Delmar / Thomson Learning.

Kruckenberg, D. W., & Williams, D. L. (1980). What research has to say—attitudes toward experiential program. *The Agricultural Education Magazine, 52*(11), 14–15.

Lee, Jasper S. (2000). *Program planning guide for agriscience and technology education,* 2nd ed. Danville, IL: Interstate Publishers, Inc.

Long, G. A. (1984). SOEP—making a good tool better. *The Agricultural Education Magazine, 56*(11), 4–22.

Martensen, J. C., & Foster, R. M. (1980). The blackfoot story—how cooperative education meets the needs of agricultural industry. *The Agricultural Education Magazine, 52*(9), 12–14.

Martin, R. A. (1984). Interfacing vocational agriculture with business. *The Agricultural Education Magazine, 56*(8), 3–16.

Watman, M., & Raymond, G. (1991). Supervised experience for everyone. *The Agricultural Education Magazine, 64*(2), 6–7.

18. THE PLAN IN BUSINESS OCCUPATIONS

Missouri University, Instructional Materials Laboratory. (1990). *Supervised business experience handbook.* Columbia: Author. (ERIC Document Reproduction Service No. ED325703)

Saunders, A. D. (1987). Individualized instruction in information processing. *Business Education Forum, 42*(1), 18.

Tomal, A. (1992). Bring performance appraisals into the classroom. *Business Education Forum, 47*(2), 17–19.

19. THE PLAN IN FAMILY AND CONSUMER SCIENCES OCCUPATIONS

Carson, L. (1990). Cooperative learning in the home economics classroom. *Journal of Home Economics, 82*(4), 37–41.

Guilinger, J. (1987). The potential of advisory councils. *Vocational Education Journal, 62*(1), 11.

Kolde, R. F. (1986). Success through partnerships. *VocEd, 78*(1), 61, 78.

Martin, R. A. (1987). Preparing teachers for advisory committee work. *Vocational Education Journal, 62*(1), 19.

Parks, D., & Henderson, G. H. (1987). Making the most of advisory groups. *Vocational Education Journal, 62*(1), 20.

Texas Tech University, Home Economics Curriculum Center. (1998). *Home economics career preparation handbook.* Austin: Author. (ERIC Document Reproduction Service No. ED439278)

Wehmeier, R. (1987). What do advisory groups really do? *Vocational Education Journal, 62*(1), 17.

Womble, M. N., Adams, E., Stiff-Gohdes, W. L. (2000). Business and marketing education programs in Georgia: Focus groups examine issues for program reform. *The Delta Pi Epsilon Journal, 42(1),* 38–63.

20. THE PLAN IN HEALTH OCCUPATIONS

Arkansas State Department of Education, Division of Vocational-Technical and Adult Education. (1985). *Health occupations education curriculum project: Final report.* Little Rock: Author.

Humbert, J. T., & Woloszyk, C. A. (1983). Essential duties and tasks performed by cooperative education coordinators. *Workplace Education, 1(4),* 14–15.

Knapp, A. H. (1987). Test your advisory group savvy. *Vocational Education Journal, 62(1),* 23.

Lozada, M. (1995). The health occupations boom. *Vocational Education Journal, 70(6),* 34.

Martin, R. A. (1987). Preparing teachers for advisory committee work. *Vocational Education Journal, 62(1),* 19.

Ohio State Department of Education, Division of Vocational and Adult Education. (1997). *Diversified cooperative training: Diversified cooperative health occupations manual of operation.* Columbus: Author. (ERIC Document Reproduction Service No. ED419104)

Ross, S., & Marriner, A. (1985). Cooperative education: Experience-based learning. *Nursing Outlook, 33(4),* 177–180.

Snell, M. (1981). A comparison of employers of cooperative work experience education programs on the secondary and post-secondary education levels. *Journal of Cooperative Education, 17(2),* 20–25.

Stitt, T., & Stitt, B. (1983). The legality of the training agreement: Implications for the vocational coordinator. *Journal of Cooperative Education, 19(2),* 83–90.

21. THE PLAN IN MARKETING OCCUPATIONS

Adams, E., Womble, M. N., & Jones, K. H. (2000). Marketing education students' perceptions toward marketing education courses. *Journal of Career and Technical Education, 17(1),* 46–63.

Asselin, S. B. (1983). MDE teachers' contribution to the IEP. *Marketing & Distributive Educator's Digest, 9(1),* 22.

Gleason, J. R. (1991, November). Tech-Prep: Threat or opportunity? *Perspectives on Marketing, 7(2),* 7.

Gleason, J. R. (1992, November). SCANS, tech-prep, apprenticeship, and more. *Perspectives on Marketing, 8(2),* 4–5.

Goins, K. (1995). Marketing education: What can we do? *Vocational Education Journal, 70(1),* 34.

James, R. L. (1991, March). Developing a marketing plan for your marketing program. *Perspectives on Marketing, 6(4),* 13–14.

Lynch, R. L. (1991, March). In search of excellence in marketing education. *Perspectives on Marketing, 6(4),* 7–10.

One of vo-tech's best and brightest (1995). *Vocational Education Journal, 70(3),* 16.

Palmieri, D. F. (1993, January). Using research to define program relevance in marketing education. *Perspectives on Marketing, 8(3),* 10, 16.

Price, W. T., Jr. (1982). Making marketing and distributive education programs accessible to the handicapped. *Marketing & Distributive Educator's Digest, 7(2),* 16.

Ruhland, S. K., & Wilkinson, R. E. (1995). *Marketing and cooperative education administrative handbook.* Columbia: Missouri University, Instructional Materials Laboratory. (ERIC Document Reproduction Service No. ED384803)

Schwartz, S. L. (1991, March). Marketing education's role in student retention. *Perspectives on Marketing, 6(4),* 11–12.

Smith, C. L., & Rojewski, J. W. (1993, September). School-to-work transition programs. *Perspectives on Marketing, 9(1),* 13–14.

Snyder, D. H. (1990, November). Developing a successful marketing option for two-year associate degree business programs. *Perspectives on Marketing, 6(2),* 3–4.

Stone, J. R., III. (1992, March). Myths and realities: The untold story of vocational and marketing education. *Perspectives on Marketing, 7(4),* 5–6.

Wentland, D. (1991, May). Managing the image of marketing education. *Perspectives on Marketing, 6(5),* 7–8.

22. THE PLAN IN TRADE AND INDUSTRIAL OCCUPATIONS

Charbonneau, W. (1986). How the job training partnership act works. *Industrial Education, 75(6),* 24.

International Technology Education Association. (2000). *Standards for technological literacy: Content for study of technology.* Reston, VA: Author.

Knapp, A. H. (1987). Test your advisory group savvy. *Vocational Education Journal, 62(1),* 23.

Martin, R. A. (1987). Preparing teachers for advisory committee work. *Vocational Education Journal, 62(1),* 19.

Parks, D. L., & Henderson, G. H. (1987). Making the most of advisory groups. *Vocational Education Journal, 62(1),* 20.

Schreiber, A. P., & Walter, R. P. (1995). Three states tackle school-to-work reform. *Vocational Education Journal, 70(3),* 41.

Thiers, N. (1996). A good trade. *Vocational Education Journal, 71(3),* 32.

Thiers, N. (1996). Technical support. *Vocational Education Journal, 71(4),* 42.

Index

Index

A

Academic sanction 305

Accountability (*See* Evaluation)

Accrediting associations 280

Administrative laws 294

Adult Workforce Development 133

    evaluating achievement 145

    financing 142

    involvement of community 142

    organizing 139

    securing instructors 143

Advisory committee 105, 243

Affective domain 176

Age certificate 309

Agricultural occupations, the plan in 315

    desired learning on the job 323

    instructional materials 319

        sources 322

    laboratories and classrooms 321

    occupations 317

    plans 318

    related instruction 319

    student application 326

    student evaluation 326

    student organization (National FFA Organization) 253, 257

    training agreement 327

    training plan 330

    training profile 324

    training stations 326

Alternating cooperative experiences 151

Alumni groups 249

American Society for Training and Development 98

American Association of Family and Consumer Sciences 98

Annual contract 95

Annual report 95

Apprentice 19

Apprenticeship 10, 19, 21

Area Redevelopment Act 49

Area vocational school/center 45

Association for Career and Technical Education 98

Assured jobs program 39

At-risk populations 31

B

Bilingual education 59

BPA (Business Professionals Association) 253, 254, 363

Business occupations, the plan in 335

    instructional materials, sources of 370

    instructional objectives 342

    laboratories and classrooms 355

    new business education curriculum 337

    occupations 337

    personal data sheet 360

    plans 344, 363

    related instruction 343

    selecting student-learners 357

    student application 358

    student evaluation 365, 369

    student organization

        Business Professionals Association 253, 254, 263

        Future Business Leaders of America 253, 256, 363

    training stations 359

C

Calendar of events 110

Career Education Act 46

Career-technical student organization 251

Carl D. Perkins Vocational and Applied Technology Education Act of 1990 52

Carl D. Perkins Vocational and Technical Education Act of 1998 52

Carl D. Perkins Vocational Education Act of 1984 51

Changing occupations 31

Chapter work activities 260

Clusteral concept 34

Cognitive domain 175

Combination plans   79

Community survey   106

Competencies   36, 74, 171
   career development   36
   job adjustment   36
   job development   36
   job intelligence   36

Competitive events/activities   262

Comprehensive Employment Training Act (CETA)   54

Cooperative education   16
   comparison to work experience   17
   key elements   16
   plan of instruction   73
   program or plan   75

Cooperative Education Association   98

Cooperative Models   71

Cooperative occupational education plans   16, 17, 23, 24

Cooperative plan   71
   basic facts   16
   characteristics   76
   combination   78
      diversified   79
      sectional   79
   specialized   78
   vocational instruction   73

Coordination   77, 117
   activities   117, 118
   in a new plan   208
   in apprenticeship   130
   recurring   242
   relation with unions   244
   summer months   242

Coordinator   83
   activities   117
   responsibilities at the post-secondary and collegiate levels   147, 153
   responsibilities at the secondary level   115
   (See also Teacher-coordinator)

Correlated instruction   15, 217

Correlation   171, 217

Cost-benefit studies   288

Counselor-coordinator   87

Critical factor technique   280

Cycling instruction   178

D

Debriefing   156

DECA (Distributive Education Clubs of America)   253, 255, 430, 436

Demographic trends   30

Dignity of work   225

Directed occupational experience   92

Disabled, individuals who are   57

Disadvantaged, individuals who are   58

Diversified occupations plans   449

Due process   296

E

Education for All Handicapped Children Act   57

Employer-employee appreciation events   247

Employment certificate   309

Engaged-time analysis   283

Equal educational opportunity   301

Evaluation   273
   cost-benefit studies   288
   critical factor   280
   follow-up studies   284
   internal   280
   process   276, 280
   product   280
   program reviews   285
   types   277
      labor market behavior   284
      learning-time analysis   283
      value added assessment   289

Experiential learning   7, 11

Exploratory work experience   10, 11, 12

Extended contract   242

F

Faculty cooperative advisor   153

Family and consumer sciences occupations, the plan in   388
   curriculum   401
   desired learning on the job   402
   instructional materials, sources of   401
   laboratories and classrooms   400
   occupations   399
   student evaluation   408
   student organization (Family, Career, and Community Leaders of America)   253, 255, 402
   training stations   400

Farm-out system   225

FBLA (Future Business Leaders of America)   253, 256, 363

Federal Revenue Act of 1978   53

Feedback   227

FFA (National FFA Organization)   253, 257

FCCLA (Family, Career, and Community Leaders of America)   253, 255, 402

Follow-up studies   284
   (See also Evaluation)

G

General occupational competencies   171

General related instruction   173

General work experience   10, 13, 17

George-Barden Act   48

George-Deen Act   48

Goals 2000: Educate America Act   64

H

Hazardous occupations   309

Health occupations, the plan in   371
   curriculum   375
   desired learning on the job   379
   instructional materials, sources of   379
   instructional procedures   376

job opportunities   374
laboratories and classrooms   378
licensure   378
occupations   374
plans   375
related instruction   377
student evaluation   381
student organization (Health Occupations Student Organization)   257, 379
training stations   380
HOSA (Health Occupations Student Organization)   257, 379

I

IEP   182
Initiating a plan   101
in loco parentis   302
In-school instruction   172, 174
(*See also* Related instruction)
Internal evaluation   280
International Technology Education Association   99
Internships (practicums)   7, 18, 151

J

Job adjustment competency   36
Job development   193
Job development competency   36
Job intelligence competency   36
Job Training Partnership Act (JTPA)   55

K

K–12 organization   45

L

Learning agreement or contract   159
Learning-Time analysis   283
Learning manager   92
Legal aspects   291
classroom behavior   294, 301
constitutional rights   297

federal role   295
federal system   295
minimum wage   309
responsibility of teachers   294
Local reports   95

M

Mandatory evaluation   277
Manpower Development and Training Act (MDTA) of 1962   46
Marketing Education Association   98
Marketing occupations, the plan in   411
curriculum   414, 415, 424–428
desired learning on the job   430
employment levels   428
goals   413
instructional materials, sources of   437
laboratories and classrooms   429
occupations   425
related instruction   420
sequence of courses   416
student evaluation   430, 431
student organization   430, 436
training stations   413
Methods of instruction   76
Monthly enrollment report   95
Morrill Act   44
Multicultural learning environment   184

N

National Association of Agricultural Educators   98
National Association of Health Occupations Teachers   98
National Association of Trade and Industrial Education   98
National Defense Education Act   48
New occupation   37
Non-additive credits   158

O

Occupational competencies   36
Occupations, changing nature of   29, 31
Open-interviewing   156
Operational experiences   150
Orientation experiences   150

P

Perkins, Carl D., Vocational and Applied Technology Education Act of 1990   52
Perkins, Carl D., Vocational Education Act of 1984   51
Perkins, Carl D., Vocational and Technical Education Act of 1998   152
Plan, initiating   101
Post-secondary and collegiate cooperative education   147
coordinator responsibilities   147, 153
employer responsibilities   154
evaluation of students   156
faculty advisor responsibilities   153
granting academic credit   158
internships   151
placement of students   155
professional cooperative organizations   162
student responsibilities   154
writing cooperative syllabus   158
Preparatory course   76
Preparatory instruction   138
Press, radio, and TV   128
Private Industry Council (PIC)   55
Process-oriented evaluation   281
Product-oriented evaluation   280
Professionalism   96
Professional organizations   97
Program review   277
Psychomotor domain   176
Public policy   41
Public relations   247

R

Rating sheets   230

Recurring activities   242

Reimbursement policies   130

Related class   220

Related instruction   169, 172
    basic learning tasks   176
    bilingual and multicultural   184
    career development approach
      174
    educational objectives   175
    equipment and facilities   189
    in-school   174
    learning theory   171
    materials   185
    nature of   172
    realistic   220
    sequence of   177
    special needs populations   178

Released-time work experience   12

Reports
    local   95
    state   95

S

School-to-Work Opportunities Act of
    1994   63

Sectional cooperative plan   79

Seminars   141

Service delivery areas (SDA's)   55

Sex equity   61

Smith-Hughes Act   48

Specialized cooperative education plans
    78

Special needs populations
    legislation for   57
    planning related instruction for
      178

Specific job competencies   172

Specific occupational competencies
    176

Sponsor development   235

Sponsors   76
    (See also Training station supervi-
      sor)

SSLO   202

State reports   95

Steering committee   104

Step-by-step training plan   201

Student-learner   76

Student organizations   251
    benefits   258
    competition   262
    cooperation   262
    coordinating with instruction   265
    evaluating the program   270
    financing chapter activities   268
    guiding the chapter   265
    legal responsibilities   270
    objectives   258
    organizing the chapter   265

T

Targeted job-tax credit   194

Teacher-coordinator   77
    basic role   85
    characteristics   90
    competencies needed   91
    contract   88
    educational preparation   93
    image   87
    job   85, 88
    schedule   117
    tasks   88
    types   86
    (See also Coordinator)

Technology Student Association
    (TSA)   253, 258

Third-party evaluation   279

Time-on-task analysis   283

Time-use analysis   283

Trade and industrial occupations   447
    apprenticeship   451
    curriculum   453
    desired learning on the job   458
    functions approach   450
    instructional materials, sources of
      457
    laboratories and classrooms   457
    occupations   450
    plans   451
    related instruction   453
      objectives   455
    scope of instructional program
      450
    student evaluation   459
    student organization
      (SkillsUSA–VICA)   253, 258,
      458
    training stations   451

Training agreement   201, 210

Training plan
    in behavioral terms   202
    involving students   219
    preparing   201

Training stations   76, 191
    as instructional laboratories   191
    criteria for selecting   195
    coordinating   165
    job development   193
    organizing   199
    setting work schedules   200
    specific training   201
    training agreement   210
    training plan   208

Training station sponsors   76, 205
    (See also Training station supervi-
      sor)

Training station supervisor   205, 235
    (See also Training station sponsors)

TSA (Technology Student Associa-
    tion)   253, 258

U

Union relationships (See Coordination)
    244

Upgrading   138

V

Value-added assessment   289

VICA (SkillsUSA–VICA)   253, 258,
    458

W

Weekly training station report   220

Work exploration   197

Workforce Investment Act   55